Computer Graphics for Engineers

Computer Graphics for Engineers

Bruce R. Dewey
University of Wyoming

HARPER & ROW, PUBLISHERS, New York
Cambridge, Philadelphia, San Francisco, Washington,
London, Mexico City, São Paulo, Singapore, Sydney

Sponsoring Editor: Cliff Robichaud
Project Editor: Steven Pisano
Text Design Adaptation and Cover Design: Maria Carella
Cover and Text Art: University of Wyoming Engineering Science Interactive Graphics Laboratory
Production Manager: Jeanie Berke
Production Assistant: Beth Ackerman
Compositor: Science Press
Printer and Binder: R. R. Donnelley & Sons Company
Cover Printer: The Lehigh Press, Inc.

Computer Graphics for Engineers
Copyright © 1988 by Harper & Row, Publishers, Inc.

All rights reserved. Printed in the United States of America. No part of this book may be used or reproduced in any manner whatsoever without written permission, except in the case of brief quotations embodied in critical articles and reviews. For information address Harper & Row, Publishers, Inc., 10 East 53rd Street, New York, NY 10022-5299.

Library of Congress Cataloging-in-Publication Data

Dewey, Bruce R.
Computer graphics for engineers.

Bibliography: p.
Includes index.
1. Computer graphics. 2. Engineering design—
Data processing. I. Title.
T385.D48 1988 006.6 87-14978
ISBN 0-06-041670-X

87 88 89 90 9 8 7 6 5 4 3 2 1

Contents in Brief

Contents in Detail

Preface

Digital computing is revolutionizing information transfer in all engineering activities. Word processing, computer graphics, and expert systems are now at the forefront of the revolution. Computer graphics is an important communications tool for all engineers to deal with plans, diagrams, schematics, and other illustrations. The quality and speed of the engineering process is greatly enhanced by the use of computer-aided engineering (CAE). While specific CAE software varies widely from one engineering discipline to another, computer graphics is always an essential part. Intended for all fields of engineering, this text integrates computer graphics and geometric modeling into professional practice. In engineering applications, the *picture* is *not* the final result. Instead, the *model* is the result and the picture is only a means for communicating information about the model. This distinction justifies a somewhat different coverage of computer graphics for engineers, one with less emphasis on picture rendition and more emphasis on the theory of modeling.

The organization into four parts reflects four different but related areas of computer graphics. Part I discusses the role of computer graphics in CAE and provides some introductory general information about hardware and software. This should be helpful for readers to understand some fundamental concepts before proceeding to specific applications. Part II treats viewing and modeling transformations. Part III treats the geometry of curves, surfaces, and solids. From this background, methods for computing some common geometric properties are developed. Part IV extends the foregoing material into chapter-length surveys of the three principal computer graphics applications utilized by engineers—presentation graphics, computer-aided drafting, and solid modeling. The topics in Part IV are presented in order of historical development and relative amount of use.

Presentation graphics has very wide application in engineering, since it deals with plots, charts, graphs and other tools for the communication of data. Techniques for smoothing and fitting data curves are based on the mathematical methods developed in Part III which pertain to curvilinear graphics and surfaces. The manipulation of three-dimensional plots is treated by the Part II material on transformations.

Computer-aided drafting, or CAD, systems automate the creation and modification of engineering drawings. In addition to providing large productivity gains in the design process, CAD systems can perform many other useful functions. Among these are data base management, reduction of paper use and storage, development of cost data, and instant communication of design information. While the usual CAD system requires no programming language skill for its operation, an understanding of the transformations of Part II promotes effective use of the system.

Solid modeling systems automate the generation and manipulation of points, curves, surfaces, and solids. In addition to use in conceptual design and in the preparation of analysis models, solid modeling also plays an important role in the broad area of computer-integrated manufacturing. All of the concepts developed in Parts II and III find application in solid modeling.

From the foregoing, it is seen that while the chapters of Part IV are more or less independent of each other, all have aspects which depend on some of the principles in the rest of the book. Because various existing computer graphics systems have such great diversity, the material has to be supplemented and interpreted in terms of the user's own system. In contrast, the material in Parts II and III is of general applicability and is central to underlying principles used in all computer graphics systems.

Concepts are introduced with simple examples and short homework problems which do not require the use of a computer. However, there is no substitute for the hands-on experience gained in doing projects such as those suggested at the ends of the chapters. Some of the projects are oriented toward design. For project work, documentation for the user's own system should be used to supplement the text. In addition to the book references cited, periodicals in computer graphics should be consulted for up-to-date information.

As preparation for a course using this text, a student is expected to be proficient in a high-level programming language (e.g., BASIC, PASCAL, or FORTRAN-77), and to be familiar with calculus, matrix, and vector algebra. For background, the dynamics course included in most engineering curricula provides valuable insight into three-dimensional (3-D) transformations and vector-valued parametric functions. The background already acquired by engineering students means, for example, that 3-D work can be approached from the beginning. The material in this text supplements more traditional courses such as finite element analysis, kinematics, robotics, and image processing. More material is included here than would ordinarily be used in a one-semester course. The latter part of Chapter 2 and Part III may be omitted for a general, application-oriented course. For a course that is more theoretically oriented, all or selected topics of Part IV may be omitted.

The book is intended for use with *generic* computers and graphics software. Although the brief programming examples in this text are in FORTRAN-77, which is widely used for engineering applications, users of BASIC or PASCAL should be able to follow the code. To reduce the programming burden, either a graphics tools subroutine library or a BASIC system with embedded graphics calls is strongly recommended. In addition, it is highly desirable to have appropriate software to support presentation graphics, computer-aided drafting, and solid modeling for the use of material in Part IV. The old maxim about reinventing the wheel applies to computer graphics software as well: Make maximum use of previously developed routines and application programs.

A graphics tools subroutine library, such as the one described in Appendix B, can effect significant reduction in programming effort for most applications. Many universities, including the author's, have developed their own library. Commercially distributed libraries are available for mainframes, minicomputers, and microcomputers. Examples include Tektronix IGL and Plot-10 GKS, DI-3000, Template and Dimension-GKS, and GW-Core. These libraries provide graphics subroutines to be called from user programs as well as necessary system-level routines.

The capabilities needed for Part IV are widely available. Presentation graphics software includes the graphics in integrated spreadsheet programs such as LOTUS 1-2-3 and EXCEL, many stand-alone microcomputer "graphics" packages, as well as CA-DISSPLA, TELL-A-GRAF, SAS-GRAPH, and extensions to the graphics tools systems listed above. There are a large number of microcomputer-based CAD systems, including AUTOCAD, CADKEY, VersaCAD, and many others. Minicomputer CAD software includes CADAM, Anvil-4000, and PAFEC-DOGS. Many of the popular professional computer graphics systems have capabilities for both CAD and solid modeling. Specialized solid-modeling software includes CATIA, Euclid, GEOMOD, Romulus, PADL, MOVIE-BYU, and PATRAN. The foregoing lists of software are merely examples and have no pretext of completeness, rating, or endorsement.

It may be of interest to note that computers were used to greatest possible advantage in the preparation of this text. The manuscript, including equations, was entirely written and edited on a microcomputer-based word processing system. All of the line drawings were computer-produced with a pen plotter. The color examples have been produced either from screen photographs of graphics workstations or on a computer graphics film recorder.

My privilege of working during the late 1970s in England with engineering software developers Richard Henshell, Keith Shaw, Ian McKenzie, Tony Christian, and others at Pafec Ltd. provided a valuable start in shaping the content of the text. In more recent years, discussions with C. H. Hamilton of PDA Engineering and John Steadman of the University of Wyoming have produced many of the ideas found in the text.

I appreciate the comments and suggestions received from the students who studied from the evolving versions of the manuscript at the University of Wyoming. Two students from the early days, Ron Wendland and Edward Kline, helped start development of UWLIB, the graphics tools subroutine library that has been the basis for much of the work. Many other people, particularly Glenn Sanders, have contributed routines and testing over the past four years.

The continuing help of the University of Wyoming Engineering Science Interactive Graphics Laboratory manager, Cheryl Hilman, has been invaluable in many ways. Two UW engineering students, Terry L. Smith and Daniel A. Hendrickson, expertly produced most of the black-and-white art work for the book, the major part of which was done with a CAD system, and the rest of which was done with various presentation graphics systems. Thomas L. Lynch, a UW computer science student, did most of the color work, including the cover. Other student contributions include a color plate which is the work of Guy Dear and an extensive presentation graphics package developed by David Boll which has been used to produce some of the illustrations. James E. Peterson helped with checking the solutions to the problems. In addition to contributing examples of professional computer graphics work, Martin Marietta, PDA Engineering, CA-ISSCO, Pafec, Inc., and Tektronix, Inc. have helped support our computer graphics laboratory.

Two engineering faculty, Sally Steadman of the University of Wyoming and Wayne C. Dowling of Iowa State University, have made detailed suggestions on algorithms and developments, have contributed problems and projects, and have provided extensive reviews of the manuscript. Sally Steadman is also a key software developer in our laboratory.

It is a pleasure to acknowledge the reviewers: Mac Casale of the University of California, Irvine, and PDA Engineering; Roy Scott Hickman of the University of California, Santa Barbara; Michael B. McGrath of the Colorado School of Mines; and Osama Soliman of the University of Tennessee.

A very special note of appreciation is extended to my wife Marilyn. Her help with many of the aspects involved in manuscript preparation, as well as her cheerful support, encouragement, and good advice are gratefully acknowledged.

Bruce R. Dewey

1
Systems

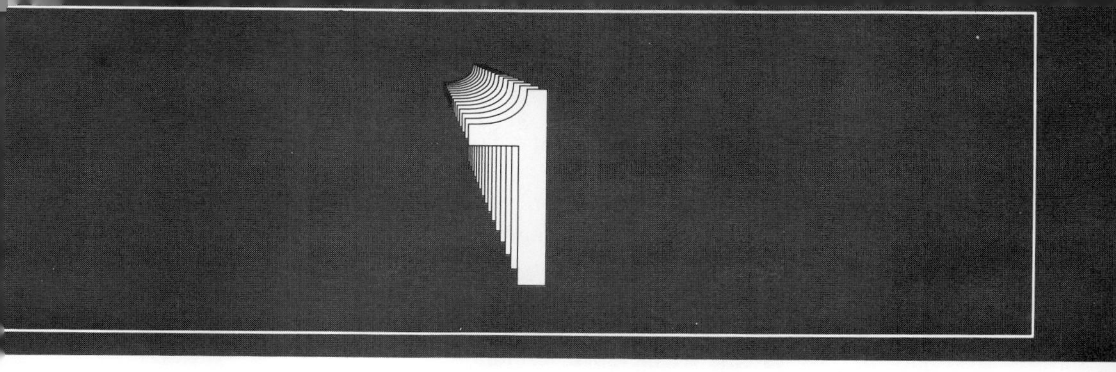

Introduction to Hardware

An alphabet soup of computer acronyms dances before the eyes of today's engineers—CAD, CAM, CIM, CAE, AI, DOS, APT, CPM, PERT, CADD, PL1, CSG, and many, many others. The hundreds of such acronyms are just one indication of how widespread the use of digital computing has become in all of engineering. Computer systems crunch numbers and draw pictures with amazing speed, freeing the engineer from drudgery and permitting work at a higher cognitive level.

1.1 COMPUTER-AIDED ENGINEERING

Computer-aided engineering (CAE) encompasses all design and production activities in engineering. CAE includes drafting, analysis, synthesis, modeling, testing, word processing, data acquisition, real-time control, numerical computation, and graphics.

Early use of the computer in engineering involved analysis or "number crunching," where specialists solved large systems of equations in fluid mechanics, aerodynamics, electrical networks, stress analysis, and the like—work that before would have been impossible. Development of computer graphics was driven by the need for understanding and communicating numerical data.

Today, there are computers and workstations with excellent and easy-to-use software systems for virtually any CAE task. Computer graphics, a well-developed technology that is central to all CAE, can be classified according to three principal functions:

Data presentation graphics accomplishes communication of ideas and numerical data to others. In addition to two-dimensional data plots, data may be presented with three-dimensional surface plots, contour plots, pie charts, maps, and bar charts.

Computer-aided drafting (CAD) prepares, copies, and modifies the line work, text, notes, and dimensions of engineering drawings. A characteristic of CAD systems is the dimensionally accurate data base, which allows production of accurately scaled drawings. When CAD systems are customized for various disciplines, different libraries of standard symbols, dimensioning features, and links to analysis systems are included. Commonly, but inaccurately, the acronym CAD is used for "computer-aided design." However, most so-called CAD systems really do only drafting; systems that do more design functions are known by other acronyms such as CADD for "computer-aided drafting and design," CIM for "computer-integrated manufacturing," or simply CAE.

Solid modeling is used in conceptual studies and layout of systems in space. Here, geometric data are described in such a way that other operations (e.g., determination of enclosed volume) are possible. A distinction between CAD and solid modeling can be made by the way the two different systems treat data. In CAD, operations (e.g., draw a straight line from point A to point B) and parameter setting (e.g., make the line green) characterize the data structure. In solid modeling, vertices, edges, surfaces, and volumes are created, manipulated, and modified in three dimensions.

Most engineering computer graphics systems can readily be classified as presentation graphics, CAD, or solid modeling, although there may be overlap between these categories. A CAD system almost always has features associated with a solid modeling system. A CAD or a solid modeling system may be used to prepare presentation graphics. An application such as a flowchart may be produced on either a CAD system or a presentation graphics system. An analysis system, such as one for finite element stress analysis, may have features of solid modeling for preparing input data and of presentation graphics for showing results. A comprehensive system for electrical circuit design may have CAD features for the input of schematic diagrams and for the preparation of printed circuit board artwork, and presentation graphics for the output of simulation results.

Computer graphics is the common element in all CAE. Even though CAE applications have wide diversity, the embedded computer graphics systems have remarkable similarity. Just how an engineering organization gets CAE in place is an inexact, evolutionary process. Whereas in the early days of CAE a preference existed for "turnkey" systems, where one vendor integrated all hardware and software, the contemporary approach is the in-house integration of CAE systems with a view toward sharing data and standardizing equipment as much as possible. Although, ideally, software should be identified before hardware is selected, it often happens that existing hardware resources dictate the choice of new software.

Even though there is a wide variety of computer graphics hardware, basic aspects of display technology are similar. The brief descriptions that follow are intended to survey the basic distinctions among various types of hardware.

1.2 WORKSTATION DISPLAY TECHNOLOGY

The term *graphics workstation* refers to computer equipment used for the input and output of graphics information. Workstations have internal processing capability as well as access to a remote computer or other workstations for augmentation of processing, enhancing display capabilities, or sharing data. In contrast, a *terminal* is an input-output device of limited capability where virtually all processing is done by a remote computer. For cost effectiveness, high performance, and ease of expansion, the trend is away from communication links and shared computer resources toward location of processors and storage in workstations.

Display technologies found in graphics workstations may be classified according to the support of interactive output, hard-copy output, and interactive input. The brief descriptions that follow illustrate some important features of interest to users of display technology. Detailed information is available in published literature[14,39] and in brochures and manuals available from various manufacturers.

1.3 INTERACTIVE OUTPUT

Almost all interactive output in computer graphics is on some type of cathode ray tube (CRT) display. As shown schematically in Figure 1.1, a CRT directs a stream of electrons to a screen, which is coated inside with a phosphor. Three technologies, *stroke-refresh, direct-view storage tube* (DVST), and *raster scan,* differ in how images are created on the screen. While all three technologies have certain advantages and drawbacks, raster scan has become the most popular.

The stroke-refresh display technology was used in the earliest interactive computer graphics systems, where the display was made from an oscilloscope

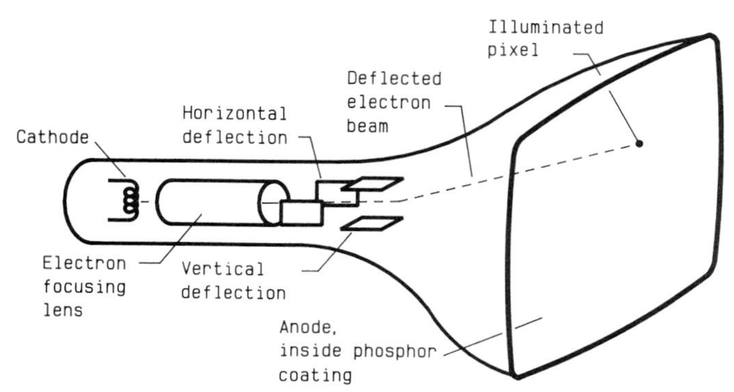

Figure 1.1 Cathode ray tube. Electrons from the heated cathode are attracted to the positively charged anode. The phosphor coating is selected for color and persistence characteristics. Elements within the CRT focus and steer the beam under processor control.

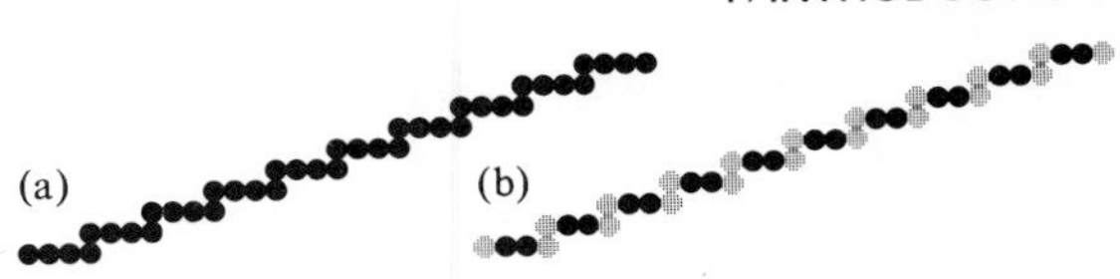

Figure 1.2 Rasterized line segment or "vector." Antialiasing circuitry, incorporated in some display systems, improves the appearance of an oblique line by partially illuminating the pixels that represent an abrupt change in a horizontal scan line. (a) Aliased; (b) antialiased.

driven by digital-to-analog converters for the horizontal and vertical deflection systems. The inside of the glass screen is coated with a very-short-persistence phosphor, and the position of the beam is continuously variable, because the voltage driving the deflection is continuously variable. Thus, very smooth lines in any direction may be displayed. The short persistence of the phosphor means that the picture must be redrawn (refreshed) several times per second. If the refresh rate is less than 30 times per second or so, the image appears to flicker.

Advantages of stroke-refresh displays include their "vector" nature (meaning that straight-line segments are produced by merely connecting two end points) and high resolution. The rapid drawing and redrawing makes these displays ideal for dynamically changing images. However, stroke-refresh displays are relatively expensive because of the inherent processing complexity. Since it is not uncommon to have an image with 4000 to 6000 vectors, it can be seen that the controller should be capable of processing 180,000 (6000 × 30) vectors per second or more.

Introduction of the DVST in 1968 provided a simpler and far less expensive graphics display technology. The phosphor in the DVST is bistable, having long-persistence and short-persistence modes. Vectors written on the screen remain until erasure by a flood of electrons, which causes the phosphors to return to the dark state. Since the screen itself stores the image and each vector must be processed only once, the processor in the terminal is relatively simple. Among the

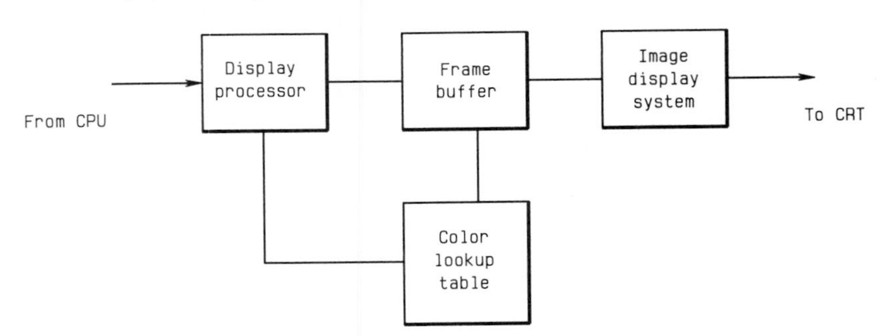

Figure 1.3 Display controller for raster scan system. The display processor acts on commands from the central processor unit (CPU), generating vectors and text as well as addressing the frame buffer and color table memories. Many other functions can be added to display processors, including shading, clipping, and image transformation.

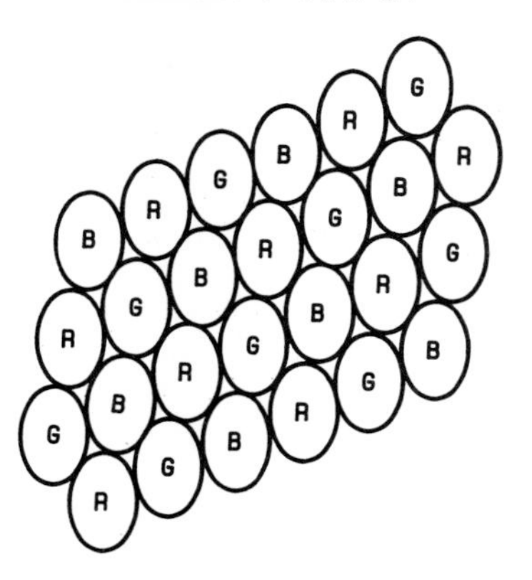

Figure 1.4 Schematic of screen in a color raster system.

disadvantages are lack of color (green is the only color of practical bistable phosphors), a relatively dim screen, and lack of animation capability.

The terms *calligraphic,* which refers to line drawing, and *random scan* are sometimes applied to the stroke-refresh and DVST displays, which draw lines between arbitrary points on the screen. In contrast, the row-by-row production of information typifies a *raster scan* device. This is essentially a point-plotting device, since it displays line segments with the dots (pixels) available from the raster, as shown in Figure 1.2. The points are illuminated in a fixed order, proceeding from left to right and top to bottom as the electron gun scans the screen. Line segments (or vectors) other than perfectly horizontal or vertical ones have "jaggies" or stairstepping, although some hardware does "antialiasing," Figure 1.2(b), to smooth the appearance of diagonal lines.

An *interlaced* display, like a home television, scans every other row of the screen in 1/60 second, then scans the skipped rows, completing all of the rows in 1/30 second. A *noninterlaced* display scans the rows in order, taking 1/60 second for a complete repetition. The latter technology—60 Hz noninterlaced—is preferred for professional use because it gives better picture clarity and less flicker. A schematic diagram of the display controller for a raster scan display system is shown in Figure 1.3.

Among the advantages of raster-scan displays are versatility and low cost. Features such as shading, a wide range of colors, and the possibility of animation have made raster-scan displays the most popular type. Color displays are produced by using screen phosphors for red, green, and blue (R, G, and B) as shown in Figure 1.4. The relative amounts of R, G, and B in each pixel are controlled by the memory area in the workstation called the color lookup table. More information on the use of color is given in Section 2.7.

1.4 HARD-COPY OUTPUT

Like output displays, devices for recording graphical information on paper and film can be classified as random-scan or raster-scan devices.

Even though the pen plotter is the earliest type of hard-copy output device, it is still very widely used. The random-scan process is accomplished by moving a pen on the paper under computer control. The pen is "down" to draw and "up" to be repositioned without drawing. Relative motion between pen and paper may be accomplished in two different ways, as shown in Figure 1.5.

The random-scan display technology is also used in computer output microfilm (COM) recorders, which provide for focusing the CRT image onto photographic film. Best-quality images from COM recorders result from using a monochromatic (white) calligraphic display. Colors, separated into red, green, and blue in software, are recorded one at a time with the insertion of appropriately colored filters between the screen and the film.

Simple film recorders and laser (or electrophotographic), dot-matrix, electrostatic, thermal, and ink-jet printer-plotters use raster-scan technology. The more expensive devices in this category, such as color electrostatic plotters, produce large continuous tone images of high quality. Processing is minimized if a "screen dump" is used, that is, if there is a one-to-one copy of the pixels on the screen to the paper. Typical screen resolutions are 100 dots per inch or less, which results in noticeable jaggies in hard copies. In contrast, high-resolution raster-scan hard-copy devices need their own processors to resolve the 300 dots per inch or more needed to avoid jaggies.

The inks and dyes used to make hard copies do not perfectly match the colors produced on CRT screens. Furthermore, the black background on the CRT and the white background on paper make striking differences in the appearance of images.

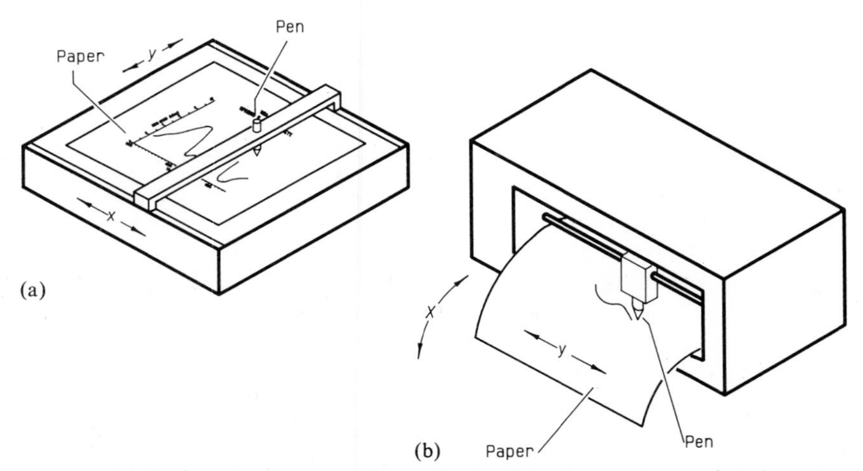

Figure 1.5 Schematic diagram of pen plotter. Stepping motors under computer control accomplish x and y motion. A solenoid positions pen "up" or pen "down." Colors and pen size are changed by swapping pens from a storage area (not shown). (a) Pen is moved in x and y directions; (b) pen is moved in y direction and paper is moved in x direction.

1.5 INTERACTIVE INPUT

Graphic and alphanumeric input occurs through the *logical input devices* associated with a workstation. The term logical input device describes an interaction technique as opposed to specific hardware. This concept provides maximum versatility and portability of graphics applications programs, since the range of usable equipment is greatly expanded. The six classes of logical input, as defined by the Graphical Kernel System (GKS), which is discussed in Section 2.10, are locator, stroke, valuator, choice, pick, and string. User inputs should always be acknowledged by the occurrence of some visual or audible action.

Locator

User specification of a point on the display surface or on an auxiliary tablet is done with the locator. If the display surface is the screen, a cross hair or other cursor can be manipulated with one or more of the hardware devices shown in Figure 1.6.

One use of the locator is in copying drawings that exist on paper with a *digitizing tablet.* The material to be copied is secured to a surface with an embedded sensing system, such as a grid of wires. Points, selected by use of an electrically actuated hand-held puck or stylus, are converted to coordinate data with an analog-to-digital converter. Although a digitizing tablet can support a mouse function to move a cursor on the screen, the mouse does not have the positional accuracy to digitize drawings.

A *light pen* can be used as a locator, although it is more commonly associated with pick, as described below. Essentially, a light pen is a fast-response

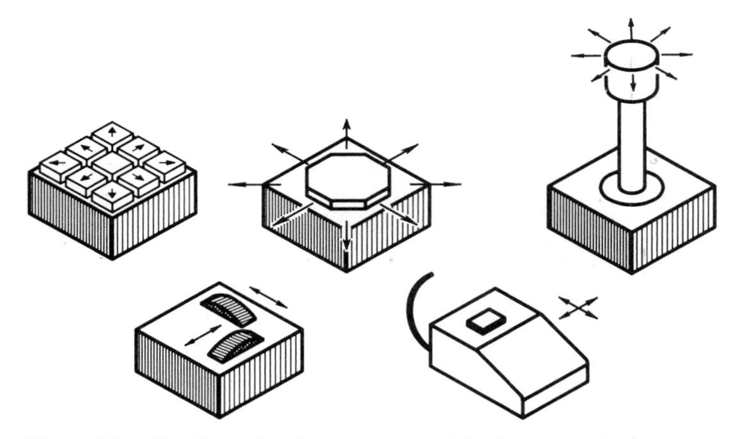

Figure 1.6 Hardware implementations of the locator include cursor keys, joydisk, joystick, thumbwheels, and mouse. To send the location to the processor, the space bar, some other designated key on the keyboard, or a button is depressed.

photomultiplier tube connected to timing circuitry. In use, the light pen is held against the screen and actuated by pressing a button. Because each pixel on the screen is written at a discrete time, timing information can be used to compute the horizontal and vertical coordinates.

Stroke

The stroke logical input is very much like the locator except that a continuous stream of data is input to the application program. A tablet is the most practicable device for stroke input. Designed to capture the dynamics of the operator's input, the stroke input usually is implemented to sample up to a predetermined maximum number of points at even time increments. Stroke is probably the least used of the standard logical inputs.

Valuator

This device allows the user to generate a floating-point number, which is returned to the application program. Common on microcomputers is the "game paddle," a potentiometer that generates a number for program functions. In interactive graphics, the valuator might be used to select the viewing angle for a three-dimensional image or vary an input parameter for a simulation.

Choice

In contrast to the valuator, the choice (or button) logical input device returns a single integer to the application program. A user's access to choice input in most systems includes selecting a menu item with the screen cursor, pressing a function key, or typing an integer number on the keyboard. Some specialized computer graphics systems have an auxiliary keyboard on which the buttons have functions assigned under program control.

Some pucks for digitizing tablets and some mice have more than one button that can be used for choice input commands assigned under program control. Screen menus or tablet menus for selection of drawing commands are the common way to give the user choice input in CAD systems. Choice input commands can also be given with instructions typed according to some specified format. However, menu choices provide an easily learned user interface, because a special command language does not have to be mastered.

Pick

Pick is used to select segments that appear on the screen. A segment can be a point, a line, a group of lines, a certain area, or any other cohesive part of a displayed image. User feedback is provided by flashing or highlighting the segment picked. It turns out that the pick logical input is invoked whenever a screen menu choice is made with the graphics cursor. Some of the hardware input devices used for locator can also be used for pick, with program logic determining the segment nearest the located point. When the search for the nearest segment is implemented in software, it can be computationally intensive.

A light pen coupled with a stroke-refresh display is often used for implementation of the pick logical input function. Because a stroke-refresh system rapidly repeats the display of an image segment by segment, a signal from the photomultiplier tube indicates that the segment under the pen is currently being redisplayed and is the one to be "picked."

Ivan Sutherland, the acknowledged "inventor" of computer graphics, developed a stroke-refresh terminal with a light pen in the early 1960s. Today, light pens are becoming less popular because of the high cost of stroke-refresh hardware and the user fatigue associated with using the pen for many hours continuously.

String

String, which is entry of information from the typewriter keyboard, is the usual way to send characters to the application program. Character strings define text on drawings, file names, and commands.

It is common to emulate the other logical inputs with use of the keyboard. For example, the screen cursor can be moved with any software-selected keys to emulate locator. Valuator and choice, which are associated with the input of a number, can be emulated with a keyboard simply by typing in a number. Pick, often associated with selection of command options, can also be done with typed inputs.

1.6 CLOSURE

A diverse range of computer graphics hardware exists. Technology is steadily providing improved equipment at more cost-effective prices. Computer graphics periodicals and trade meetings are good ways to keep up with the changes.

PROBLEMS

1.1. The common asynchronous serial transmission protocol used between host computers and many graphics workstations is rated in baud or bits per second. Assume that the command from the host to the workstation to draw a vector requires five 8-bit bytes. What is the maximum number of vectors per second that can be drawn if the communication rate is 9600 baud?

1.2. Of the logical input devices, stroke finds the least use. Propose some reasons for this.

1.3. Explain how a tablet can be used for (a) locator, (b) stroke, (c) valuator, (d) choice, (e) string input.

1.4. A light pen is to be used as a locator on a noninterlaced 60-Hz raster device with a screen resolution of 1280×1024 pixels. What timing accuracy, in microseconds, is required?

PROJECT

1.1. Prepare an inventory of the computer graphics workstation hardware available in your laboratory. List the important technical characteristics, including

Communication speed with host (if any).

Resolution of CRT screens (displayed rows and columns).

Number of colors simultaneously displayable and number of available colors in the palette for CRT displays.

Logical input devices supported.

Resolution specification of digitizers.

Possible sizes of hard copy.

Number of different pens and resolution of pen plotters.

Cost information.

Introduction to Graphics Systems

A computer graphics system contains a great quantity of software. This software may be classified as (1) computer operating systems, (2) utilities, (3) drivers for graphics workstations, and (4) applications programs.

An operating system is the basic software that controls computer functions. Utilities include text editors, data base managers, interpreters, compilers, and linkage editors. In-depth acquaintance with operating systems and utilities is generally necessary for engineers developing computer graphics systems, but not for users. Machine-specific operating systems require software developed on one computer system to be converted to run on another. Operating system standardization, such as that provided with the UNIX and MS-DOS/PC-DOS operating systems, provides many advantages to developers and users alike.

Drivers, specialized codes used to control graphics devices, also lack standardization. De facto standards have evolved from the drivers for such popular equipment as Hewlett-Packard plotters and the Tektronix 4010, a DVST terminal dating back to the 1960s. Standardization has addressed applications programming and the use of *metafiles* for machine-independent interchange of graphics information.

Applications programming standards specify how basic graphics actions are incorporated in programs written in high-level languages like FORTRAN-77. Examples of graphics actions are definition of coordinate systems, output and input primitives, and rendition of images. There are three recognized programming systems: Core, Graphical Kernel System (GKS), and Programmer's Hierarchical Interactive Graphics Standard (PHIGS). For various reasons, however, much of the programming for graphics applications is not written in strict concurrence with one of these systems.

This chapter treats basic concepts used in applications programs, such as generation of lines and other graphic entities, adjustment of images to the work surface, removal of hidden lines and surfaces, control of color and shading, and animation.

2.1 DISPLAY GENERATION

A minimum understanding of how computer graphics workstations generate images is helpful for all users.

The most basic concept in computer graphics software is drawing a point, which is located by its horizontal and vertical coordinates. A straight line is an array of points. By hand, a line is drawn on paper with (1) a *move* of the pencil to the starting point without touching the paper and (2) a *draw* to the stopping point with the pencil in contact with the paper. Move and draw graphics is also basic to computer-generated images, where the coordinates of the two end points are contained in the calls to move and draw.

Coordinate Systems

Engineers are familiar with various coordinate systems, such as Cartesian, polar, and spherical coordinates. For computer graphics, Cartesian coordinates are the only practical system; usually, coordinates in other systems are changed to Cartesian coordinates before any transformation or other manipulation is done.

Coordinates can be classified as *device coordinates* or *world coordinates*. The conversion between device (or screen) coordinates and world (or user) coordinates is done with the windowing transformation described in Section 2.4. The convention for device coordinates (u, v) is shown in Figure 2.1.

It should be noted that screen coordinates are two-dimensional (2-D), and by convention always in the u-v plane. In most cases, u-v coordinates accepted by workstations are in integer format. Since world coordinates are three-dimensional (3-D), any image on the screen is simply an *orthographic projection*. An orthographic projection into the x-y plane of an object described

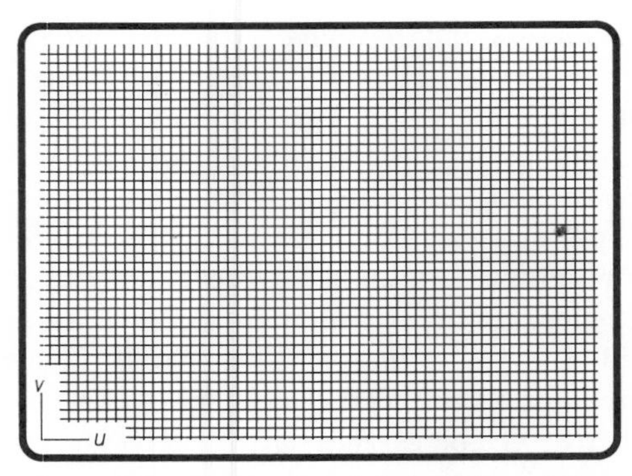

Figure 2.1 Conventional designation of Cartesian screen coordinates. In some systems the origin is at the upper left corner.

in world (x, y, z) coordinates is trivially constructed by simply disregarding the z coordinates. Similarly, an orthographic projection into the y-z plane is constructed by disregarding the x coordinates.

Pixels and Resolution

The smallest addressable cell on the computer graphics screen is called a *pixel,* or "picture element." In a raster terminal a pixel is simply one of the "dots" making up the image, while in a DVST or stroke-refresh display a pixel is the smallest spot that can be addressed and displayed.

A *bit map* is a pixel-by-pixel description of an image. Each pixel may have a specified state, "on" or "off," or specified intensity or color. In most raster graphics displays, the bit map is stored in a *frame buffer* or *display memory.* The resolution of a raster graphics screen is given by the number of rows and the number of columns of pixels; for other workstations resolution measures how precisely a point is addressed.

The generation of *dot matrix* text is a good illustration of how pixels may be bit-mapped to generate images. A typical alphanumeric display screen generates 80 columns of 6-pixel-wide character cells with 480 pixels horizontally on the screen. Simple characters require 5 pixels horizontally, plus a 1-pixel gap between characters to ensure readability. Typically, uppercase characters are 7 pixels high and the complete character cell is 10 pixels high. Higher-quality text than that shown in Figure 2.2 can be generated if more pixels are allocated for each character. For compatibility with plotters and other stroke-type devices, raster graphics terminals internally generate the bit maps from lines, characters, and other primitives.

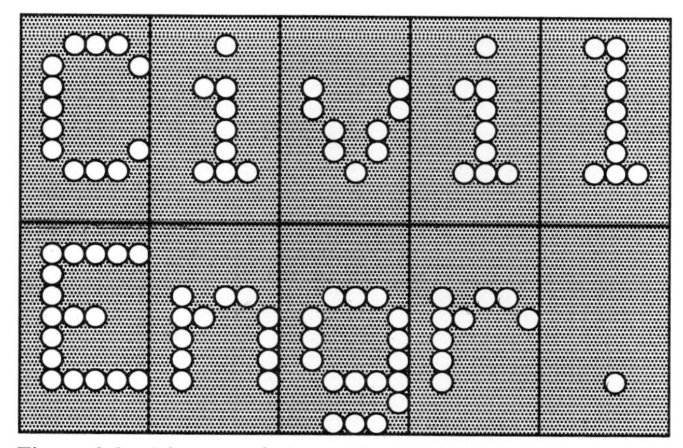

Figure 2.2 Bit map of simple dot matrix text using 6 × 10 character cells on a monochromatic display. In the frame buffer, black pixels are set to 0 and white pixels are set to 1. With nonproportional spacing, the space between narrow letters ("i" and "l") is large compared to that between most letters.

Vector Line Generation

Storage of graphics images and programming complexity are greatly reduced by use of *vector graphics,* in which line segments are defined by their end points. For example, a line that bit-maps into 200 pixels can be described as a vector with only two data. Most applications employ vector graphics for simplicity, economy, speed, and transportability. In DVSTs, stoke-refresh terminals, and pen plotters, vectors can be displayed directly. However, in raster terminals and dot matrix plotters, vectors are converted to bit-map form, or "rasterized," by the internal display processor of the graphics workstation, which runs an algorithm known as the digital differential analyzer.[29,39]

In an applications program, creation of a vector involves one or two instructions. A common scheme is to "move" to one end of the vector and then "draw" with a straight line to the other end. In the case of a pen plotter, the move instruction is done with the pen "up" and the draw with the pen "down." If another vector is to be drawn starting where the first vector ended, all that is necessary is another draw instruction. Thus, a circle that is drawn with, say, 60 chords (straight line segments) requires 61 commands (one move and 60 draws).

Although one can generate an image with a pixel-by-pixel description, the move and draw instructions of vector graphics provide relief from very tedious programming. The output primitives described in the next section represent an even higher level of function than move and draw.

2.2 OUTPUT PRIMITIVES

Graphics output can be accomplished with the four standard output primitives—*line, polygon, marker,* and *text*. Intelligent workstations create these and sometimes other primitives from instruction codes that contain a minimum amount of information. Besides these four, there are two less frequently used output primitives. The *cell array* permits direct specification of the bit map for any arbitrary rectangular region. The *generalized drawing primitive* provides a means of accessing a nonstandard capability such as a circle or interpolating curve. The standard output primitives represent higher-level commands than the move and draw commands described in the preceding section.

Line

As an output primitive, *line* means a straight line segment, the shortest distance between two points. Three or more points connected with straight line segments define a *polyline,* which is a GKS (Section 2.10) output primitive.

For convenience in manipulation, points will be described in matrix notation. A *points matrix* for one point gives the Cartesian coordinates of the point as a row matrix or vector,

$$[\mathbf{P}] = [x \quad y \quad z] \tag{2.1}$$

In the case of n points being stored in a points matrix, the convention is expanded to an $n \times 3$ points matrix,

$$[\mathbf{P}] = \begin{bmatrix} x_1 & y_1 & z_1 \\ x_2 & y_2 & z_2 \\ \cdot & \cdot \; \cdot \; \cdot & \cdot \\ x_n & y_n & z_n \end{bmatrix} \tag{2.2}$$

For 2-D graphics, conventionally all the z's are zero. Thus, a 2-D points matrix might have the dimension $n \times 2$.

Some graphics systems avoid the use of doubly dimensioned arrays for points matrices. In this case, the matrix of Eq. (2.2) is partitioned into three column matrics,

$$[\mathbf{P}] = [[\mathbf{X}] \quad [\mathbf{Y}] \quad [\mathbf{Z}]] \tag{2.3}$$

where [**X**], [**Y**], [**Z**] each have the single dimension of n. For compactness and convenience, the doubly dimensioned points matrix convention will be used subsequently in this text.

The n-row points matrix is convenient for definition of the *polyline* produced by a move to the point of the first row of [**P**] and subsequent draws to the rest of the points.

Lines and polylines do not necessarily have to be drawn with solid lines. Various broken-line styles such as those shown in Figure 2.3 are typical of attributes supported by intelligent workstations. Broken lines can be produced by using an attribute tagged to the line as well as by creating and storing separate line segments. Obviously, using attributes saves a great deal of overhead.

Figure 2.3 Typical line types. Numbered styles are GKS standards.

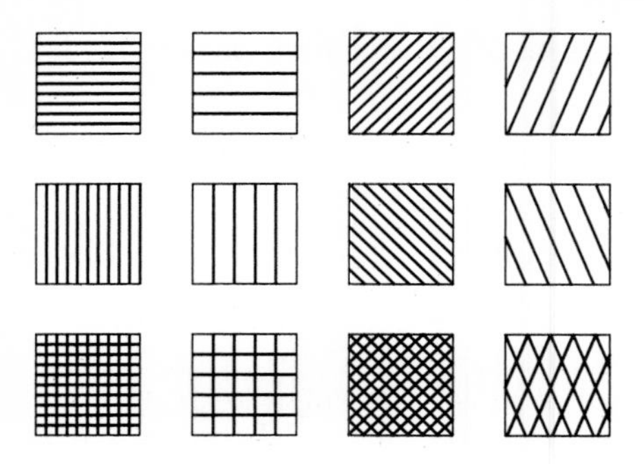

Figure 2.4 Sample hatch patterns available for panel filling in GKS.

Polygon

Logically, a polyline can be extended to a *polygon* or *panel* by closing the figure with one additional draw back to the first point in [**P**]. Because a polygon defines a bounded area, a "filled panel" primitive is easily implemented. The panel filling built into intelligent workstations has attributes covering a range of colors and patterns. The set of hatch patterns in Figure 2.4 is typical of those available for panel filling.

A *tessellated* or *faceted* 3-D surface can be built from polygons with shared edges and vertices. The resulting surface is a *polyhedron* if it completely encloses a volume. The surface of a cube, for example, is formed by six squares, which have 12 shared edges and eight shared vertices. One method of representing a tessellated surface uses two matrices: the points matrix, Eq. (2.2), specifies the vertices, and a *topology matrix* lists the vertices bounding each of the polygons. Examples of a points matrix and a topology matrix for a tetrahedron are shown in Figure 2.5.

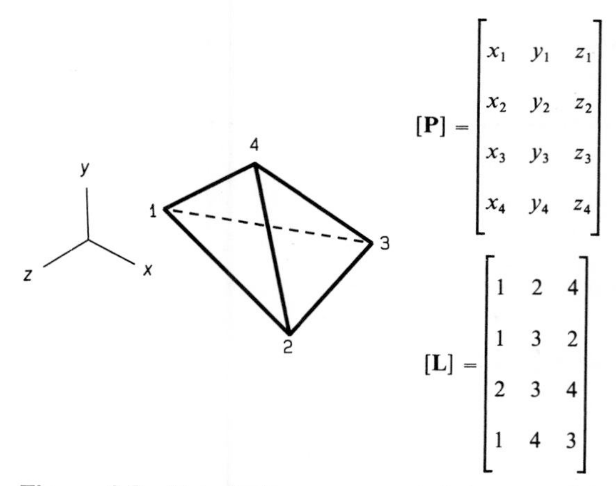

$$[\mathbf{P}] = \begin{bmatrix} x_1 & y_1 & z_1 \\ x_2 & y_2 & z_2 \\ x_3 & y_3 & z_3 \\ x_4 & y_4 & z_4 \end{bmatrix}$$

$$[\mathbf{L}] = \begin{bmatrix} 1 & 2 & 4 \\ 1 & 3 & 2 \\ 2 & 3 & 4 \\ 1 & 4 & 3 \end{bmatrix}$$

Figure 2.5 Tetrahedron represented by a points matrix and a topology matrix. Arbitrarily, the vertices in the topology matrix are numbered in counterclockwise order as viewed from the outside.

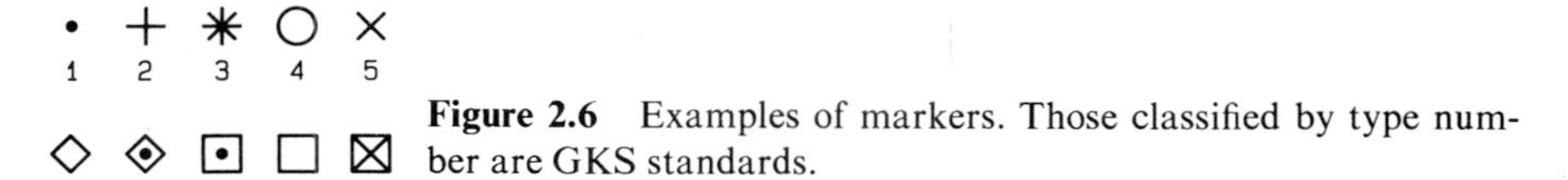

Figure 2.6 Examples of markers. Those classified by type number are GKS standards.

Marker

The *marker* is convenient for drawing data points in presentation graphics. The GKS output primitive is the *polymarker,* which places the same type of marker at all points listed in a points matrix. A selection of standard markers is shown in Figure 2.6.

If the marker is supported in the workstation intelligence, the system sets the desired attributes and executes a move to the desired position to render a marker.

Text

Whether or not the text is generated in software furnishes one way to classify text. *Software text* is generated stroke by stroke in the processor under program control. A relatively large amount of data is processed and transmitted in producing software text.

Hardware text is generated from instructions in the workstation memory. Hardware text may be permanently stored in the workstation memory or, for more versatility, may be either downloaded from the processor or read from plug-in modules. After the necessary attributes are set, only one byte (seven or eight bits) is needed to describe a character. For reference, text bit codes are given in Appendix D. The use of hardware text speeds up drawing by reducing

Figure 2.7 Examples of proportionally spaced stroke text in order of increasing complexity. The terms simplex, duplex, and triplex suggest the number of strokes in each line making up the characters. The exact names and styles of text fonts vary from one computer graphics system to another.

FUTURA
SERIF
FASHON
LOGOI
SWISSL
SWISSM
SWISSB

Figure 2.8 Examples of shaded fonts. The names and fonts are from the proprietary DISSPLA library produced by ISSCO, San Diego, California.

the amount of processing and transmitting data. Hardware text often produces the most legible characters in raster displays, since the text is bit-mapped exactly to the pixel spacing.

The term *font* refers to the style of text. Fonts of *stroke text,* Figure 2.7, are formed by line drawing. *Shaded* or *filled* text fonts, Figure 2.8, are drawn in outline and filled in the same way as polygons. The exact names applied to different fonts vary from one graphics system to another. The hardware text often found in workstations is similar to the Cartographic Utility font seen in Figure 2.7.

For engineering work, elaborate text fonts should be avoided. Font choice depends on final size, quality level, and speed of production. Text formed by a minimum number of line segments, as in the Cartographic Utility font, is satisfactory in small sizes. Text with more line segments, such as the Simplex font, is a better choice in medium sizes. Larger text may appear too light unless a font with extra line segments, such as the Duplex font, is used. Shaded fonts are an optimum choice for large text of highest quality.

Three distinct levels of *text precision* may be specified: string, character, and stroke. In choosing among these, the inevitable trade-off between quality and speed exists.

1. *String precision* permits fastest generation, since only the starting location of text is specified. For simplicity, string precision text is usually not proportionally spaced (e.g., Figure 2.2).
2. *Character precision* allows character-by-character control of text placement. Text may be proportionally spaced.
3. *Stroke precision* can produce text of the highest quality. Each character is generated exactly in software, with full control of all the attributes discussed below.

Expansion is the ratio of the width to the height of text. This ratio is smaller for "compressed" text, which permits more characters in a string of given height and length, and larger for "expanded" text. The closeness of one character to another in a string is controlled by *spacing*. The direction of the text line can be set so that it is horizontal, vertical, or at any arbitrary angle. The text slant is defined as the angle between the edges of a character such as "H" and a line perpendicular to the text line. Normal text has a slant of 0° and the usual italic text has a slant of about 20°. The position of the point that specifies the location of text relative to the text line is *alignment*. Possible specifications include

> Horizontal alignment: where the text position is located by the left end (normal), center, or right end of the string or character.
>
> Vertical alignment: where the text position is located by the top, the center, or the bottom of the string.

Most engineering computer graphics systems provide string precision hardware text in a simple font that gives clear, readable results. Users of these systems will be concerned only with the size and alignment of text.

2.3 WINDOW AND VIEWPORT

One of the functions of a graphics driver program is the conversion from world coordinates to device coordinates. Program control of the size and location of an image on the display surface is accomplished by the window and viewport.

The *window* specifies the minimum and maximum values of world coordinates that are to be displayed. The *viewport* is the rectangular area on the screen specified in device coordinates where the window is displayed. The default viewport is the entire screen. The relation between window and viewport is shown in Figure 2.9, where the world coordinates are in the *x*-*y* coordinate system and the device coordinates are in the *u*-*v* coordinate system.

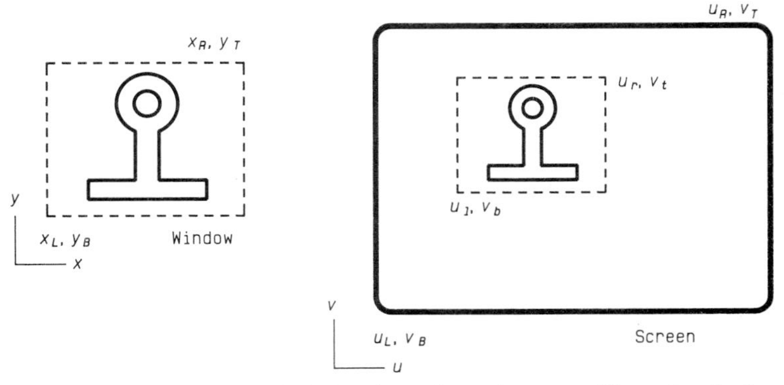

Figure 2.9 Window and device-dependent viewport. Since the device surface is bounded by u_L, v_B and u_R, v_T, the viewport must be in the range $u_L \le u_l < u_r \le u_R$ and $v_B \le v_b < v_t \le v_T$. In many systems $u_L = v_B = 0$.

The transformation between world coordinates and device coordinates in terms of the window and viewport as shown in Figure 2.9 can be written as

$$u = \frac{u_r - u_l}{x_R - x_L}(x - x_L) + u_l$$

and (2.4)

$$v = \frac{v_t - v_b}{y_T - y_B}(y - y_B) + v_b$$

Since, for display of any one image, the viewport and window do not change, it is convenient to collect the constant terms in Eq. (2.4) and compute them just once, so that the transformation is

$$u = ax + b \quad \text{and} \quad v = cy + d \tag{2.5}$$

Use of these transformations is *device-dependent,* which means that the user must specify the viewport in terms of the screen coordinates of a particular terminal or plotter.

For transportability between various workstations, standards (see Section 2.10) specify that all information including the viewport be in *normalized device coordinates* (NDC). In NDC the viewport is bounded according to $0 \leq u_l < u_r \leq 1$ and $0 \leq v_b < v_t \leq 1$. Thus, Eqs. (2.4) produce an intermediate result labeled u_i, v_i, where $0 \leq u_i \leq 1$ and $0 \leq v_i \leq 1$, instead of u, v, which are bounded by the device-dependent range of screen addresses. Conversion from NDC to actual device coordinates is done by the additional step

$$\begin{aligned} u &= u_i(u_R - u_L) + u_L \\ v &= v_i(v_T - v_B) + v_B \end{aligned} \tag{2.6}$$

which may be simplified for many devices where $u_L = v_B = 0$. The extra step involved in the use of NDC is shown in Figure 2.10.

Special attention should be given to the proper aspect ratio of the horizontal and vertical dimensions of windows and device-dependent viewports if images are not to be distorted. This requires that the window and the viewport be in the same proportion. If the viewport is the whole screen, the width-to-height ratios of the window and screen must be equal. Commonly, this aspect ratio is 4 horizontal to 3 vertical. Distortion of the aspect ratio is unacceptable for engineering drawings; for example, circles appear as ellipses. However, for data presentation plots, the window and viewport may have different proportions to fit the image into the desired space.

Conventionally, device-independent viewports produce undistorted images if the window and the viewport (in NDC) are both square. Since this produces an aspect ratio of 1 to 1, the final mapping to device coordinates must be done with a modified version of Eq. (2.6),

$$u = u_i v_T \quad \text{and} \quad v = v_i v_T \tag{2.7}$$

where $u_L = v_B = 0$ and $u_R \geq v_T$. Note that using Eq. (2.7), instead of Eq. (2.6), will utilize only a square region on the screen.

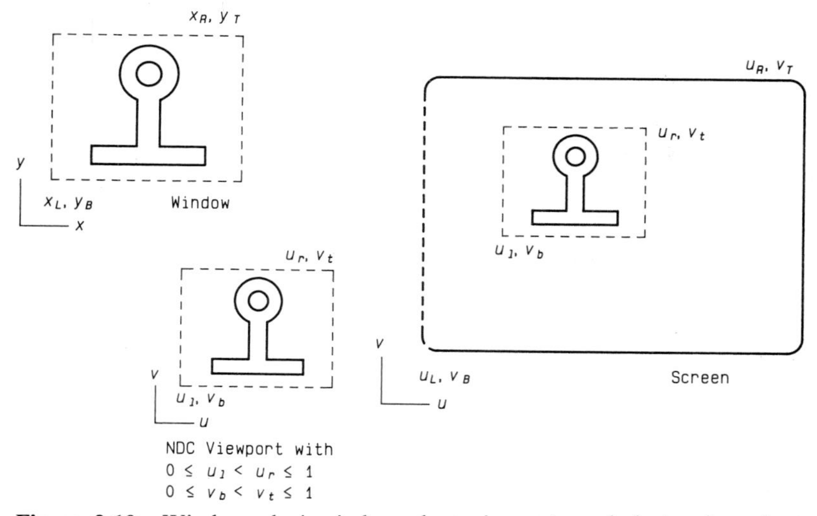

Figure 2.10 Window, device-independent viewport, and device-dependent viewport. The intermediate image in NDC is converted for display on a specific device with Eq. (2.6).

One solution for the apparent inconsistency between Eqs. (2.6) and (2.7) is to make the ranges of vertical and horizontal screen addresses on the device equal. Because the screen is wider than it is high, points in the shaded region of Figure 2.11 can be addressed but not displayed.

With some systems, attempting a plot device coordinates that are out of range may lead to unpredictable results. Clipping, described in the next section, eliminates any such portions of images.

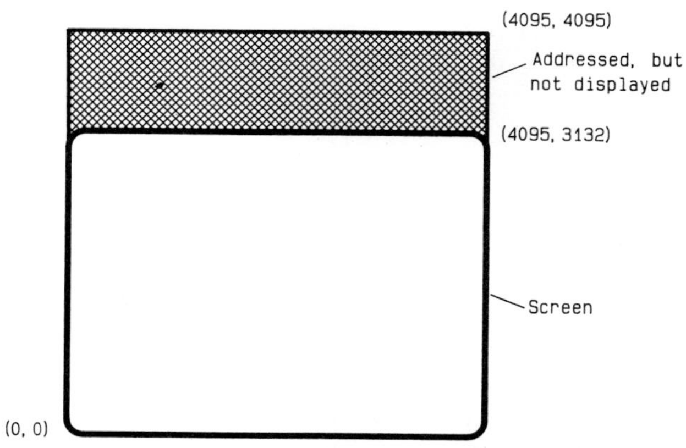

Figure 2.11 Example of device coordinates that address a square region, with only a rectangular subregion being displayed.

EXAMPLE 2.1

A 150 mm × 120 mm printed circuit board (PCB) is to be displayed on a screen with an aspect ratio of 4 horizontal to 3 vertical. It is desired to have the image centered on the screen with the 120 mm dimension vertical and at least 20 mm empty space on all sides. Furthermore, the image is to be as large as possible while keeping the aspect ratio undistorted. Determine the proper size for the window, if the viewport defaults to the whole screen and the lower left corner of the image of the PCB is located at (0, 0).

Solution. The aspect ratio of the entire drawing with the required edge space is $(150 + 20 + 20)/(120 + 20 + 20) = 1.19$. Since this is less than 4/3, the vertical dimension of the PCB controls the placement of the window. It follows that $y_B = -20$ and $y_T = 140$, where the y coordinate of the bottom of the image is 0. For no distortion, the height of the window just determined, 160, dictates that the width is $(4/3) \times 160 = 213.3$. To center the image horizontally, the space on either side must be $(213.3 - 150)/2 = 31.7$. If the x coordinate of the left side of the image is to be 0, the window coordinates are $x_L = -31.7$ and $x_R = (213.3 - 31.7) = 181.6$. The display will appear as illustrated.

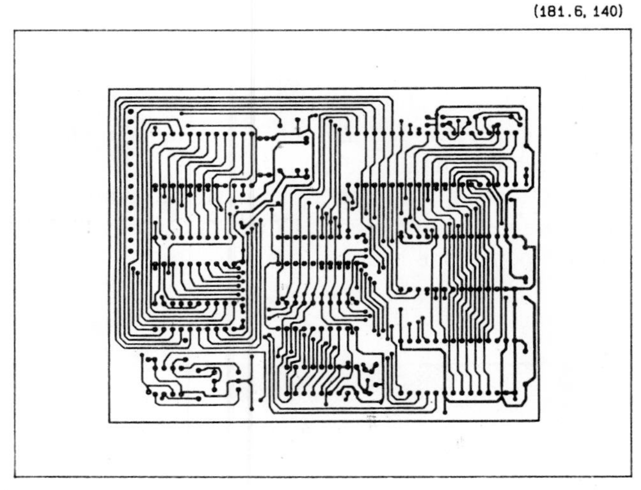

2.4 CLIPPING

When portions of the image have coordinates that are outside the window, clipping must be invoked to prevent error. Clipping algorithms, which can be done either in hardware or in software, must be efficient because every part of the image is processed. Some display hardware has built-in clipping that handles the task automatically. In case the display does not have built-in clipping, it should be provided in software. (See calls to CLIP and NOCLIP in Appendix B.) To speed up program execution, software clipping may be turned off when not needed, such as in a fully debugged program.

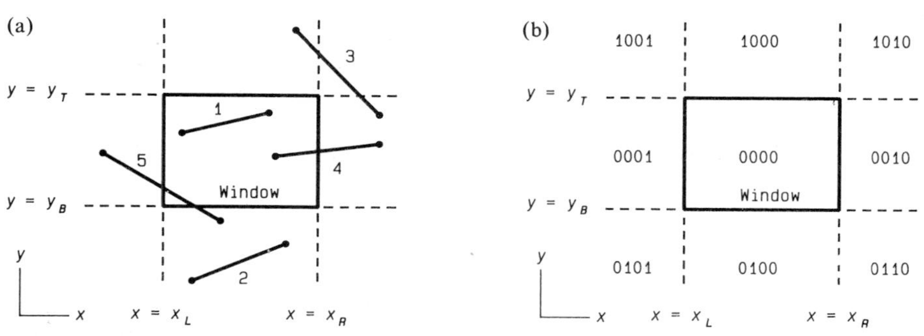

Figure 2.12 Two-dimensional line clipping to a window. (a) Example line segments; (b) region code or end point bit code.

Different clipping algorithms deal with points, lines, and perhaps polygons. Point clipping is trivial, since all that is necessary is testing to see that a particular point is not in the window. Line clipping must be able to display the part of the line that is in the window. Although polygons can be clipped as a collection of lines, additional provisions must be made for filled polygons, which may become open as a result of the clipping. Thus, if filled polygons are not being used, a polygon clipping algorithm is unnecessary.

An extension of the concept of clipping at window edges is volume clipping. Here, a region in 3-D space is truncated by six planes, which can be visualized as the six faces of a box. Clipping in three dimensions is particularly useful when part of a model must be removed to permit viewing of details that are obscured.

Line Clipping

One of the most widely used methods for clipping lines in a window is the Cohen-Sutherland[39,49] algorithm. For illustration, a selection of numbered line segments is shown in Figure 2.12(a). The area containing the window is divided into nine regions with the codes in Figure 2.12(b).

A four-digit region code is assigned to each point, where the bit positions are numbered from right to left and the use of each bit is

bit 1—left

bit 2—right

bit 3—below

bit 4—above

If the point is within the region, the bit is set to 1; otherwise the bit is set to 0. For the lines of Figure 2.12(a), the visibility tests are given in Table 2.1.

From Table 2.1 it can immediately be concluded that if the logical AND† of the end points is not zero, the line must be completely invisible. Furthermore, if *both* end point codes are zero the line is completely visible. If only *one* end point is zero, the line is partially visible. Otherwise, no conclusions can be drawn without further processing.

†The logical AND states that 1 AND 0 = 0, 0 AND 1 = 0, 0 AND 0 = 0, but 1 AND 1 = 1.

TABLE 2.1 VISIBILITY TEST BIT CODES FOR IMAGE IN FIGURE 2.12

Line	End point bit codes		Logical AND of end points	Conclusion from end point codes
1	0000	0000	0000	Completely visible
2	0100	0100	0100	Completely invisible
3	1000	0010	0000	None
4	0000	0010	0000	Partially visible
5	0001	0100	0000	None

For the lines which are not completely invisible or visible as indicated by the end point tests, the intersections with the window boundaries (or window boundaries extended) must be found. If the end points of the line in question are denoted by (x_0, y_0) and (x_1, y_1) the slope m of the line is

$$m = \frac{y_1 - y_0}{x_1 - x_0} \tag{2.8}$$

If the slope is infinite, the line is vertical, with the result that only the top and bottom of the window are checked for an intersection. Similarly, if the slope is zero, only the left and right sides of the window are checked. The intersections are

$$\begin{array}{lll} x = x_L & y = m(x_L - x_0) + y_0 & \text{Left side} \\ x = x_R & y = m(x_R - x_0) + y_0 & \text{Right side} \\ x = x_0 + \dfrac{y_B - y_0}{m} & y = y_B & \text{Bottom} \\ x = x_0 + \dfrac{y_T - y_0}{m} & y = y_T & \text{Top} \end{array} \tag{2.9}$$

where x_L, x_R, y_B, and y_T are the left, right, bottom, and top coordinates of the window, respectively.

With the intersections known, the original line segment can be divided into two or three new segments, which are treated separately. The end point codes completely specify whether or not each line segment is visible, so the logical AND does not have to be done.

Polygon Clipping

When a closed polygon is clipped, it is necessary that the polygon remain closed. An example of a polygon to be clipped is shown in Figure 2.13. The Sutherland-Hodgman algorithm[39,49] clips to one edge at a time, creating intermediate polygons, until the final result is reached.

For the polygon in Figure 2.13, clipping might start on the left edge with the determination of the intersection points a and c by means of Eq. (2.9). This defines the intermediate polygon *aBCc,* which is next clipped at the top edge to determine points b and e. The resulting polygon, *aBbe,* lies completely within the window and may subsequently be filled to give the expected result.

Volume Clipping

The 2-D Cohen-Sutherland algorithm for line clipping can be extended[39,49] to limit the display of images from a 3-D data base. The third dimension means that

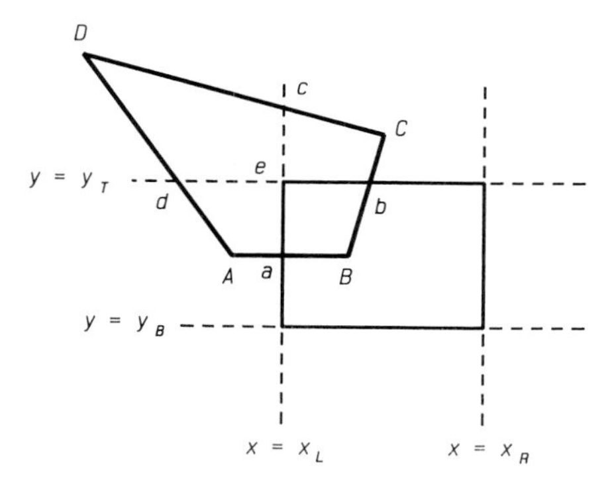

Figure 2.13 Polygon clipping.

line segments connect points in x-y-z space. It is conventional in computer graphics that the z axis is perpendicular to the screen, so that the clipping planes are easily related to a graphics terminal. By clipping with hither and yon clipping planes, Figure 2.14, the user can reduce the complexity of the displayed image. Volume clipping is useful for viewing portions of 3-D regions, where the image of the whole region may be too cluttered.

For three dimensions, a six-bit end point code is needed. A bit is set to 1 if the corresponding coordinate of the end point is out of the volume and to 0 if it is inside. The order of bits is arbitrarily selected as (1) left of the volume, (2) right of the volume, (3) below the volume, (4) above the volume, (5) in front of the volume, and (6) behind the volume. The method is essentially similar to that used in 2-D line clipping.

Clipping a perspective projection requires special consideration. One simple way to handle perspective projections is to clip to planes of constant, x, y,

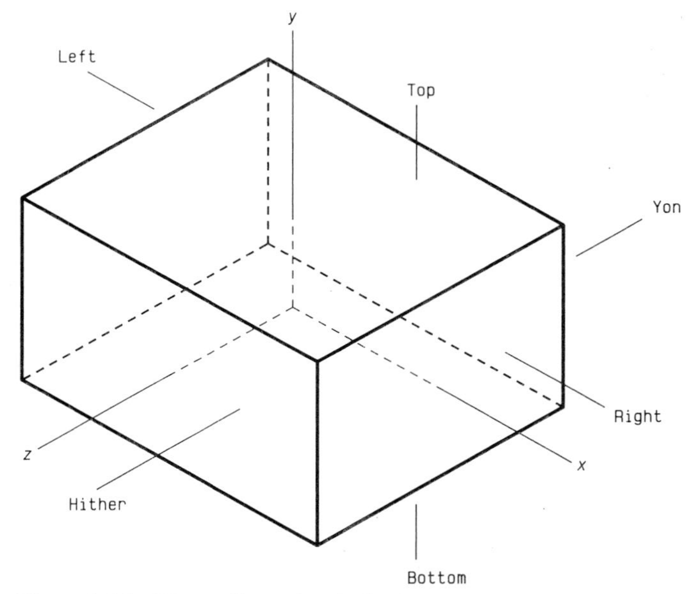

Figure 2.14 Three-dimensional clipping planes.

or z before applying the perspective transformation described in Section 4.7. This simple method leaves an unfilled region around the edges. A more comprehensive method involves clipping to a frustum of a pyramid. Another strategy that may be satisfactory is to do the hither and yon clipping in three dimensions, followed by 2-D clipping of the displayed image to the window.

2.5 HIDDEN-LINE AND HIDDEN-SURFACE REMOVAL

Displaying only the visible lines and surfaces in the rendering of an image considerably enhances clarity and impact. *Hidden-surface removal* algorithms prevent display of any part of a surface that is obscured from view; *hidden-line removal* algorithms prevent display of line segments that are obscured. Although surfaces are displayed by drawing lines, the apparent overlap in the meaning of the two terms is resolved by the way in which actual algorithms process the information.

Algorithms for removal of hidden lines and surfaces depend on two properties: the *depth* of any part of the object and the *coherence* of the parts of the object being displayed. A simple illustration of depth and coherence is shown in Figure 2.15. The depth, or distance away from the observer, is determined through *geometric sorting,* which assigns priorities to the surfaces, edges, or points making up the object. When two or more parts of an image occupy the same space on the display surface, the one closest to the viewer is visible. Sorting every pixel in an object leads to an effective, although computationally intensive, hidden-surface removal algorithm. With the use of coherence, which is the tendency for parts of the scene to be locally constant, larger entities can be considered in the sort, which leads to faster execution.

Figure 2.15 The coherence of the three objects—fence, tree, and house—reduces the sorting required for hidden-surface removal.

Algorithms can be classified[29,39] as *object space* or *image space.* Object-space algorithms, which work in the 3-D world coordinate system, deal with the geometric relationships among the objects in the scene. Image space algorithms, which work in the screen coordinate system, deal with the displayed image, possibly pixel by pixel. Some algorithms are a combination of both classifications.

For simplicity in what follows, objects will be assumed to be bounded by flat faces and straight edges without discontinuities. Various hidden-surface algorithms determine the orientation and location of the faces and check for conflicts and intersections with other faces. Thus, the coherent nature of faces and edges is important for the proper functioning of hidden-line and hidden-surface algorithms.

Unfortunately, the most general algorithms for hidden-line and hidden-surface removal are generally complex and long. In applications that involve interactivity or animation, hidden-line or hidden-surface elimination may be unacceptably slow, which is a problem that can be remedied by use of faster processors and special-purpose algorithms. There is no one best algorithm for hidden-line and hidden-surface removal, as attested to by the attention paid this problem in the literature.[23,29,39,49]

In this section, three very simple but useful algorithms are described. It should be noted that the following algorithms (1) may be suitable only for certain limited applications, (2) may not always produce perfect removal of hidden lines and surfaces, and (3) may not be as efficient as other algorithms.

Masking

The *masking* or *floating horizon* technique works in image space to remove hidden lines.[38,49] It works well for 3-D surface plots (Section 8.7) and other images where a large number of line segments make up a single coherent surface. As illustrated in Figure 2.16, relatively long line segments should be converted to two or four short ones for proper functioning of the algorithm.

Masking depends on (1) sorting the line segments in the order starting *closest* to the viewer and (2) processing the line segments in sorted order, where visibility exists only for segments that lie outside a region bounded, or "masked," by previously plotted lines. The speed of the algorithm is inversely proportional to the fineness of the mask and the amount of interpolation used for partially hidden line segments.

In implementing this algorithm, it is convenient to imagine the display surface being divided into some large number, say 100, or vertical columns or strips that define the sides of the mask. Two arrays, one for the maximum values of y in each column and one for the minimum values of y in each column, are used to define the top and bottom of the mask. The algorithm is illustrated in Figure 2.16.

As noted, relatively long line segments that cross the mask should be divided into shorter segments for processing. Line segments with both end points obscured by the mask are not displayed and no further action is necessary. Line segments with both end points visible are displayed and the mask is updated. Line

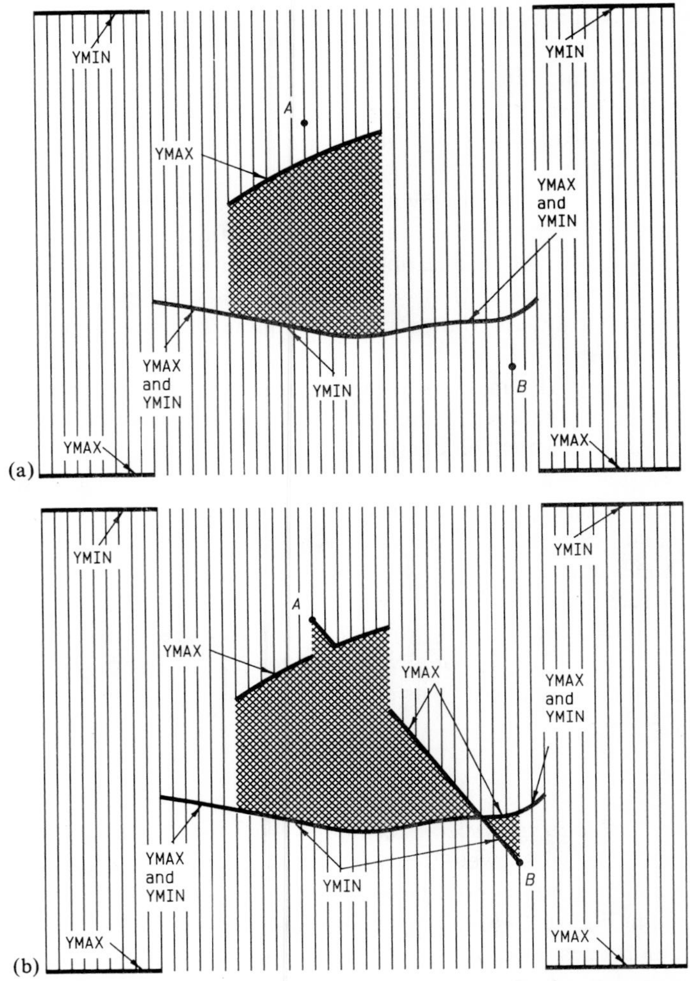

Figure 2.16 Masking algorithm. The 50 values in the array YMAX(I), I = 1, 50, are initialized to an arbitrarily small value (shown as the bottom of the window) and the 50 values of YMIN(I) are initialized to an arbitrarily large value (top of the window). As lines are processed, the mask is defined by YMAX and YMIN. (a) Values of YMAX and YMIN existing before line *AB* is processed. (b) A long line segment such as *AB* that crosses the mask should be changed into shorter segments for processing. The new range of the mask is shown.

segments that pass from a visible point to an invisible point or vice versa are subdivided with the following algorithm:

1. Designate the visible end point $\mathbf{P}_v$ and the invisible endpoint $\mathbf{P}_i$.
2. Set $\mathbf{P}_0 = \mathbf{P}_v$ and $\mathbf{P}_1 = \mathbf{P}_i$.
3. Determine the midpoint of the line segment, $\mathbf{P}_m = (\mathbf{P}_0 + \mathbf{P}_1)/2$.

4. Test for visibility of $\mathbf{P}_m$.
   - If $\mathbf{P}_m$ is visible, set $\mathbf{P}_0 = \mathbf{P}_m$.
   - If $\mathbf{P}_m$ is not visible, set $\mathbf{P}_1 = \mathbf{P}_m$.
5. Repeat steps 3 and 4 an arbitrary number of times.
6. Draw a line segment between $\mathbf{P}_v$ and the last value of $\mathbf{P}_0$. (Use of $\mathbf{P}_0$ leaves a gap between the end of the plotted line and the mask.)

The accuracy of the result can be improved by increasing the number of bisections of partially hidden line segments and by using more strips in the mask. However, this improvement is at the expense of computing time and storage. Line segment length and strip width have a practical lower limit, which is dictated by the resolution of the display device. A coarse mask and a small number of bisections is a good choice for interactive applications where speed is needed. Refinement is warranted for final-quality presentation plots.

EXAMPLE 2.2

An image consists of two polygons and a curve which are in conflict. Polygon A is known to be closer to the viewer than polygon B, which in turn is closer than curve C. The x-y coordinates of the vertices of A and B and the equation of curve C along with the resulting figures before hidden lines are removed are shown in the accompanying illustration. Use the masking technique in a program to render the image with hidden lines removed.

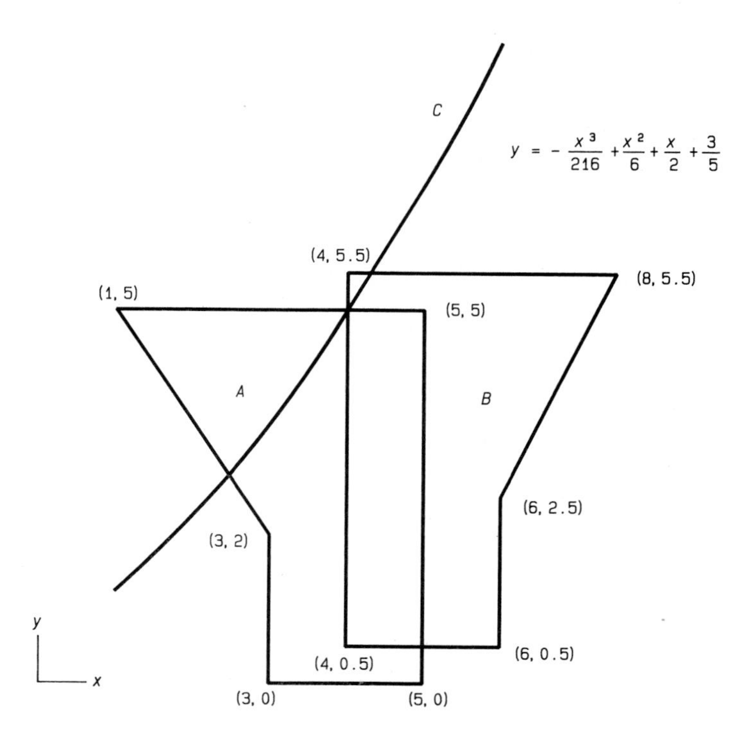

Solution. A sorting algorithm is unnecessary because the priority of the objects is already known. The FORTRAN-77 program that follows contains subroutines to construct the mask and to test the visibility of each point. A rather coarse mask, 65 vertical strips, is used for fast execution. Curve C will be approximated by 30 line segments on the interval $1 \leq x \leq 6$. The number of bisections for partially hidden lines is set at 8 for above-average quality.

```
C
C        PROGRAM POLY
C
       DIMENSION PA(6,2), PB(6,2)
       COMMON /AMASK/ YMAX(65),YMIN(65),XL,XH,NSTOT
C
C    PA AND PB - THE VERTICES (X,Y) OF POLYGONS A AND B
C    YMAX AND YMIN - MAX AND MIN VALUES FOR MASK
C    DX - INCREMENT FOR CURVE C
C    XL,XH,YL,YH - THE GRAPHICS DEVICE WINDOW SIZE
C    NSTOT - TOTAL NUMBER OF VERTICAL STRIPS IN THE MASK
C    LACC - NUMBER OF BISECTIONS FOR VISIBLE LINE SEGMENT
C    IFLAG - VISIBILITY FLAG -> 0 = NOT VISIBLE,   1 = VISIBLE
C          IFLAG0 - FLAG FOR FIRST POINT (X0,Y0) OF LINE SEGMENT
C          IFLAG1 - FLAG FOR SECOND POINT (X1,Y1) OF LINE SEGMENT
C    DEFINE INITIAL DATA VALUES AND EQUATION OF CURVE C
C    POLYGON DATA MUST BE IN THE ORDER P(1,1), P(2,1), ETC.
C    FIRST POINT OF POLYGON IS REPEATED TO CLOSE POLYGONS A AND B
C
       DATA PA/1.0,3.0,3.0,5.0,5.0,1.0, 5.0,2.0,0.0,0.0,5.0,5.0/
       DATA PB/4.0,4.0,6.0,6.0,8.0,4.0, 5.5,0.5,0.5,2.5,5.5,5.5/
20     DATA YMAX/65*-1.0E20/, YMIN/65*1.0E20/, NSTOT/65/, LACC/8/
       DATA XL,XH,YL,YH/0.0,10.0, -0.5,7.0/
C
       CURVE(X) = -X**3/216. + X**2/6. + X/2. + 3./5.
C
C    INITIALIZE GRAPHICS (SEE APPENDIX B FOR GRAPHICS SUBROUTINES)
C
       CALL GRINIT (4107,7550,1)
       CALL WINDOW (XL,XH,YL,YH)
       CALL NEWPAG
C
C    PROCESS POLYGON A, WHICH CAN'T BE HIDDEN, BUT IS STILL MASKED
C
       CALL MOVE(PA(1,1),PA(1,2))
       DO 100 I=2,6
          CALL DRAW(PA(I,1),PA(I,2))
          CALL MASK(PA(I-1,1),PA(I-1,2),PA(I,1),PA(I,2))
100    CONTINUE
C
C    PROCESS POLYGON B, WHICH MAY HAVE HIDDEN LINES
```

```
C
      CALL MOVE(PB(1,1),PB(1,2))
      CALL VISIBL(PB(1,1),PB(1,2),IFLAG0)
      DO 200 I=2,6
         CALL VISIBL(PB(I,1),PB(I,2),IFLAG1)
         IF (IFLAG0 .EQ. 1 .AND. IFLAG1 .EQ. 1) THEN
            CALL DRAW(PB(I,1),PB(I,2))
            CALL MASK(PB(I-1,1),PB(I-1,2),PB(I,1),PB(I,2))
         ELSE IF (IFLAG0 .EQ. 1 .OR. IFLAG1 .EQ. 1) THEN
            CALL DRAWLN(PB(I-1,1),PB(I-1,2),PB(I,1),PB(I,2),
     &                  LACC,IFLAG0,IFLAG1)
            CALL MASK(PB(I-1,1),PB(I-1,2),PB(I,1),PB(I,2))
         END IF
         IFLAG0 = IFLAG1
200   CONTINUE
C
C   GENERATE CURVE C, WHICH MAY HAVE HIDDEN LINE SEGMENTS
C   SINCE THIS IS DONE LAST, THE MASK IS NOT UPDATED
C
      DX = 1.0/6.0
      X0 = 1.0
      Y0 = CURVE(X0)
      CALL MOVE(X0,Y0)
      CALL VISIBL(X0,Y0,IFLAG0)
      DO 300 I=1,30
         X1 = X0+DX
         Y1 = CURVE(X1)
         CALL VISIBL(X1,Y1,IFLAG1)
         IF (IFLAG0.EQ.1 .AND. IFLAG1.EQ.1) THEN
            CALL DRAW(X1,Y1)
         ELSE IF (IFLAG0.EQ.1 .OR. IFLAG1.EQ.1) THEN
            CALL DRAWLN(X0,Y0,X1,Y1,LACC,IFLAG0,IFLAG1)
         END IF
         X0 = X1
         Y0 = Y1
         IFLAG0 = IFLAG1
300   CONTINUE
      CALL GRSTOP
      STOP
      END
C
C------------------------------------------------------------
C
      SUBROUTINE MASK (XX0,YY0,XX1,YY1)
C
C   CONSTRUCTS MASK BOUNDED BY YMAX AND YMIN FOR HIDDEN LINES
C
      COMMON /AMASK/ YMAX(65), YMIN(65), XL, XH, NSTOT
C
C   CREATE TEMPORARY NAMES FOR XX0, XX1, YY0, YY1.  ORDER L TO R
C   NS0 - BEGINNING VERTICAL STRIP NUMBER
C   NS1 - ENDING VERTICAL STRIP
```

```
C
      CALL ORDER(XX0,YY0,XX1,YY1,X0,Y0,X1,Y1)
      NS0 = (X0-XL)*NSTOT/(XH-XL) + 1
      NS1 = (X1-XL)*NSTOT/(XH-XL) + 1
      IF (NS0 .LT. 1) NS0 = 1
      IF (NS1 .GT. NSTOT) NS1 = NSTOT
      IF (NS0 .EQ. NS1) GO TO 200
C
C   BRANCH FOR NS0, NS1 IN DIFFERENT STRIPS
C   LINEARLY INTERPOLATE BETWEEN NS0 AND NS1
C
      DO 100 N=NS0,NS1
         Y = Y0 + (Y1-Y0)*(N-NS0)/(NS1-NS0)
         IF (Y .GT. YMAX(N)) YMAX(N) = Y
         IF (Y .LT. YMIN(N)) YMIN(N) = Y
100   CONTINUE
      RETURN
C
C   BRANCH FOR NS0 AND NS1 IN SAME STRIP
C
200   IF (Y0 .GT. YMAX(NS0)) YMAX(NS0) = Y0
      IF (Y0 .LT. YMIN(NS0)) YMIN(NS0) = Y0
      IF (Y1 .GT. YMAX(NS0)) YMAX(NS0) = Y1
      IF (Y1 .LT. YMIN(NS0)) YMIN(NS0) = Y1
      RETURN
      END
C
C-----------------------------------------------------------------
C
      SUBROUTINE VISIBL(X,Y,IFLAG)
C
C   CHECKS VISIBILITY     IFLAG = 0  -> NOT VISIBLE
C   OF A POINT            IFLAG = 1  -> VISIBLE
C
      COMMON /AMASK/ YMAX(65), YMIN(65), XL, XH, NSTOT
      IFLAG = 0
C
C     DETERMINE STRIP NUMBER IN MASK AND CHECK IF VISIBLE
C
      NS = NSTOT*(X-XL)/(XH-XL) + 1.0
      IF (NS .GT. NSTOT) NS = NSTOT
      IF (NS .LT. 1) NS = 1
      IF (Y .GE. YMAX(NS)) IFLAG = 1
      IF (Y .LE. YMIN(NS)) IFLAG = 1
      RETURN
      END
C
C-----------------------------------------------------------------
C
      SUBROUTINE ORDER(XX0,YY0,XX1,YY1,X0,Y0,X1,Y1)
C
C  PLACES POINTS 0 AND 1 IN LEFT-TO-RIGHT ORDER
```

```
C
      IF (XX0 .GT. XX1) GO TO 200
100   X0 = XX0
      Y0 = YY0
      X1 = XX1
      Y1 = YY1
      RETURN
C
200   X0 = XX1
      Y0 = YY1
      X1 = XX0
      Y1 = YY0
      RETURN
      END
C
C-----------------------------------------------------------------
C
      SUBROUTINE DRAWLN(XA,YA,XB,YB,LACC,IFLAG0,IFLAG1)
C
C   THIS ROUTINE ASSIGNS A VISIBLE LINE END TO X0, Y0.
C   THE OTHER END OF THE VISIBLE LINE IS FOUND BY BINARY SEARCH,
C   AND THE LINE IS DRAWN.  INCREASE  LACC  FOR MORE ACCURACY.
C
C   ASSIGN DUMMY VARIABLES.  X0 AND Y0 ARE ALWAYS VISIBLE
C
      IF (IFLAG0 .EQ. 1) THEN
         X0 = XA
         Y0 = YA
         X1 = XB
         Y1 = YB
      ELSE
         X0 = XB
         Y0 = YB
         X1 = XA
         Y1 = YA
      END IF
C
C   TRIM LINE WITH BINARY SEARCH
C
      DO 100 I=1,LACC
         X = (X0+X1)/2.
         Y = (Y0+Y1)/2.
         CALL VISIBL(X,Y,IFLAG)
         IF (IFLAG .EQ. 0) THEN
            X1 = X
            Y1 = Y
         ELSE
            X0 = X
            Y0 = Y
         END IF
100   CONTINUE
C
```

```
C     DRAW VISIBLE LINE SEGMENT
C
      IF (IFLAG0.EQ.1 .AND. IFLAG1.EQ.0)THEN
         CALL DRAW(X0,Y0)
      ELSE IF (IFLAG0.EQ.0 .AND. IFLAG1.EQ.1) THEN
         CALL MOVE(X0,Y0)
         CALL DRAW(XB,YB)
      END IF
      RETURN
      END
```

The resulting figure shows a small, but acceptable, gap near the bottom between polygon A and polygon B. In this example, it should be noted that if the processing of B started at the point (4, 5.5) and proceeded clockwise around the figure, a problem with vertical lines that lie in one strip would appear. Because of the order of processing, the entire vertical line from (4, 0.5) to (4, 5.5) would be masked. The subroutines in this example are used in Example 8.4 for construction of a surface that is a function of two variables.

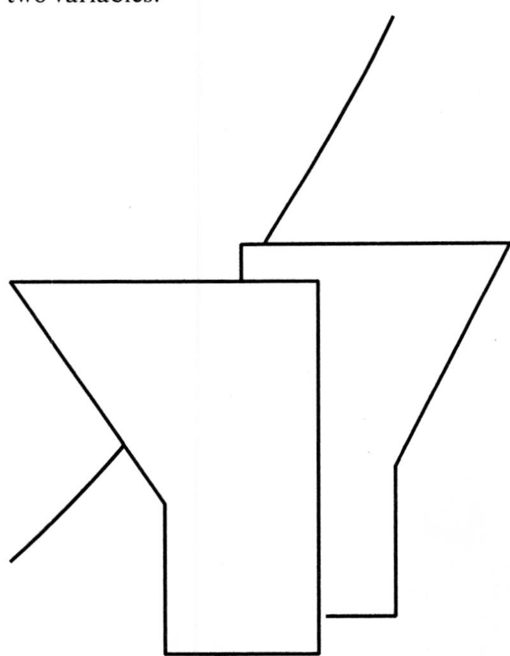

Back Side Elimination

Polygonal objects or tessellated surfaces are frequently found in engineering applications. Back side elimination checks the orientation of each of the flat coherent faces bounding the model and determines whether it faces away from or toward the viewer. It works in object space. No sorting is needed unless the object is concave (has depressions) or there are multiple objects.

The orientation of a face is given by a vector that is perpendicular to the face. Such a normal vector **N** is computed from the cross product of any two vectors **a** and **b** lying within the face,

$$\mathbf{N} = \mathbf{a} \times \mathbf{b} \tag{2.10}$$

If is it not known whether this normal is outward or inward, the following test may be applied to convex polyhedra. (Convex means that a line drawn between any two vertices of the object lies entirely within the object.) A third vector **c** is selected which goes from any point on the questionable face to any other point on the convex surface or within the object. If the dot product

$$\mathbf{c} \cdot (\mathbf{a} \times \mathbf{b}) \text{ or } \mathbf{c} \cdot \mathbf{N} \begin{cases} >0 & \mathbf{a} \times \mathbf{b} \text{ (or } \mathbf{N}\text{) is an inward normal} \\ =0 & \mathbf{c} \text{ lies in the plane of } \mathbf{a} \text{ and } \mathbf{b} \\ <0 & \mathbf{a} \times \mathbf{b} \text{ (or } \mathbf{N}\text{) is an outward normal} \end{cases} \tag{2.11}$$

The dot product determines the cosine of the angle between **c** and **N**. When the angle between **c** and **N** is greater than 90°, the cosine is negative, and **c** and **N** go in opposite directions from the plane. If it is found that **N** is inward, the sign of **N** is changed to make it outward. The test in Eq. (2.11) may be skipped if it is already known that **N** is outward.

When the outward normal points toward the viewer, the face is visible. If a vector toward the viewer from any point on the questionable plane face is designated as **M**, the dot product test states that if

$$\mathbf{M} \cdot \mathbf{N} \begin{cases} >0 & \text{the face is visible} \\ =0 & \text{the face appears as an edge} \\ <0 & \text{the face is not visible} \end{cases} \tag{2.12}$$

where **N** is an *outward* normal to the surface.

Usually, tessellated surfaces or polyhedra are described with vertices, which in turn are convenient for defining normals. On a typical face, Figure 2.17, it is convenient to denote the three vertices as $\mathbf{P}_i$, $\mathbf{P}_j$, and $\mathbf{P}_k$. If the vertices are not regularly numbered, another vertex in the solid $\mathbf{P}_c = [P_{cx}\ P_{cy}\ \ P_{cz}]$ is needed to test whether the normal is outward or inward. The vectors **a** and **b** become $\mathbf{P}_j - \mathbf{P}_i$ and $\mathbf{P}_k - \mathbf{P}_i$, respectively. The vector **c** is $\mathbf{P}_c - \mathbf{P}_i$. Substitution of the foregoing expressions for **a**, **b**, **c** into Eq. (2.11) (with the sign changed so that positive denotes an outward normal) yields

$$k_1 = -\mathbf{c} \cdot (\mathbf{a} \times \mathbf{b}) = -\begin{vmatrix} (P_{cx} - P_{ix}) & (P_{cy} - P_{iy}) & (P_{cz} - P_{iz}) \\ (P_{jx} - P_{ix}) & (P_{jy} - P_{iy}) & (P_{jz} - P_{iz}) \\ (P_{kx} - P_{ix}) & (P_{ky} - P_{iy}) & (P_{kz} - P_{iz}) \end{vmatrix} \tag{2.13}$$

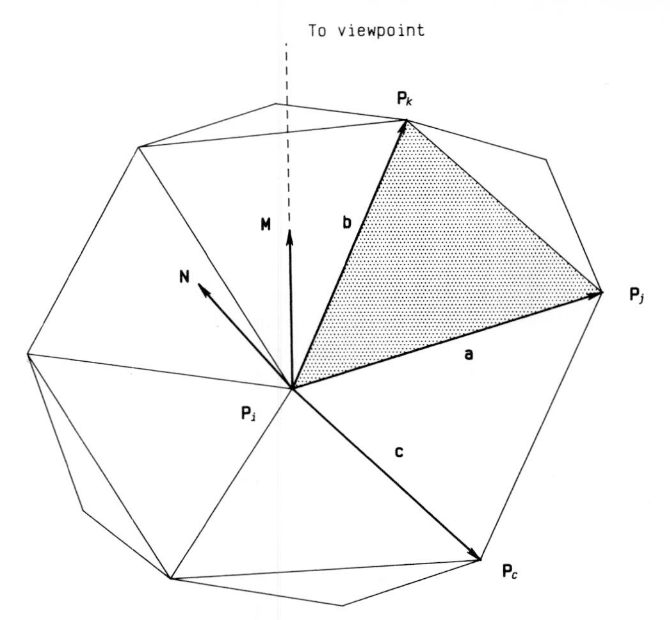

Figure 2.17 Example of tessellated surface with hidden surfaces removed.

If the three vertices are ordered in advance so as to produce an outward normal, the sign of k_1 is known to be positive and this test can be skipped.

The vector **M** from the questionable face to the viewer is $(\mathbf{P}_m - \mathbf{P}_i)$. Combination of Eqs. (2.10) and (2.12) yields

$$k_2 = (\mathbf{P}_m - \mathbf{P}_i) \cdot (\mathbf{a} \times \mathbf{b}) = \begin{vmatrix} (P_{mx} - P_{ix}) & (P_{my} - P_{iy}) & (P_{mz} - P_{iz}) \\ (P_{jx} - P_{ix}) & (P_{jy} - P_{iy}) & (P_{jz} - P_{iz}) \\ (P_{kx} - P_{ix}) & (P_{ky} - P_{iy}) & (P_{kz} - P_{iz}) \end{vmatrix} \tag{2.14}$$

Finally, visibility can be determined by the product of the two scalars resulting from Eqs. (2.13) and (2.14). If

$$k_1 k_2 \begin{cases} >0 & \text{the face is visible} \\ =0 & \text{the face appears as an edge} \\ <0 & \text{the face is invisible} \end{cases} \tag{2.15}$$

The principle employed here is part of the Roberts algorithm,[39,49] which treats the visibility of several complex objects. Here, elimination of the back sides is followed by tests for the priority of the remaining faces. The process of determining normals to surfaces is also used in light-source shading, as seen in Section 2.7.

EXAMPLE 2.3

The four vertices of a tetrahedron are given by

$$[\mathbf{P}] = \begin{bmatrix} 0 & 0 & 0 \\ 2 & 1 & 0 \\ 1 & 2 & 1 \\ 0 & 1 & 1 \end{bmatrix}$$

Determine whether the face defined by the first three vertices is visible to a viewer at [4 2 4].

Solution. The given values $\mathbf{P}_i = \mathbf{P}_1 = (0, 0, 0)$, $\mathbf{P}_j = \mathbf{P}_2 = (2, 1, 0)$, $\mathbf{P}_k = \mathbf{P}_3 = (1, 2, 1)$, $\mathbf{P}_c = \mathbf{P}_4 = (0, 1, 1)$, and $\mathbf{P}_m = (4, 2, 4)$ are substituted into Eqs. (2.13) and (2.14) to give

$$k_1 = -\begin{vmatrix} 0 & 1 & 1 \\ 2 & 1 & 0 \\ 1 & 2 & 1 \end{vmatrix} = -1$$

and

$$k_2 = \begin{vmatrix} 4 & 2 & 4 \\ 2 & 1 & 0 \\ 1 & 2 & 1 \end{vmatrix} = +12$$

The negative sign of the product $(-1)(+12)$ indicates that the face in question is hidden. A plot of the tetrahedron as seen from the view point with the hidden lines dashed, shown in the accompanying figure, verifies the result.

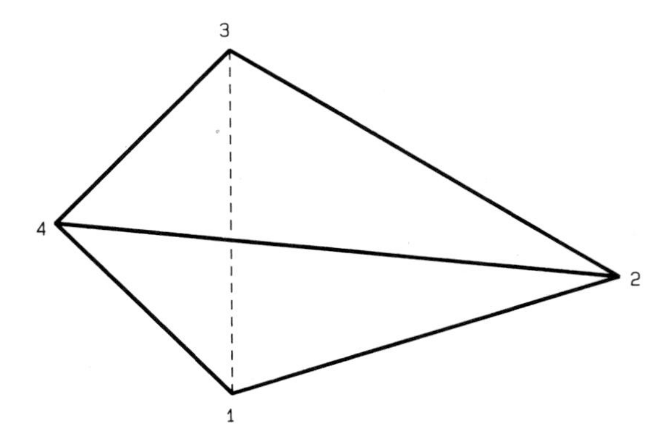

Priority Fill and *z*-Buffer

In raster graphics devices, the frame buffer, Figure 1.3, is overwritten whenever new information is inserted. Thus, when a filled polygon (Section 2.2) is drawn on a raster screen, any existing information in that location is replaced. The easy updating of the frame buffer provides a simple and fast method for elimination of hidden surfaces. Algorithms for polygon filling in software[49] are available for systems lacking hardware polygon filling.

The *priority fill* or *painter's algorithm* operates in image space with the scheme that items *farthest* away from the viewer are rendered first. Items closer to the viewer are said to have priority and simply replace the existing information in that location in the frame buffer. This is the same as an artist's technique, where the background is painted first, followed by addition of foreground objects with opaque paint. Obviously, this algorithm cannot be used on film recorders, pen plotters, storage terminals, and so forth.

Flat polygonal surfaces simplify coherence and sorting. All that is necessary in priority fill is that the polygons are displayed in order starting with the farthest and finishing with the closest. When one polygon partially obscures another, no computation is needed for the intersection problem. The following are possible different ways to do geometric sorting for the priority filling algorithm:

1. Sort the polygons using the mean depth of each. In the usual case of an observer being located normal to the x-y plane, it is necessary to compute just the z coordinate of the mean depth,

$$z_a = \frac{1}{n} \sum_{i=1}^{n} z_i \tag{2.16}$$

   where n is the number of vertices.
2. Sort the polygons using the nearest vertex of each. A sort on the z coordinates of each polygon is required before the sort on all the polygons is done.
3. Sort the polygons using the farthest vertex (least z) of each. Again, an additional sort is required.

Because it is faster to compute the mean depth of polygons than to do the extra sorting, the first scheme is the one most often incorporated in engineering interactive graphics applications. The back-to-front order of rendition in the priority fill is an advantage, because the user can see the hidden parts of the model before they are covered. Unfortunately, such a simple sort does not always work, and users understand that an occasional glitch may occur. Figure 2.18 shows a simple example with three overlapping polygons that the simple algorithms listed above cannot resolve.

A sort of the polygons by two of the methods followed by a comparison of the two lists identifies possible problems due to overlapping and intersecting polygons. Often, such problems can be resolved by subdividing the affected polygons. A more detailed algorithm that does handle conflicts not resolved in the initial priority sort can be found in Rogers.[49] Refinements such as these, of course, slow down the execution of the hidden-surface removal.

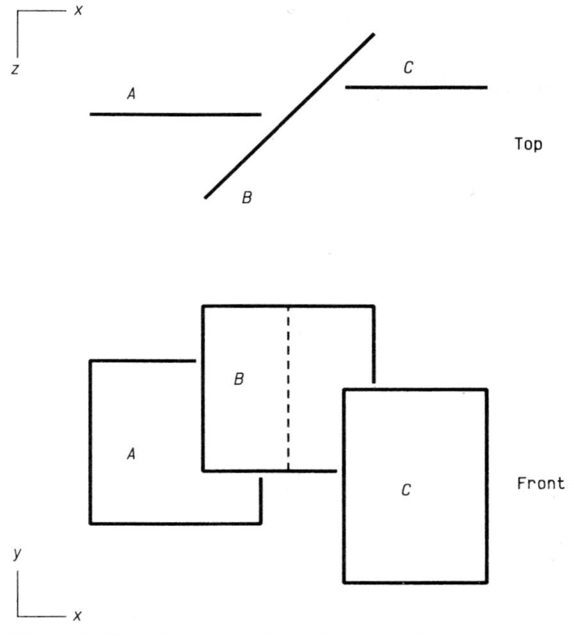

Figure 2.18 Orthographic projections of three overlapping polygons shown with correct order of rendition *ABC*. All three polygons are perpendicular to the x-z plane to simplify the illustration. The front view is the view seen by an observer located on the positive z axis. A problem that arises over how the priority can be based on vertex z coordinates can be resolved by dividing *B* into two polygons as shown by the dashed line.

When the polygons are reduced down to pixels, the priority fill algorithm is the *z-buffer* algorithm.[29,49] The z-buffer contains the depth z of each pixel. As each new pixel is processed for display, its depth is compared to the depth corresponding to the current display that is already in the z-buffer. If the new pixel has a larger z, it is passed to the frame buffer and the z-buffer is updated. The significant extra memory (one word for each pixel location) required to support this algorithm can be packaged in the graphics workstation, which gives "hardware" hidden-surface capability. Mapping coarser locations, such as groups of 4 by 4 pixels, significantly reduces the memory requirement at the expense of image quality.

The priority fill and z-buffer hidden-surface removal techniques work very well and are among the many advantages available to users of raster graphics display screens.

EXAMPLE 2.4

The model with five quadrilaterals (polygons with four vertices) shown as a wireframe in orthographic projection is to be rendered using the priority fill algorithm to remove hidden surfaces. The viewpoint is to be located in front of the model. Develop a program that sorts the polygons and renders the image on a raster display.

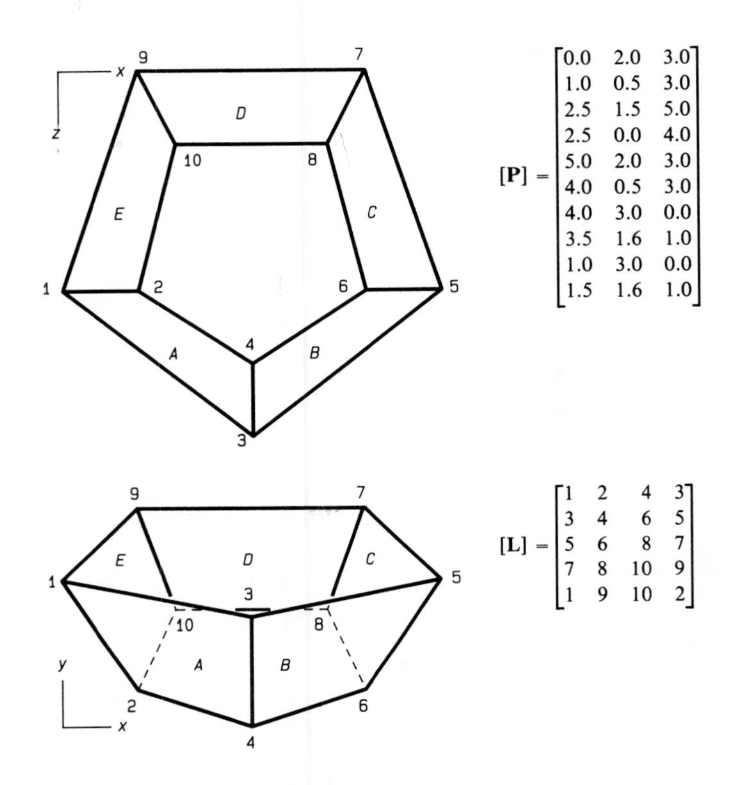

$$[\mathbf{P}] = \begin{bmatrix} 0.0 & 2.0 & 3.0 \\ 1.0 & 0.5 & 3.0 \\ 2.5 & 1.5 & 5.0 \\ 2.5 & 0.0 & 4.0 \\ 5.0 & 2.0 & 3.0 \\ 4.0 & 0.5 & 3.0 \\ 4.0 & 3.0 & 0.0 \\ 3.5 & 1.6 & 1.0 \\ 1.0 & 3.0 & 0.0 \\ 1.5 & 1.6 & 1.0 \end{bmatrix}$$

$$[\mathbf{L}] = \begin{bmatrix} 1 & 2 & 4 & 3 \\ 3 & 4 & 6 & 5 \\ 5 & 6 & 8 & 7 \\ 7 & 8 & 10 & 9 \\ 1 & 9 & 10 & 2 \end{bmatrix}$$

Solution. The FORTRAN-77 program that follows sorts the quadrilaterals according to the z-coordinates of the centroids and renders the resulting image using the graphics tools subroutines described in Appendix C. Note: the x-y-z coordinates of the vertices in the **[P]** matrix of the program appear in numerical order, and the topology list in the **[L]** matrix appears in the alphabetic order of the quadrilaterals.

```
C
C  PRIORITY SORT OF POLYGONS FOR PAINTERS ALGORITHM
C
      COMMON /GEOM/ P(10,3),L(5,4),LTEMP(4),NVER
C
C   NPOL - NUMBER OF POLYGONS TO BE PROCESSED (=5)
C   NP - NUMBER OF POINTS (=10)
C   NVER - NUMBER OF VERTICES PER POLYGON (=4)
C   L(NPOL,NVER) - LIST OF TOPOLOGIES FOR POLYGONS
C   P(NP,3) - POINTS MATRIX    X,Y,Z OF VERTICES
C   LS - ROW OF L WHICH CONTAINS THE FARTHEST (SMALLEST Z) CENTROID
C   ZS - Z OF FARTHEST CENTROID FOUND IN J-TH PASS
```

```
C   ZSC - VALUE OF CENTROID Z FOR COMPARISON
C
      DATA L/1,3,5,7,1, 2,4,6,8,9, 4,6,8,10,10, 3,5,7,9,2/
      DATA P/0.0,1.0,2.5,2.5,5.0,4.0,4.0,3.5,1.0,1.5,
     1       2.0,0.5,1.5,0.0,2.0,0.5,3.0,1.6,3.0,1.6,
     2       3.0,3.0,5.0,4.0,3.0,3.0,0.0,1.0,0.0,1.0/
C
      DATA NPOL/5/,NVER/4/
C
C   SELECTION SORT FOR PLOTTING PRIORITY.  PLOT POLYGONS
C    IN ORDER STARTING WITH THE ONE HAVING THE FARTHEST,
C    I.E. THE SMALLEST Z, CENTROID.
C
      DO 200 I=1,NPOL-1
         LS = I
         CALL CENTR(LS,ZS)
         DO 100 J=I+1,NPOL
            CALL CENTR(J,ZSC)
            IF(ZSC.LT.ZS) THEN
               LS = J
               ZS = ZSC
            ENDIF
100      CONTINUE
C
C   SWAP ORDER OF VERTICES
C
         DO 180 J=1,NVER
            LTEMP(J) = L(LS,J)
            L(LS,J) = L(I,J)
            L(I,J) = LTEMP(J)
180      CONTINUE
200   CONTINUE
C
C   PLOT TETRAHEDRON WITH FILLED PANELS
C
      CALL GRINIT(4107,0,0)
      CALL WINDOW(-2.0,8.4,-1.0,7.0)
      CALL FILPAN(10)
      DO 300 I=1,NPOL
         CALL BEGIN(P(L(I,1),1),P(L(I,1),2),1)
         DO 250 J=2,NVER
            CALL DRAW(P(L(I,J),1),P(L(I,J),2))
250      CONTINUE
         CALL ENDPAN
300   CONTINUE
      CALL GRSTOP
      STOP
      END
```

```
C
C-----------------------------------------------------------------------
C
      SUBROUTINE CENTR(LROW,ZC)
      COMMON /GEOM/ P(10,3),L(5,4),LTEMP(4),NVER
C
C    FINDS CENTROID OF POLYGON WITH NVER VERTICES
C
      ZC = 0.0
      DO 100 J=1,NVER
         ZC = ZC + P(L(LROW,J),3)
100   CONTINUE
      ZC = ZC/NVER
      RETURN
      END
```

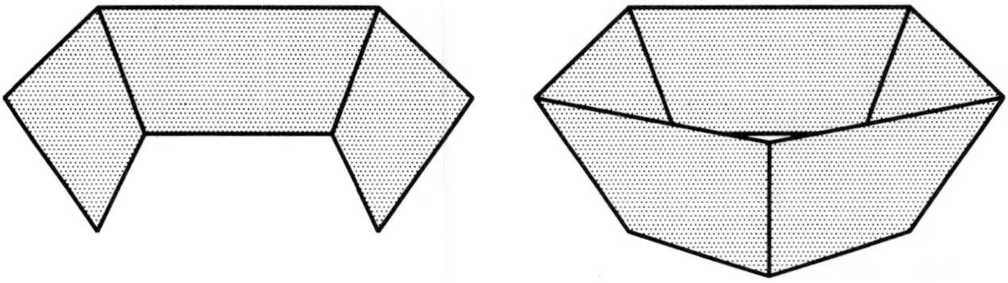

An intermediate image showing the first three polygons and the final image are shown. The program typically runs less than a second or two. Models with hundreds of polygons can still be rendered rapidly enough with this routine for interactive use.

2.6 COLOR IN COMPUTER GRAPHICS

Color adds visual appeal and information content to images. Understanding color theory is important for using color computer graphics advantageously. The simplest way to specify color is with common names such as "orange" and "blue"—ideal for pen plotters, where the names are the colors of the pens. Color CRTs provide a wide range of colors, controlled by user-selected parameters such as hue, lightness, and saturation.

A color name is usually associated with *hue,* which can be expressed numerically as the predominant wavelength of the color. Some plots of "greens" with different spectral content are shown in Figure 2.19. Another way to attach a numerical value to hue is the color wheel, where blue is arbitrarily set at 0° and a spectral order is followed around the circle.

The purity or vividness of a color is given by *saturation.* Subdued or grayish colors are said to be unsaturated. Fully saturated colors are said to be *monochro-*

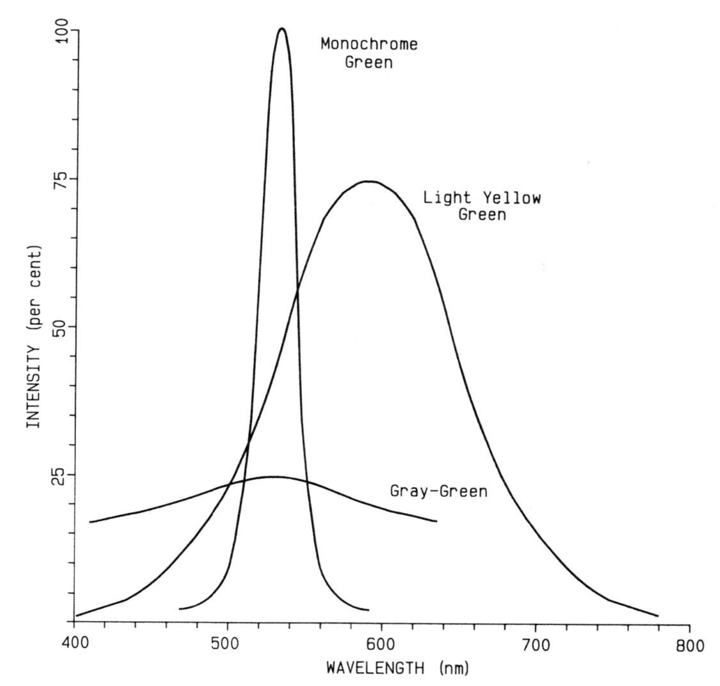

Figure 2.19 Spectral content of example "green" colors.

matic, which means that the spectrum is composed of just one wavelength. Conversely, when all wavelengths are present in equal amounts, the light is white or gray and completely unsaturated. The continuum of saturation can be expressed as a percentage in the range 0 to 100%.

The third parameter for color specification is *lightness,* which measures brightness or luminance. Lightness is specified as a percentage from 0 to 100%, where the continuum goes from black to white. Colors are most vivid at a lightness of 50%, where higher values produce pastel colors and lower values produce dull colors. The 3-D space defined by hue, lightness, and saturation (*HLS*) can be visualized as a region bounded by two cones as shown in Figure 2.20 and in Plate 2.1.

Standardization and control of color in computer graphics are complicated by the fact that additive color theory applies to CRTs, whereas subtractive color theory applies to hard-copy technology. In additive theory, light from two or more sources is combined by wave superposition. In subtractive theory, the properties of surfaces control reflection and absorption of various wavelengths. Frequency content from illuminant sources is controlled by color filters or by using emitters of known characteristics; the color properties of surfaces are controlled by coating them with colorants such as dyes and pigments. Additive theory and subtractive theory are compared in Plate 2.2.

The primaries in additive color theory are the illuminants red, green, and blue. Equal parts of red and green add to produce yellow; green and blue add to

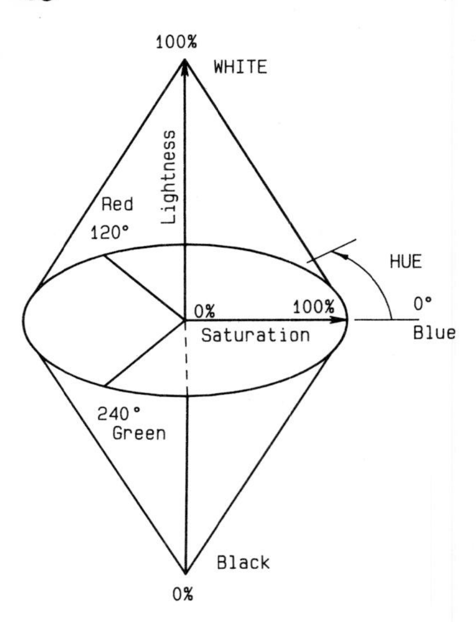

Figure 2.20 The *HLS* (hue-lightness-saturation) color system coordinates. See Plate 2.1.

produce cyan; blue and red add to produce magenta. Equal parts of all three illuminants produce white, which is light with equal amounts of all wavelengths.

The primaries in subtractive color theory are the colorants cyan, magenta, and yellow. These colorants selectively absorb the wavelengths of light other than the hue itself. Combination of the colorants causes even more wavelengths to be absorbed. Equal parts of cyan and magenta make blue; magenta and yellow make red; yellow and cyan make green. Equal parts of all three colorants produce black, which is the absence of all color.

H. Grassmann, a nineteenth century physicist, demonstrated[21,33] that three stimuli—hue, lightness, and saturation, or red, green, and blue—are needed to produce the range of recognized colors. In 1931 the Commission Internationale de l'Éclairage (CIE) applied Grassmann's observations to color standards for combination of illuminants. The CIE standard[33] is based on three hypothetical sources of specified spectral content known as the *standard tristimulus observer functions*. The relative proportions of these three functions are known as *chromaticity coordinates*. Designated as x, y, and z, the chromaticity coordinates can be combined to match the hue and saturation of any monochromatic color, as shown in Figure 2.21.

The data for all combinations of the chromaticity coordinates are represented in Plate 2.3 for an intermediate value of lightness. In viewing this plate, one must realize that printing inks fall far short of producing the full range of visible color. With this plate, the x and y coordinates can be related to a particular color. Because the sum $x + y + z = 1$, specifying z is redundant. The curved boundary represents the fully saturated monochromatic colors. Combinations of the chromaticity coordinates outside the colored area produce invisible radiation. The straight line along the bottom—the magentas—represents color formed by combination of red and blue wavelengths. The diagram has been calibrated so that white is at the point (0.33, 0.33), which indicates equal

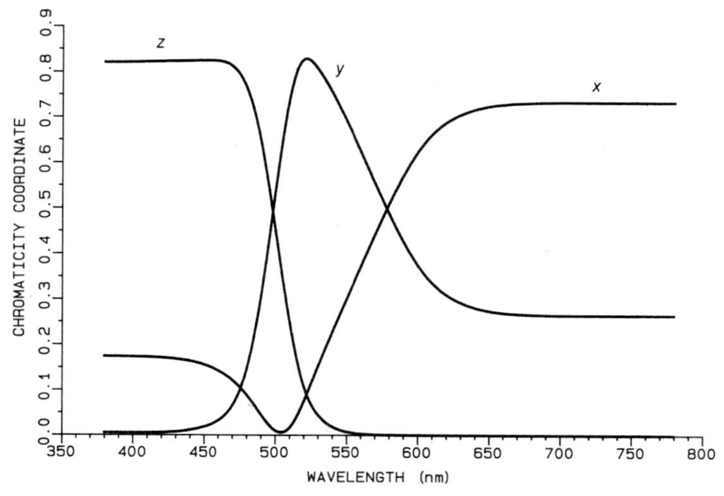

Figure 2.21 Combination of CIE chromaticity coordinates to form monochromatic colors. Note that the sum of x, y, and z is always 1.

amounts of x, y, and z. When lightness is changed, x and y stay constant. For increased lightness, the chromaticity diagram in Plate 2.3 would have a larger white area in the center.

The spectrum of the illuminants from the phosphors used in color CRT monitors is shown in Figure 2.22. In terms of the CIE standard, the red, green, and blue produced by color monitors[14] typically have chromaticity coefficients of (0.628, 0.346), (0.268, 0.588), and (0.150, 0.070), respectively. In Figure 2.23 these coordinates are marked R, G, and B.

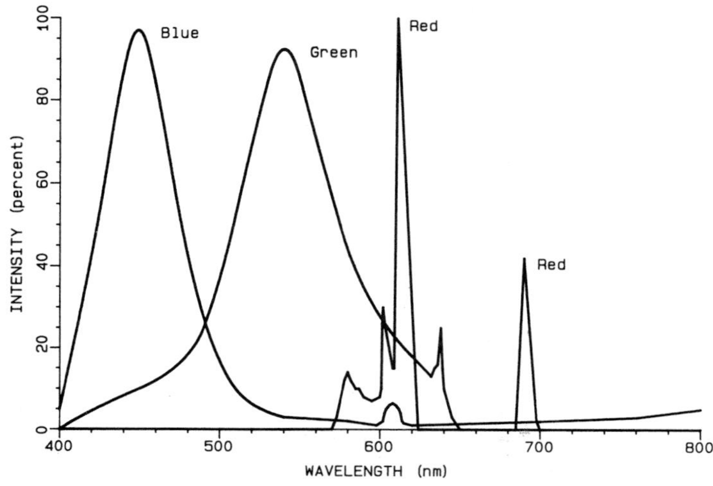

Figure 2.22 Spectral content of red, green, and blue phosphors found in typical color CRT monitors used for computer graphics display.

Figure 2.23 Interpretation of colors in CRT displays from the CIE chromaticity diagram shown in Plate 2.3. (a) Colors as produced by CRT phosphors are indicated by *R*, *G*, *B*, and *W*. (b) Hue values (0 to 360°) and saturation contours (0 to 100%) for lightness = 50%.

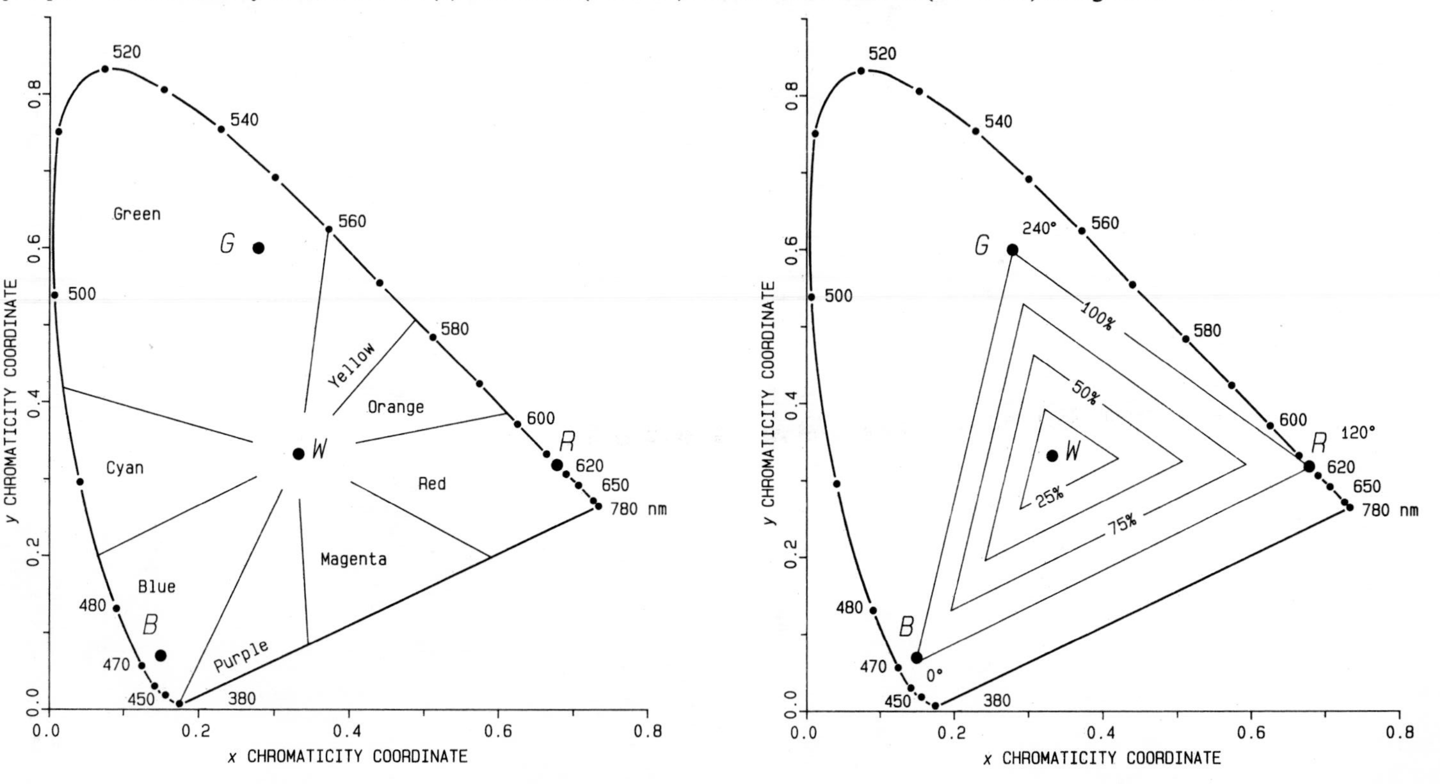

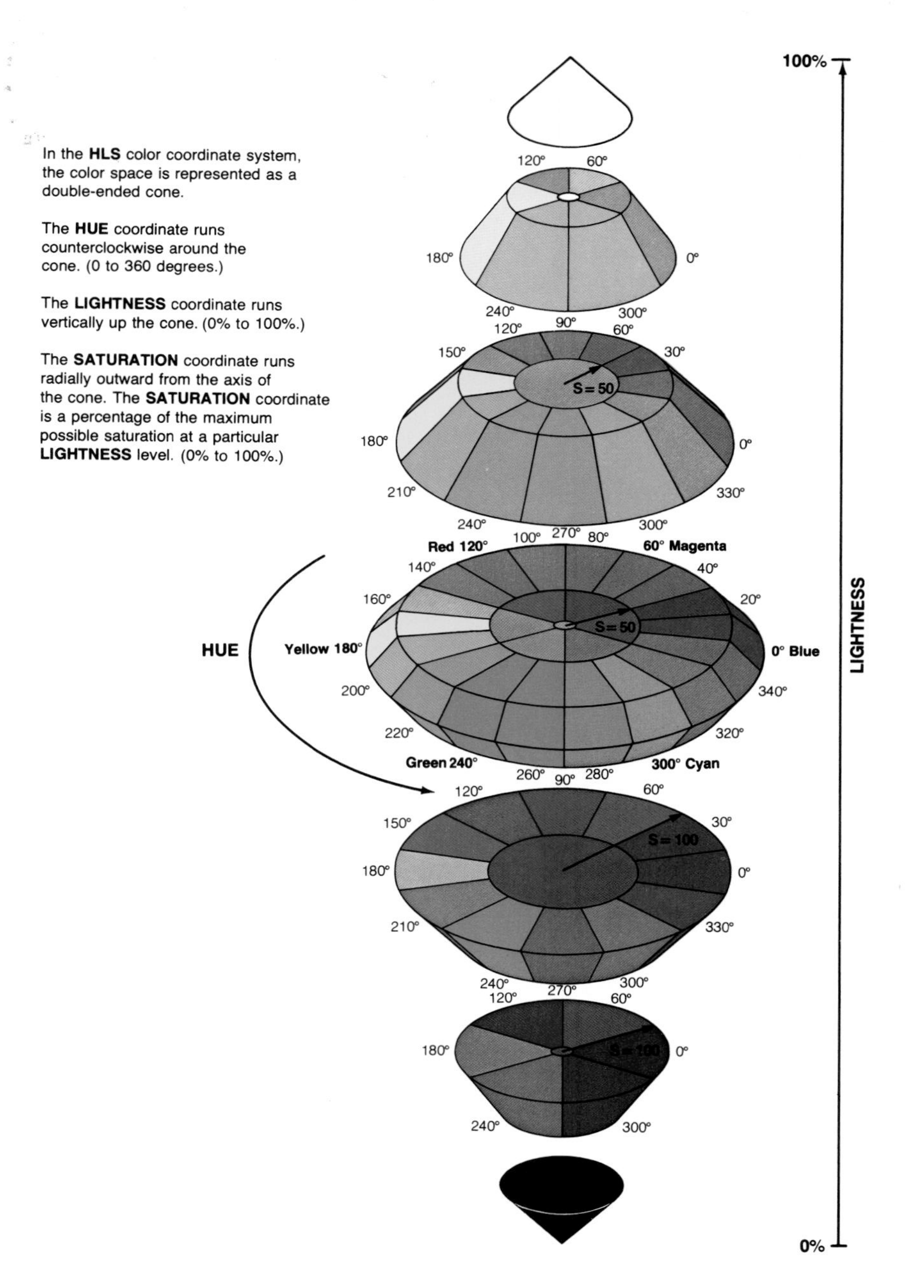

Plate 2.1 The HLS (Hue–Lightness–Saturation) color cone. (Courtesy of Tektronix, Inc.)

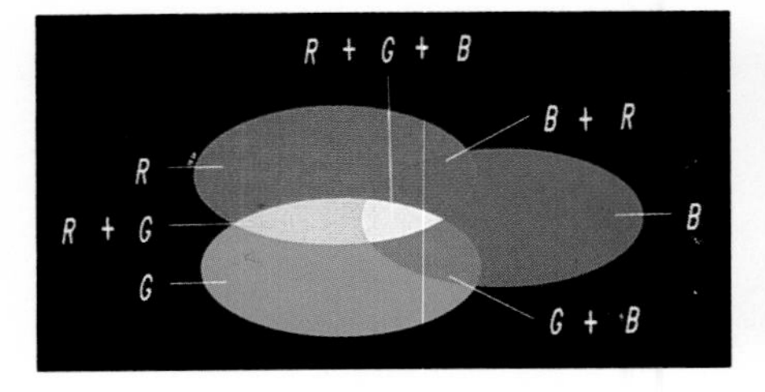

(a) Additive color theory

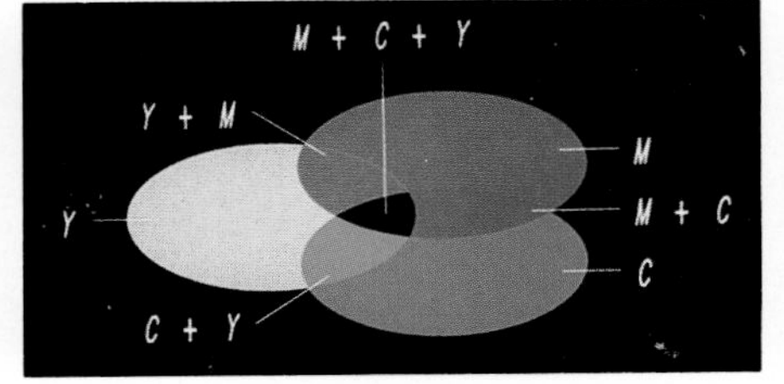

(b) Subtractive color theory

Plate 2.2 Combination of color as light *(a)* and as ink *(b)*. (Courtesy of University of Wyoming Engineering Science Interactive Graphics Laboratory.)

Plate 2.3 Approximation of chromaticity diagram. (Courtesy of General Electric Co.)

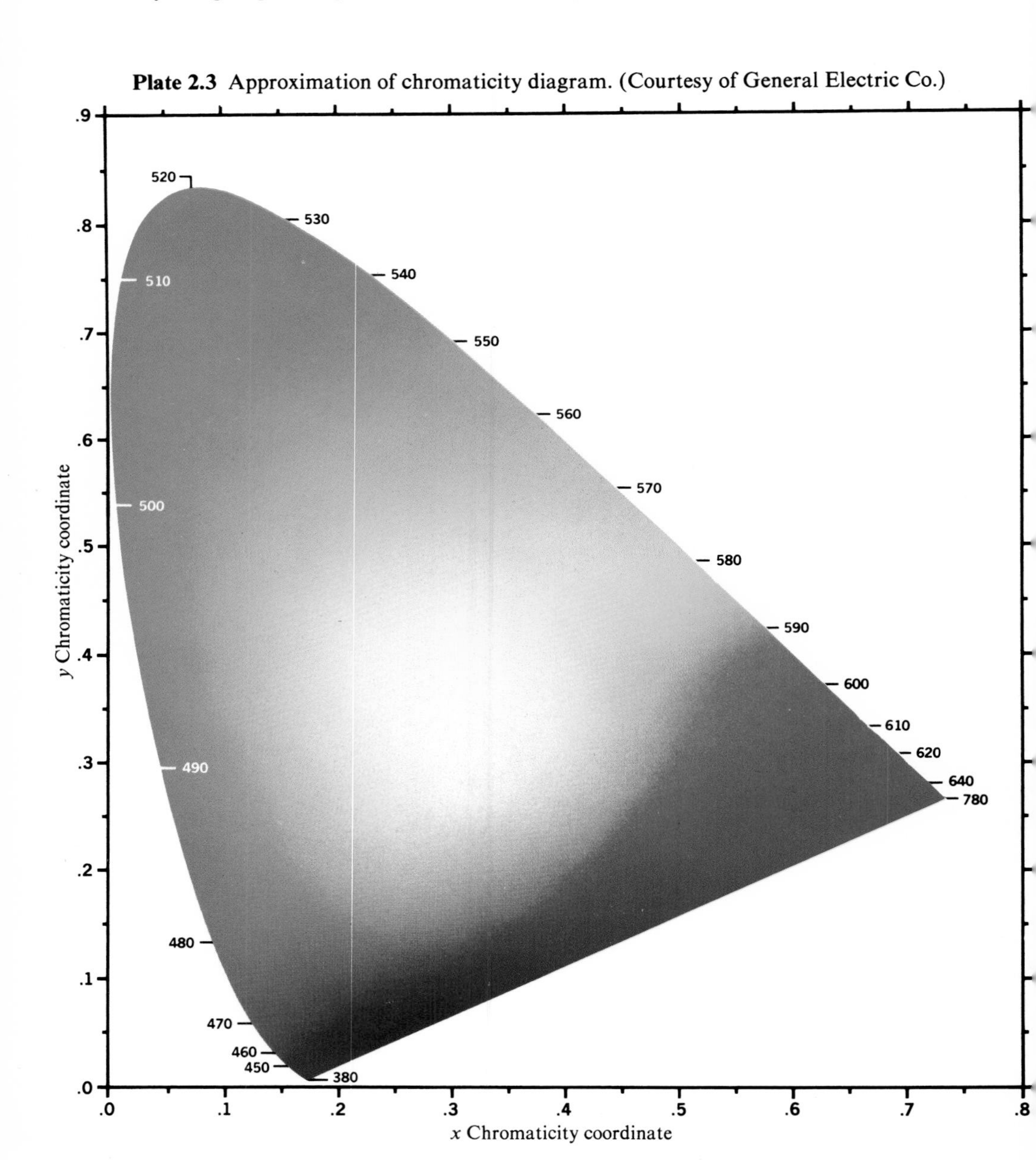

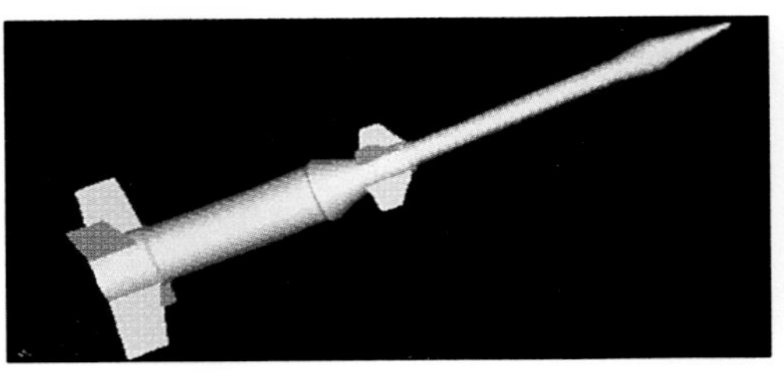

(a) Complete image

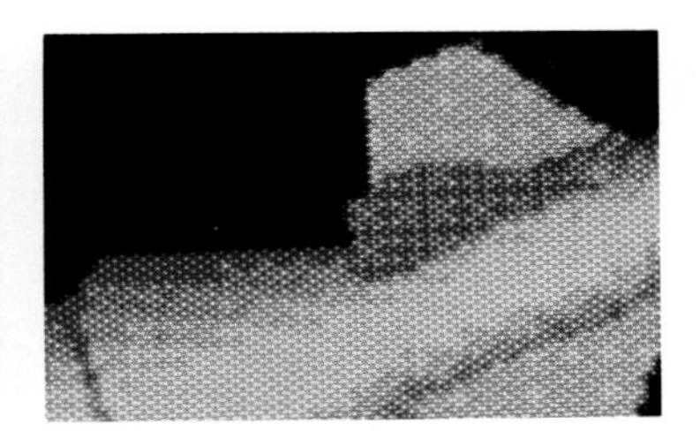

(b) Close up of center section

Plate 2.4 Use of dithering to produce shaded color image on low resolution (360 × 480) color display. Only three colors—white, cyan, and blue—are used. (University of Wyoming Engineering Science Interactive Graphics Laboratory.)

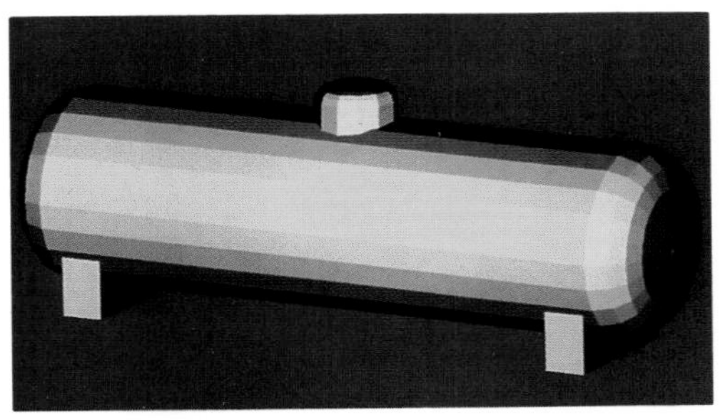

(a) Cosine shading

(b) Gouraud shading

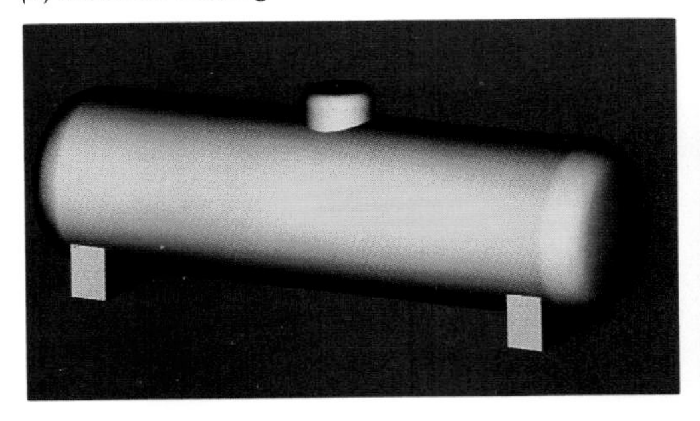

Plate 2.5 Comparison of two widely used shading algorithms. (University of Wyoming Engineering Science Interactive Graphics Laboratory.

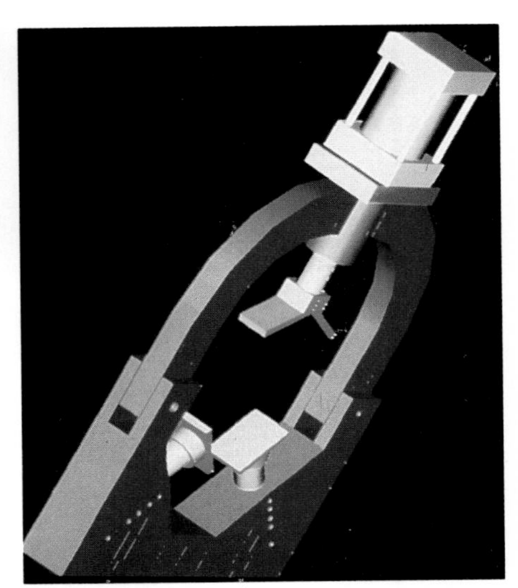

Plate 2.6 Color coding distinguishes components of precision clamping device. (University of Wyoming Engineering Science Interactive Graphics Laboratory.)

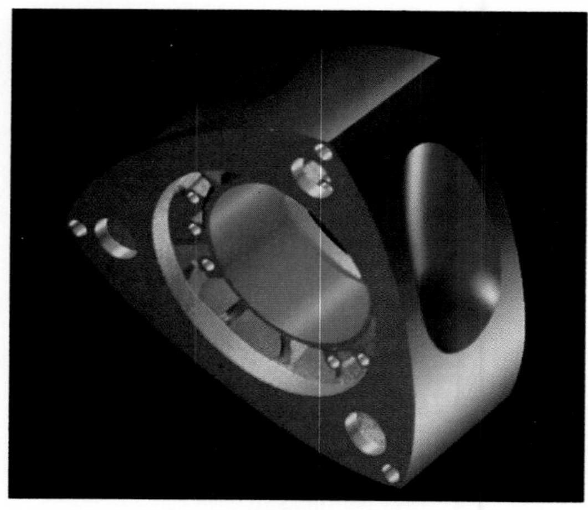

Plate 2.7 Multiple light sources and Phong shading enhance rendering of engine rotor. (Courtesy of Patran Division, PDA Engineering.)

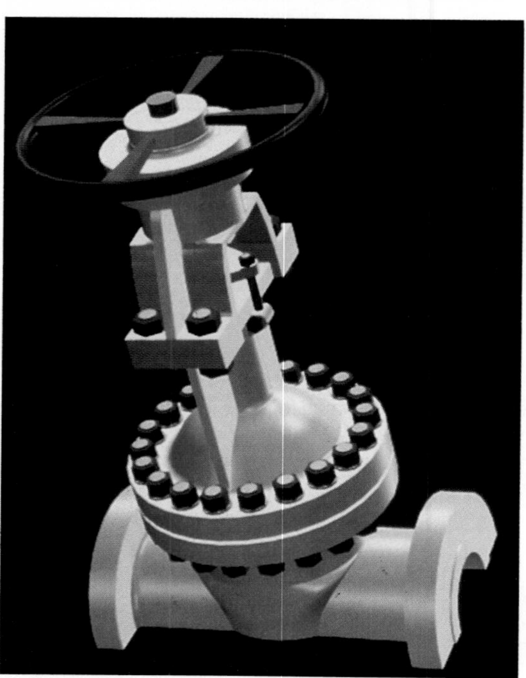

Plate 2.8 Use of multiple light sources and Phong shading to produce realism in model. (Courtesy of Patran Division, PDA Engineering.)

Plate 2.9 *(below)* Use of transparency and other advanced imaging techniques. (Courtesy of Patran Division, PDA Engineering.)

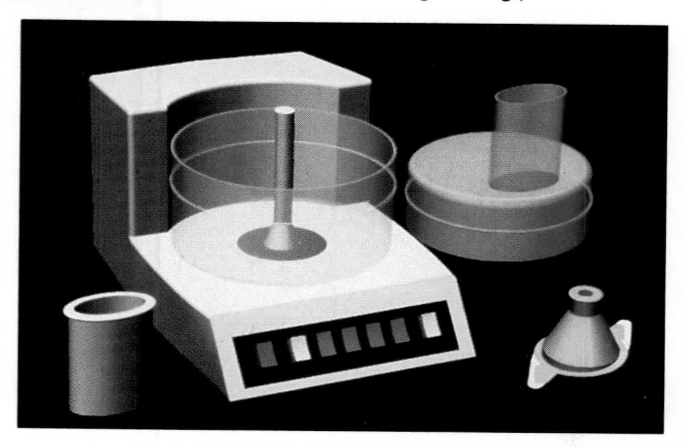

The CIE chromaticity diagram contains a great deal of information in addition to standardizing color specification. A line drawn between any two points on the diagram illustrates the additive combination of the indicated illuminants. Should the line pass through white (0.33, 0.33), colors on opposite sides of this point would be complementary. Superposition of color names on the CIE chromaticity diagram yields Figure 2.23(a). Also shown are the red, green, and blue produced by typical CRT displays. The area of Figure 2.23(b) defined by the 100% triangle with vertices at R, G, and B bounds all the possible hues on color monitors. While it may be noted that color monitors fall short of being able to produce all real-world hues, it turns out that the range of hues reproduced by printing inks is even more limited.

Some color workstations use input of *RGB* instead of *HLS*. For any chromaticity coordinate falling within the triangle of Figure 2.23(b), the proportions of red, green, and blue are simply the relative distances from the coordinate to the vertices R, G, B. A simple conversion[14] between *HLS* and *RGB* is shown in Figure 2.24.

The array of different colors that can be displayed on a workstation is known as a *palette*. The size of the palette is determined by the product of the number of steps of H, L, and S, or of R, G, and B. Thus, for example, if a workstation has 10 steps each of R, G, and B, it is said to have a palette of 1000 ($10 \times 10 \times 10$) colors. The number of colors that can be displayed at the same time is usually much smaller than the number of available colors in the palette.

The workstation *color lookup table*[14] contains a list of *HLS* or *RGB* settings for each displayable color. While the frame buffer of a simple raster display consists of one bit plane (one bit per pixel), three or more bit planes are needed for a monochromatic display with gray shades or a usable color display. Three bit planes, for example, make it possible to encode 2^3 or 8 different colors (or gray shades). The encoded color number corresponds to an entry in the color lookup table such as in Table 2.2.

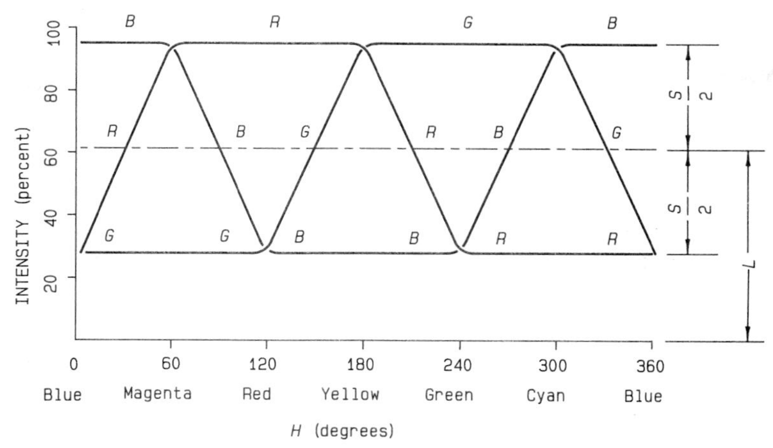

Figure 2.24 Conversion between *HLS* and *RGB*. When $L + (S/2) >$ 100%, redefine $S = 2(100 - L)$. When $L - (S/2) < 0$, redefine $S = 2L$.

TABLE 2.2 TYPICAL DEFAULT COLOR LOOKUP TABLE FOR A SYSTEM THAT DISPLAYS EIGHT COLORS SIMULTANEOUSLY

Color		Color designation system					
		Red-green-blue			Hue-lightness-saturation		
No.	Name	*R* (%)	*G* (%)	*B* (%)	*H* (deg)	*L* (%)	*S* (%)
0	Black	0	0	0	any	0	any
1	White	100	100	100	any	100	any
2	Red (bright)	100	0	0	120	50	100
3	Green (bright)	0	100	0	240	50	100
4	Blue (bright)	0	0	100	0	50	100
5	Cyan (bright)	0	100	100	300	50	100
6	Magenta (bright)	100	0	100	60	50	100
7	Yellow (bright)	100	100	0	180	50	100

A color graphics display suitable for production of realistic shaded images should be capable of displaying at least 256 colors from a palette of over 4000. A display adequate for applications that do not have shading needs only a small color lookup table and a limited palette.

As noted before, the range of colors that can be created with inks, dyes, and paints is very limited, because color is created by altering the reflected properties of light. The saturation possible from reflected colors is far less than that for illuminants. For critical applications, the best quality reproductions come from color photographic transparencies, which have about the same response as a color monitor, Figure 2.23(b).

Another way to mix colors in computer graphics, *dithering,* makes use of alternately colored pixels. With dithering, for example, an area that appears to have a pink color actually is produced by having an array of alternating red and white pixels. Dithering produces colors in a way similar to the halftone process used in printing, where color depends on the control and placement of small spots of colored ink. A sample of dithering is shown in Plate 2.4. Dithering lets a wide range of shades be produced in low-cost color terminals, which usually have a limited palette and a small color lookup table. With color hard-copy devices such as ink-jet plotters, dithering techniques greatly extend the range of colors.

Color improves the usefulness of computer-generated images in many ways. As may be observed from the color plates in this chapter and in Chapter 10, there are many ways in which color improves the clarity and appeal of computer-generated images. One of the techniques seen in the color plates, shading, involves selective variation of lightness as explained in the next section.

2.7 SHADING

Enhancement of 3-D images with shading has become a well-developed technology. The most basic shading involves assignment of lightness values for a surface illuminated with a single light source. Multiple light sources, color coding, transparency, shadowing, and texture can also be added to enhance the image rendition.

Surfaces are visible because light is reflected. The appearance of an illuminated surface depends on the color and geometry of the light source and the orientation and properties of the surface itself. One important property of a surface is its effect on the way incident light is reflected. *Diffuse* reflection means that light is scattered equally in all directions, while *specular* reflection means that the light is reflected only in one direction. Diffuse reflection models give dull, mattelike surfaces such as those in Plates 2.4 through 2.6. Specular reflection models produce shiny, mirrorlike surfaces such as those in Plates 2.7 through 2.9. The characteristics of real surfaces, of course, lie between the two extremes.

Being easier to model than specular reflection, diffuse reflection is often incorporated in engineering solid modeling systems. Here, so-called cosine shading is computed from Lambert's law, which states that the intensity of light reflected from a perfect diffuser is proportional to the cosine of the angle between the surface normal and the vector toward the light source, or

$$I = I_0 r \cos \theta = I_0 r(\mathbf{n} \cdot \mathbf{m}) \tag{2.17}$$

where I_0 is the intensity of the light source at the surface, r is the reflectivity of the surface $(0 \leq r \leq 1)$, θ is the angle of incidence, and $\mathbf{m}$ and $\mathbf{n}$ are the unit vectors shown in Figure 2.25. In more realistic situations, light from several sources combines to illuminate the image. One source is ambient light, a distributed uniform source from the surroundings. Accordingly, Lambert's law is modified to give

$$I = I_a + \sum_{i=1}^{n} I_i r \cos \theta_i = I_a + \sum_{i=1}^{n} I_i r(\mathbf{m}_i \cdot \mathbf{n}) \tag{2.18}$$

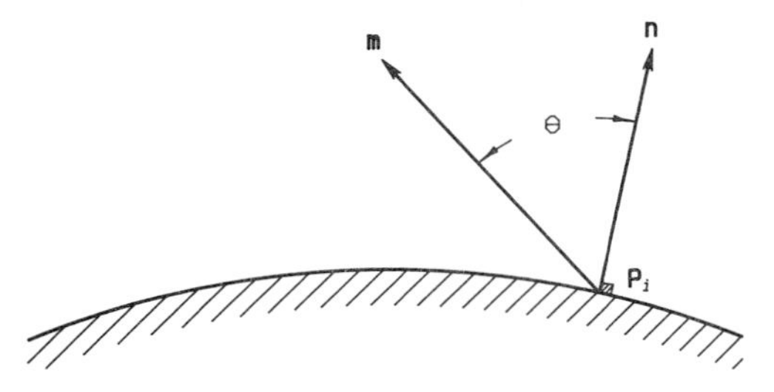

Figure 2.25 Angle of incidence θ at point $\mathbf{P}_i$ is measured between the light source and the surface normal.

where I_a is the contribution of ambient light, r is the reflectivity of the surface, and I_i are the intensities of the n individual sources oriented in the directions $\mathbf{m}_i$.

It is also known that the intensity is inversely proportional to the square of the distance between the light source and the surface. Thus, the terms for the individual sources in Eq. (2.18) can be modified accordingly if the lights are measured at the source instead of the surface.

Specialization of the foregoing for the polygonal surfaces of a faceted model is useful. The procedure is essentially similar to that in Eqs. (2.10) through (2.15). If $\mathbf{P}_i$, $\mathbf{P}_j$, $\mathbf{P}_k$ are points describing three points in the plane under consideration, the unit normal is given by

$$\mathbf{n} = \frac{(\mathbf{P}_j - \mathbf{P}_i) \times (\mathbf{P}_k - \mathbf{P}_i)}{|(\mathbf{P}_j - \mathbf{P}_i) \times (\mathbf{P}_k - \mathbf{P}_i)|} \tag{2.19}$$

For this to be the *outward* normal, the three points $\mathbf{P}_i$, $\mathbf{P}_j$, and $\mathbf{P}_k$ must be in counterclockwise order as viewed from the outside of the surface. [Alternatively, the test in Eq. (2.11) or Eq. (2.13) can be applied to determine whether the normal is outward.] The unit vector from point $\mathbf{P}_i$ on the surface to the light source is

$$\mathbf{m} = \frac{(\mathbf{P}_m - \mathbf{P}_i)}{|(\mathbf{P}_m - \mathbf{P}_i)|} \tag{2.20}$$

which can be assumed to be the same over the whole surface if the distance to the source is sufficiently large. The cosine of the angle between the two unit vectors $\mathbf{n}$ and $\mathbf{m}$ is given by the dot product $\mathbf{m} \cdot \mathbf{n}$. Thus, incorporation of Eqs. (2.19) and (2.20) into Eq. (2.17) yields for point $\mathbf{P}_i$,

$$I = \frac{I_0 r}{|\mathbf{P}_m - \mathbf{P}_i|\,|(\mathbf{P}_j - \mathbf{P}_i) \times (\mathbf{P}_k - \mathbf{P}_i)|} \begin{vmatrix} (P_{mx} - P_{ix}) & (P_{my} - P_{iy}) & (P_{mz} - P_{iz}) \\ (P_{jx} - P_{ix}) & (P_{jy} - P_{iy}) & (P_{jz} - P_{iz}) \\ (P_{kx} - P_{ix}) & (P_{ky} - P_{iy}) & (P_{kz} - P_{iz}) \end{vmatrix} \tag{2.21}$$

An example of a faceted model shaded by Eq. (2.21) is shown as cosine shading in Plate 2.5(a).

The use of cosine shading creates a "banded" appearance because the normals are discontinuous from one polygonal surface to another. Using a large number of small polygons will give the image a smoother appearance. However, Gouraud shading,[23,39] Plate 2.5(b), presents a more economical solution. In Gouraud shading, the first step is to determine the intensity at each vertex, which is based on the average of the normals of the adjacent polygons. The normals are averaged by

$$\mathbf{n}_j = \frac{\Sigma \mathbf{n}_i}{|\Sigma \mathbf{n}_i|} \tag{2.22}$$

where the $\mathbf{n}_i$ are the unit normals for the polygons sharing vertex j. Note that the denominator of Eq. (2.22) is *not* simply the number of polygons.

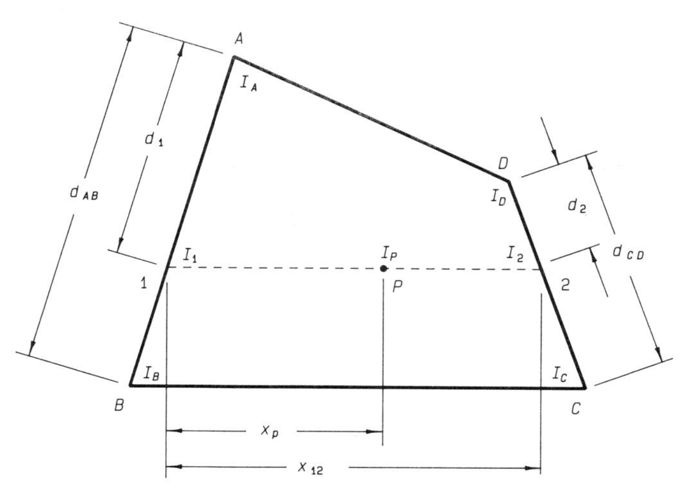

Figure 2.26 Interpolation along scan line for Gouraud shading.

Between the vertices, linear interpolation is used for the intensity of each pixel. For example, consider the point P located on a typical scan line crossing the polygon shown in Figure 2.26. Polygon $ABCD$ is illuminated so that the average intensities, computed from Eqs. (2.22), (2.17), [or (2.18)], at the vertices are I_A, I_B, I_C, and I_D. Linear interpolation along edges AB and DC yields the values

$$I_1 = \frac{(I_B - I_A)d_1}{d_{AB}} + I_A$$
$$I_2 = \frac{(I_C - I_D)d_2}{d_{CD}} + I_D \tag{2.23}$$

and it follows that the intensity at P at a distance x_P along the scan line is

$$I_P = \frac{(I_2 - I_1)x_p}{x_{12}} + I_1 \tag{2.24}$$

where x_{12} is the length of scan line crossing the polygon. Gouraud shading is unnecessary for surfaces described by the parametric geometry of Chapter 6. Such analytic surfaces have continuous curvature, and the normal at any point on the scan line can be determined from the interpolation equations that describe the surface.

Implementation of the shading is done on graphics workstations by varying the L (lightness or intensity) over an appropriate range. For pleasing results in the resulting rendition, some experimentation may be necessary. The low range of L may be all black while the higher range may be all white. Typically, a range such as $0.2 \leq L \leq 0.8$ presents a continuous tone color rendition.

Many other shading algorithms including Phong shading, which models specular reflection, are available.[23,39,49] Besides being computationally expensive, the best quality shading requires display systems with large color maps.

Dithering, shown in Plate 2.4, can be used to replace the requirement for a large color map or to do shading on monochrome displays that do not support gray scales. Plates 2.6 through 2.9 give examples of several advanced color shading techniques used to render engineering models.

2.8 SEGMENTS

Output primitives and their associated attributes may be grouped into subpictures known as *segments* for treatment as a single unit. Graphical data in stroke-refresh terminals are organized as segments. Segments may be manipulated under program control or under interactive control by the user. Some intelligent workstations provide memory for segment storage and local processing for segment operations. Alternatively, host-based segment operations are necessary to support less intelligent workstations.

Some of the formal operations with segments are[14,30]:

Creation. A program command that opens the segment also provides that the segment be identified with a number. All subsequent graphics operations such as line drawing and text are stored until a segment is closed. If desired, other existing segments can be incorporated into the open segment.

Renaming. The identification number on a segment may be changed.

Modification. Segments may undergo such transformations as change of size or change of location. The change of size is called *scaling.* Location changes can be classified as *translation,* where the segment moves a specified amount in the x and y directions, or *rotation,* where the angular location of the segment is changed.

Visibility. Whether the segment is displayed at the workstation depends on whether the segment is made visible. Segments that are set to invisible are available for reuse.

Detectability. This attribute determines whether the segment is available for pick input. A detectable but invisible segment cannot be picked.

Priority. If two or more segments overlap, priority of display and picking can be set.

Association. Segments may be tagged for use on specific workstations, allowing system-dependent segments in an environment having different kinds of workstations.

Deletion. Any previously defined segment may be deleted and its identification number may be reused.

Pan and *zoom* operations built into some workstations make use of the segment processing capabilities. In the case of pan, the visible segment(s) is translated and clipped to the window. For zoom, the visible segment(s) is scaled and clipped. Workstations with fast processing ability can do pan and zoom fast enough to give the appearance of continuous motion.

The *pick* logical input (Section 1.5) as implemented in some computer-aided drafting (CAD) systems works in conjection with segments. To be available for pick, the detectability of segments is required to be "on." When the segment is "picked" it may be highlighted by flashing or by being more brightly displayed.

Another important use of segments in CAD systems is usually known in context as *symbols*. Typically, libraries of symbols are accessed interactively by the user for incorporation into drawings. An architectural library, for example, would contain details such as plumbing fixtures and electrical symbols that could be retrieved and manipulated.

Segmentation is used well in animation, where each part of a scene that moves as a group may be defined as a segment. The changing parts are conveniently manipulated as segments while the rest of the scene remains stationary. Local storage of segments in workstations also aids animation, where each frame in an animated sequence can rapidly be made visible and invisible under program control.

2.9 ANIMATION

Motion of linkages, vibrating structures, cutting-tool paths, and other applications can be shown with the rapid display of a sequence of scenes. Simple animation can be displayed in real time on a CRT; more involved animation requires recording and playback on videotape or photographic film.

Several frames out of a sequence to animate a simple mechanism are superimposed in Figure 2.27. Because the content of each frame can be produced by locating and drawing just two vectors, personal-computer workstations can produce satisfactory animation in real time. The apparent speed of the resulting animation depends on the spacing of the scenes and the speed of execution.

The simplest way to implement real-time animation is by drawing "black" vectors to selectively erase changed parts of each scene. However, if it happens that each scene in the sequence overwrites the previous scenes, erasure is redundant. In some systems, processor speed limits the ability to do animation. In host-based systems, the data rates with serial communication to the graphics workstation are usually inadequate for animation. For such systems, parallel communication can provide the necessary speed for adequate animation quality.

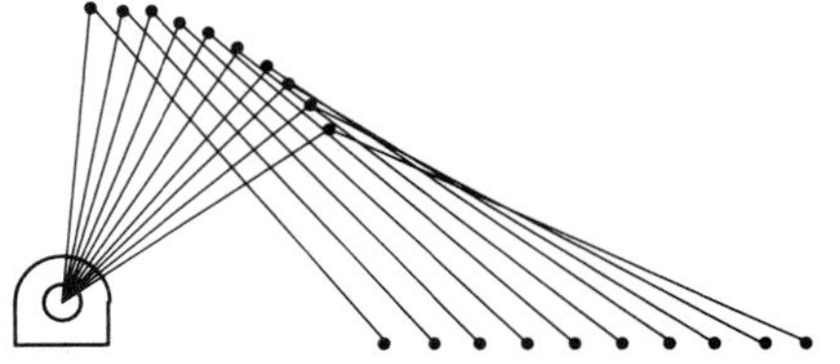

Figure 2.27 Superposition of frames for animation of slider crank mechanism.

Another strategy for improving animation is to compute and store the sequence of frames. Storage can be in memory or on disk in the computer system or as an image on a video recorder or photographic film. At playback, the frames are displayed in rapid sequence.

In animation applications, the use of segments is advantageous both for the creation of images and for storage of the sequence of frames. As each frame is prepared, unchanging material can be incorporated simply by including the appropriate segments. The changed content of each frame can be placed either by transforming existing segments or by computing new information.

Each frame, as it is prepared, can be defined as a new segment and stored in the workstation. Management of the playback of the stored segments takes place under program control. Animation is controlled by setting the visibility of each segment in the sequence "on" and then "off." Storage of segments in workstation memory is a usable animation strategy in systems where they are processor or data transmission limitations.

On the order of 10 to 30 frames per second are recommended for good animation quality. In the case of computer display, quality is improved if the old segment is turned "off" before the new one is turned "on." If hardware limits the framing rate or the total number of frames, playback from videotape or motion picture film is indicated. Special video and film recorders are available which, under computer control, record each scene automatically.

Two methods for generating the content of each frame are recognized[56]: *keyframe* or computer-assisted animation and *modeled* animation or image synthesis.

Keyframe animation uses the computer to improve drawings, to specify motion along a path, to produce "in-between" images, and to color and shade images. In practice, a series of keyframes may be supplied by the animator. Various interpolation techniques[56] are available to create the in-betweens. Choices of interpolation determine whether the motion has constant velocity, acceleration, or other spatial and temporal paths.

Modeled animation involves generating and manipulating true 3-D representations of the objects being animated. The mechanism of Figure 2.27 is a simple example of modeled animation, since the positions of the elements of the mechanism are computed from kinematics equations. In three dimensions, however, building the models is much more time-consuming than actually describing the motion.[56] Techniques for model building include photographing and digitizing real-world models (see Section 4.7) and using solid modeling (see Section 10.1). Because manual digitizing is time-consuming and tedious, mathematical description of models is gaining popularity. This is one of many areas driving the improvement of computer graphics user interfaces.

2.10 GRAPHICS STANDARDS

With standards, the problems of interchanging software between different systems, sharing computer graphics data, and interconnecting equipment are reduced. There is and will continue to be proliferation of standards that deal with various aspects of computer graphics. Setting of standards is a dynamic process; new developments and new standards go hand in hand.

Historically, de facto standards are set by one manufacturer introducing a new system and other manufacturers then following with "compatible" systems. However, as the computer graphics industry matures, nonstandard systems will find less acceptance.

Graphics standards can be placed in three broad categories:

1. Application programming conventions. Subroutines structured according to Core, GKS (Graphical Kernel System) or PHIGS (Programmer's Hierarchical Interactive Graphics Standard) are incorporated into programs to produce graphical actions.
2. Protocols for storage and transmittal of graphics data. Initial Graphics Exchange Specification (IGES) and Computer Graphics Metafile (CGM) deal with encoding images in such a way that the files can be shared among differing systems.
3. Graphics device interfaces. North American Presentation-Level Protocol Syntax (NAPLPS) defines data transmission interfaces for hardware. Computer Graphics-Virtual Device Interface (CG-VDI) deals with the interface between software and device drivers.

The relationship between the categories is shown schematically in Figure 2.28. Standards of interest to the widest range of programmers and users in engineering applications are Core, GKS, PHIGS, and IGES.

At present, there are several agencies dealing with establishment of computer graphics standards. The American National Standards Institute

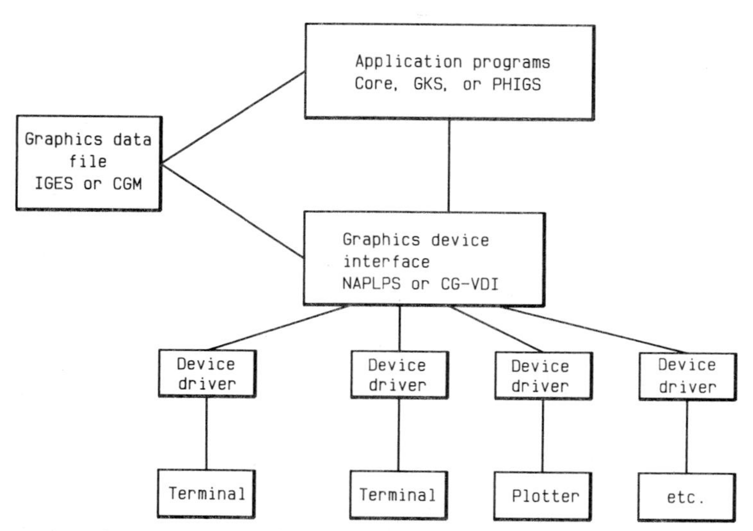

Figure 2.28 Functional division of graphics standards.

(ANSI) and its international counterpart the International Standards Organization (ISO) coordinate the standardization activities of industry and professional organizations. Computer graphics is one of the many areas in which these organizations provide standards. The first graphics standards were set by the Special Interest Group on Graphics (SIGGRAPH) of the Association for Computing Machinery (ACM). Several aerospace companies, the National Bureau of Standards (NBS), the Canadian Standards Association (CSA), and American Telephone and Telegraph (AT&T) also have roles in establishment of standards. The chief standardization agency, however, is ANSI, which has several ongoing committees involved in setting computer graphics standards.

Up-to-date information on graphic standards can be found in the periodical literature. Full published specifications are available from standardization agencies. Condensed versions of standards are available.[14,30,58]

Core, GKS, and PHIGS

The evolution of standards that deal with the application interface started with the publication of a proposed Core standard in 1977. However, a European development, GKS (2-D), became the international standard in 1986. Subsequently, the GKS-3-D and PHIGS standards have evolved. Application interface standards deal with the translation of graphics concepts into computer-usable form. Associated with these standards are "bindings" to high-level programming languages such as FORTRAN-77 and Pascal.

Conformity of graphics tools subroutine packages to a certain standard means that there is 100 percent interchangeability of the resulting code among systems. Because the full set of functionalities defined under any of the standards leads to a very large package, many subset implementations exist—both standard and nonstandard.

TABLE 2.3 CAPABILITIES IN VARIOUS OUTPUT LEVELS OF CORE IMPLEMENTATIONS

	Output levels		
Output functions	1 Basic	2 Buffered	3 Dynamic
Transformation and viewing (window, viewport, clipping)	•	•	•
Output primitives (lines, markers, text strings, polygons)	•	•	•
Primitive attributes (line style, width, color, font, etc.)	•	•	•
Temporary segments	•	•	•
Retained segments		•	•
Segment attributes (visibility, detectability)		•	•
Segment transformations			•

In Core there are two standard implementation levels for dimension (2-D and 3-D), three for output (Table 2.3), and three for input.

GKS defines a matrix of four output levels and three input levels, for a total of 12 standard implementation levels. The four output levels, with some of the distinctions between the levels, are

Level m: Subset of output primitives—polyline, polymarker, text, and fill area. Subset of attributes.

Level 0: Minimal output—all output primitives of level m plus cell array and generalized drawing primitive. Most attributes. Only one output workstation may be open.

Level 1: All the above plus the ability to use segments. More than one output workstation may be open.

Level 2: All the above plus workstation-independent segment storage facilities are allowed. This permits segments created for one workstation to be converted for display on a different workstation.

The three GKS input levels, with some of the distinctions, are

Level a: No input functions supported. Plotters and hard-copy devices fit this level.

Level b: Only request input supported. In request input, the application program asks for input and waits for its return. Included are request locator, request stroke, request valuator, request choice, and request string. Request pick is supported at output levels 1 and 2.

Level c: Request, sample, and event input support. In sample input, the value on a logical input device is taken without waiting for operator action. In event input, any input values are placed in a queue, where they are kept until needed by the application program. Sample and get pick are supported at output levels 1 and 2.

The proposed 3-D extensions to GKS involve the addition of 3-D viewing functions, optional removal of hidden lines and surfaces, and optional 3-D locator and stroke logical input devices.

PHIGS has evolved from GKS and Core. The same six logical input devices (locator, stroke, choice, valuator, string, and pick) with the same three operating modes (request, event, and sample) as in GKS are defined. The viewing functions in PHIGS are similar to those in 3-D Core.

All three systems recognize world coordinates. However, there are some differences among the systems in the acceptance of device-level coordinate systems and in the treatment of viewing operations. PHIGS offers additional levels of coordinate systems to ease programming.

With the evolution of the three standards, the number of recognized output primitives has increased. The Core primitives include line, polyline, polygon, marker, polymarker, and text. In addition to these primitives, GKS has a generalized drawing primitive and a cell-array primitive that can take advantage of new and potential hardware developments. Even more primitives have been added to PHIGS, where there are subpolygons and 3-D shapes.

Interestingly, the *move absolute, move relative, draw absolute,* and *draw relative* primitives in Core were not carried forward into GKS and PHIGS. These systems prefer higher-level functionalities.

Additional attributes (color, line width, character size, etc.) have been specified in GKS and PHIGS. Furthermore, attributes may be *bundled,* that is, grouped together and modified as a unit. Bundling permits, for example, changing the attributes pertaining to a segment without redrawing the segment.

Although the computer graphics industry recognizes the importance of portability of programs from one computer system to another, acceptance of application programming standards has been slow. New hardware, innovative user interfaces, and the desire for simplicity have all contributed to the slow progress.

IGES

The problem of transferring geometric data between different computer graphics systems has been central to the development of IGES. Files transferred under the IGES protocol[58] are used to generate engineering drawings and part description for numerically controlled machining operation.

The IGES standard specifies the form of the "neutral files" that describe an engineering drawing. IGES entities are classified according to whether they are geometry, annotation, or structure and definition.

Geometry entities include categories such as arcs, lines, planes, spline curves, and ruled surfaces. Each geometry entity must be accompanied by specified parameters. For example, the parameters for a line are the coordinates of the starting and ending points.

Annotation entities include dimensions, notes, labels, witness lines, and other features pertinent to engineering drawings. Parameters associated with a linear dimension, for example, include the text, the leaders with arrowheads, and extension lines.

Structure and definition entities define relationships among the elements of the model. Typical capabilities include description of the interrelation of the parts of the drawing, properties associated with lines and regions, and definitions of text fonts.

In a sense, an IGES file is a "picture" because it contains all the information needed to generate an engineering drawing independent of any particular type of system. The IGES system is popular because software modules for IGES preprocessing and postprocessing can be added to existing systems, and the IGES primitives are similar to those found in CAD-CAM. On the other hand, IGES files are so verbose that they will be used only when required for data transfer.

CGM, CG-VDI, and NAPLPS

Among changes in these developing standards have been the names themselves. CGM was previously called Virtual Device Metafile (VDM), and CG-VDI was previously known by the names Computer Graphics Interface (CGI) and Virtual Device Interface (VDI).

CGM is a protocol for encoding computer graphics related to Core, GKS, and PHIGS in a system-independent form. The term *metafile* refers to a device-independent protocol for storage and transmission of graphical data and related control information.

CG-VDI is an interface between the device-dependent and device-independent levels of a graphics system. Device-dependent levels pertain to the coordinates and control information needed by plotters, raster screens, and so forth. At this level coordinates may be in meters, raster units, or other units. At the device-independent level, the coordinates (NDC) are in 0-1 space.

CGM and CG-VDI share several primitives and attributes. The attribute structure parallels that of GKS. Because CG-VDI deals with the interface, it also includes provision for input functions.

NAPLPS[14,58] deals with compression of graphic data for transmission over communications systems. Both text and graphics are converted to efficient binary code under this standard, which can use narrowband communications links (e.g., telephone lines). It is a standard for sending videotext and teletext to homes, schools, and offices. NAPLPS uses *opcodes* (one-byte characters) for graphics primitives, attributes, and control functions that are executed at the receiving terminal.

2.11 CLOSURE

The evident diversity and complexity of graphics systems indicate the importance of users being familiar with the documentation on their own systems. Furthermore, advanced techniques and algorithms have been published and incorporated in graphics systems. New professional systems should be based on previously developed software, such as the excellent graphics tools packages that are available.

Gradually, standardization is being developed and implemented by suppliers and users. Although the economic benefit of interchangeability is self-evident, the desire to keep existing systems, the inflexibility toward new developments, the complexity introduced by unneeded features, and other such factors have impeded more rapid acceptance of standards.

PROBLEMS

2.1. Three edges of a 2-in. cube correspond with the three positive Cartesian coordinate axes. Determine the points matrix and the topology matrix.

2.2. An octahedron (shown in the accompanying figure) is formed from eight equilateral triangles having 30-mm edges. The figure is symmetrical with respect to the y axis, with one vertex on the origin and the opposite vertex on the positive y axis. The square formed by the other four vertices is oriented such that two of these vertices are in the x-y plane. Find the points matrix and the topology matrix.

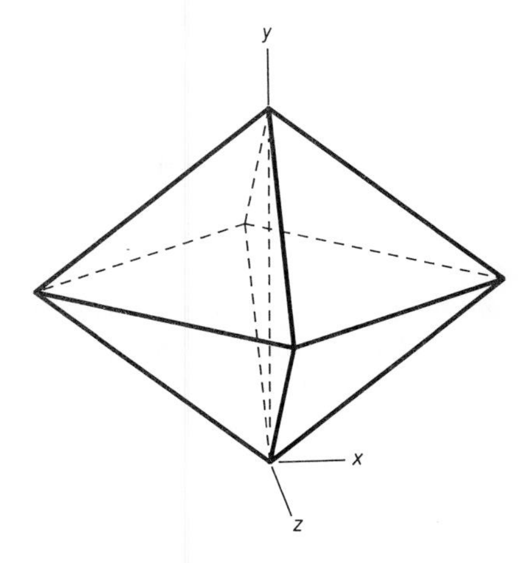

2.3. A tetrahedron is a figure composed of four triangular faces. A tetrahedron composed of four equilateral triangles with 2-in. edges is oriented such that one of the faces is in the x-y plane with one edge coincident with the y axis and such that the fourth vertex is in the positive z direction. Determine the points matrix and the topology matrix.

2.4. Combine Eqs. (2.4) and (2.6) into a single equation that describes the window-viewport transformation with the viewport defined in NDC.

2.5. For a certain display screen, the device coordinates are (0, 0) for the lower left corner and (1023, 768) for the upper right corner. The lower left corner of the window is (0.0, 0.0), and the upper right corner is (400.0, 300.0) in user coordinates. Determine the device coordinates of the point (315.2, 102.9) given in user coordinates. Assume the default viewport is used.

2.6. For the display screen and window of Problem 2.5, write the window transformation in the form of Eq. (2.5).

2.7. The display screen of a microcomputer has device coordinates of (0, 0) at the upper left corner and (279, 191) at the lower right corner. For a viewport covering the entire screen and a window with the lower left corner at (−50.0, 0.0) and the upper right corner at (50.0, 75.0), determine the window transformation in the form of Eq. (2.5).

2.8. The width of the physical screen of the microcomputer system in Problem 2.7 is 1.3 times the height. Rework Problem 2.7 to determine the window transformation so that objects are not distorted. The viewport will not use the bottom portion of the screen.

2.9. Repeat Problem 2.7, but with a viewport bounded by $u_l = 0.2$, $u_r = 0.6$, $v_t = 0.5$, and $v_b = 0.1$ in NDC. The viewport uses the whole screen.

2.10. The screen of a display is addressed in device space with the range $0 \leq u \leq 4095$, $0 \leq v \leq 4095$, where the origin is at the bottom left. The top one-fourth of the screen is addressed but not displayed. The window selected has world coordinates of (0, 0) at the bottom left and (130, 100) at the top right. Determine the transformation from screen coordinates (u, v) to world coordinates (x, y) in the form $x = Au + B$ and $y = Cv + D$.

2.11. A map covers an area 5 miles square. Distances in world coordinates are in feet. Determine the transformation equations to convert world coordinates to normalized device coordinates.

2.12. For a window defined by normalized device coordinates and a line segment that connects the points (−0.2, 0.1) and (0.9, 1.2) determine (a) the bit codes for the end points, (b) the logical AND of the end point bit codes, and (c) the intersections of the line segment with the window.

2.13. Repeat Problem 2.12 but with a line segment that connects the points (0.2, 0.1) and (0.9, 1.2).

2.14. A window is defined in normalized device coordinates. The vertices of a triangle lie at points (0.1, −0.2), (0.8, 0.5), and (1.2, 0.4). Determine the sequence followed by the Sutherland-Hodgman algorithm in clipping this triangle.

2.15. The outward normal from a face passing through the origin on a tessellated model is $\mathbf{n} = 0.333\mathbf{i} + 0.667\mathbf{j} - 0.333\mathbf{k}$. An observer can move along an axis parallel to the z axis which is defined by $x = 3, y = 4$. Determine the z location along this axis where the face changes from visible to invisible.

2.16. The coordinates of the vertices defining one face of a convex faceted solid are (0, 0, 0), (1, 1, 0), and (0, 1, 2). Determine whether this face is visible to an observer at the point (0, 1, 5), if a vertex on another face is located at (2, 2, 2).

2.17. A tetrahedron has vertices at (0, 0, 0), (1, 1, 0), (−2, 3, 1), and (2, −2, 1). Denote the vertices by the numbers A, B, C, and D in the order given so that the faces may be designated by their topology. Determine which faces are visible to an observer located at (a) (10, 0, 0), (b) (0, 10, 0), and (c) (0, 0, 10).

2.18. The *HLS* color values at a point are 210°, 50%, and 100% respectively. (a) Convert these values to the *RGB* system. (b) Estimate the values for the x and y chromaticity coordinates.

2.19. For the *RGB* color of 80%, 60%, 20%, determine the corresponding description in *HLS*.

2.20. A color is described in *HLS* by 110°, 75%, 100%. Convert this to *RGB*.

2.21. Determine the x and y chromaticity coordinates for the following colors described in *HLS:* (a) 120°, 50%, 50%, (b) 300°, 50%, 100%, and (c) 60°, 50%, 25%.

2.22. A plane surface has the normal unit vector $0.231\mathbf{i} + 0.923\mathbf{j} - 0.308\mathbf{k}$. A single light source is located in the direction given by $\mathbf{m} = 0.311\mathbf{i} + 0.545\mathbf{j} - 0.779\mathbf{k}$. If the product $I_0 r = 10.0$, determine the shaded intensity of illumination on the surface.

2.23. The three points (2, 3, 1), (0, 1, 2), and (1, 3, 0) lie on a surface. An illumination source with a value of 15 is located at a large distance from the surface such that $\mathbf{m} = (\mathbf{i} + \mathbf{j} + \mathbf{k})/\sqrt{3}$. Determine the shaded intensity on this surface if $r = 0.3$.

2.24. Rework Problem 2.22, but add a second light source with the same value, but located such that $\mathbf{m} = \mathbf{j}$. Assume the contribution due to ambient light is negligible.

2.25. The normal unit vectors of four faces that meet at one vertex are

$$\mathbf{n}_1 = 0.600\mathbf{i} + 0.800\mathbf{j} \qquad \mathbf{n}_2 = 0.707\mathbf{i} + 0.200\mathbf{j} + 0.678\mathbf{k}$$

$$\mathbf{n}_3 = 0.707\mathbf{j} + 0.707\mathbf{k} \qquad \mathbf{n}_4 = 0.800\mathbf{i} + 0.600\mathbf{k}$$

In preparation for application of the Gouraud shading model, determine the averaged normal unit vector at this vertex.

2.26. Demonstrate that the denominator of Eq. (2.22) is not equal to n, the number of polygons.

2.27. The four edges of a convex planar polygon have the lengths

$$AB = 3 \text{ in.} \qquad BC = 4 \text{ in.}$$

$$CD = 2.5 \text{ in.} \qquad DA = 5 \text{ in.}$$

Sides AB and BC are perpendicular and the scan lines are parallel to BC. The average intensities of reflected illumination at the four vertices are

$$I_A = 10.0 \qquad I_B = 5.0$$

$$I_C = 4.0 \qquad I_D = 8.0$$

Determine the reflected intensity as predicted by the Gouraud shading model at a point that is 1 in. from line AB and 2 in. from line BC.

PROJECTS

NOTE: Appropriate routines from a graphics tools subroutine library should be used to implement these projects.

2.1. Write an interactive program to place graphics text on the output display. A menu with the user choices COLOR, TEXT SIZE, GO, and QUIT should be offered before each line is typed. When GO is selected, the cursor should come up to locate the start of the text strings. Use your program to make a poster announcing a meeting.

2.2. Develop a program that draws pie charts automatically. Ensure that the data are input in percent, with the total being 100 percent. Use a library routine to draw a circle, followed by drawing straight lines from the center to the circle at the appropriate angles as computed. If time permits, include routines that bring up the cursor to locate text and permit interactive input of text.

2.3. Write a program that continuously samples the system clock in your computer and displays, in proper position, the hour, minute, and second hands of a clock face. Optionally, add the numerals for the clock face. For a host-based system, it may be preferable to sample the clock only once and draw only hour and minute hands. If it is not possible to access a system clock, you may substitute typed user input.

2.4. Develop a program that computes the position of a slider in a slider-crank mechanism (see Figure 2.27) as a function of the lengths of the links and the angle of the crank. Display a series of successive positions to animate the operation of the mechanism.

2.5. Develop a program that determines the position of the links in a four-bar mechanism as a function of the lengths of the links and the angle of the input link. Create an animated sequence that shows the full range of the mechanism.

2.6. Write a routine that does line clipping to a viewport. Demonstrate that your routine works in cases where lines are completely out of the viewport, completely in the viewport, and partially in the viewport, with one or both end points out of the viewport.

2.7. Write a routine that treats hidden surfaces by back side elimination. Test the routine with a polyhedron you select.

2.8. Work out the points and topology matrices for an octahedron you select. Determine the midpoints of the faces by averaging the coordinates of the four vertices. Write a program to sort the faces in order of increasing z, and demonstrate the priority fill algorithm on a raster device.

2.9. Develop a program that accepts legal values of *HLS* as input and computes the corresponding values of *RGB* and vice versa. Use the algorithm of Figure 2.24. Display a sample of the color if a suitable system is available.

2.10. With Lambert's law, write a program to shade the faces of an octahedron. Assume the light source is along one of the coordinate axes. Arrange the octahedron so that the four shaded faces have noticeably different reflected intensities.

2

Transformations

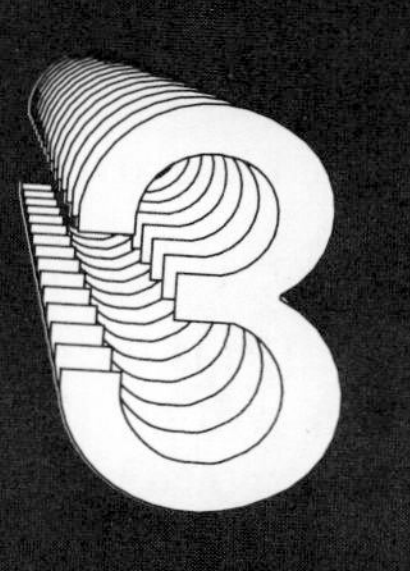

Two-Dimensional Transformations

Images of two-dimensional (2-D) objects can be displayed directly on a graphics workstation. The points matrix, Eq. (2.2), may be written with the z terms set equal to zero,

$$[\mathbf{P}] = \begin{bmatrix} x_1 & y_1 & 0 \\ x_2 & y_2 & 0 \\ \cdot & \cdot\ \cdot\ \cdot & \cdot \\ x_n & y_n & 0 \end{bmatrix} \tag{3.1}$$

The term *transformation* applies to mathematical operations that change the position of the points in various ways. Furthermore, a distinction is made between *modeling* and *viewing transformations.* A modeling transformation is used to change the actual geometry of the object, for example, make it larger. A viewing transformation is used to alter the displayed image, for example, zoom up on a small detail for a better look. In interactive graphics applications, the distinction can be drawn that a modeling transformation changes the data that describe the object, while a viewing transformation does not.

Three basic transformations can be performed on the points matrix: scaling, translation, and rotation. In scaling, the magnitudes of the coordinates change. In translation, points move in such a way that all lines connecting points retain the same direction. In rotation, points move in concentric circular arcs. These transformations may be either modeling or viewing.

Other possible transformations in two dimensions include shearing and mirror reflection. Additional transformations can be constructed by multiplying together any combination of transformations. All of the 2-D transformations find common use in computer-aided drafting (CAD) systems.

Transformation can be expressed in matrix form as

$$[\mathbf{P}^*] = [\mathbf{P}][\mathbf{T}] \tag{3.2}$$

where [**P***] is the new points matrix, [**P**] is the old points matrix, and [**T**] is the transformation matrix. In equation writing the convention is used that the left-hand side contains the result or assigned value, as in statements in programming languages.

In modeling transformations, the [**P***] matrix replaces the [**P**] matrix. In viewing transformations, the [**P***] matrix is used to display the object, and an unchanged copy of [**P**] is retained. While this is an important distinction in graphics systems, the same mathematics governs both types.

It should be noted that the matrices in this chapter, with a few exceptions, provide for three-dimensional data. It might be observed that storage and manipulation resources can be reduced through the use of matrices that do not have columns and rows for z. However, the advantage of treating 2-D as a special case of 3-D is that only one set of transformations and one type of points matrix is needed. The homogeneous coordinate system, which provides a consistent framework for all transformations, is introduced with translation and is utilized thereafter.

3.1 SCALING

Scaling can be classified as uniform or nonuniform: in uniform scaling the same change occurs in all directions; in nonuniform scaling the enlargement or contraction varies in different directions. The process is analogous to that of

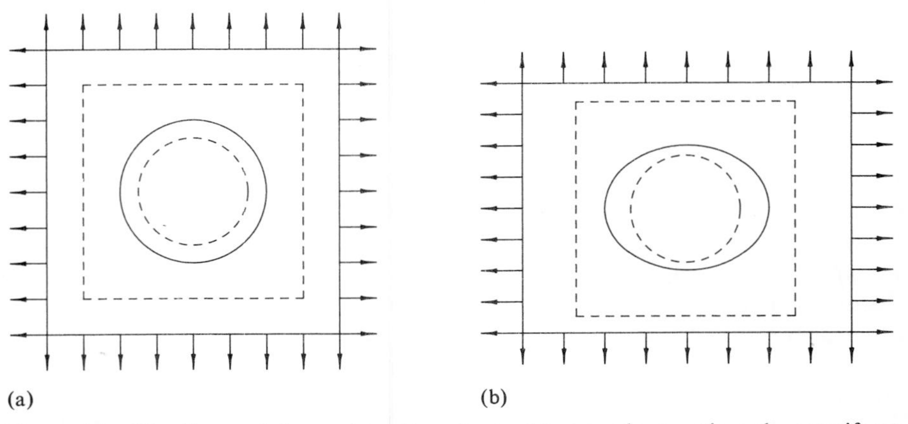

Figure 3.1 The distorted shape of a rubber sheet subject to edge tractions shows uniform and nonuniform scaling. Before loading, the sheet is square with a painted circle as indicated by the dashed outline. The center of scaling (point that does not move) is at the center of the circle. (a) Same loading on horizontal and vertical edges gives uniform scaling. (b) Different magnitudes of loading on horizontal and vertical edges give nonuniform scaling.

normal strain in the mechanics of materials. As illustrated in Figure 3.1, the distortion of a rubber sheet depends on the uniformity of the loading.

The transformation matrix for uniform scaling is simply

$$[\mathbf{T}] = s[\mathbf{I}] \tag{3.3}$$

where [**I**] is the identity matrix and s is the scale factor. This may be written out as

$$[\mathbf{T}] = \begin{bmatrix} s & 0 & 0 \\ 0 & s & 0 \\ 0 & 0 & s \end{bmatrix} \tag{3.4}$$

For a single point, the [**P**] matrix is a row vector and uniform scaling produces

$$[\mathbf{P}^*] = [x \quad y \quad 0] \begin{bmatrix} s & 0 & 0 \\ 0 & s & 0 \\ 0 & 0 & s \end{bmatrix} = [sx \quad sy \quad 0] \tag{3.5}$$

which shows that x and y have each been multiplied by the factor s. Note that the same result would occur in Eq. (3.5) if the s in row 3 of the matrix of Eq. (3.4) were 0, as long as only 2-D scaling is considered.

Whenever the diagonal terms of the matrix in Eq. (3.4) are not equal, nonuniform scaling occurs. Here there is unequal change in the vertical and horizontal dimensions, with the result, for example, that a circle becomes an ellipse. It is convenient to call the term in the first row s_x, the one in the second row s_y, and the one in the third row s_z. (Of course, s_z is redundant in 2-D transformations.) The transformation of a single point in this case is

$$[\mathbf{P}^*] = [x \quad y \quad 0] \begin{bmatrix} s_x & 0 & 0 \\ 0 & s_y & 0 \\ 0 & 0 & s_z \end{bmatrix} = [s_x x \quad s_y y \quad 0] \tag{3.6}$$

If the off-diagonal terms of [**T**] in Eq. (3.4) are nonzero, a shearing transformation occurs. More about the shearing transformation can be found as part of the six-point transformation in Section 3.6. However, an example such as the one that follows shows how off-diagonal terms in the transformation matrix produce shearing.

EXAMPLE 3.1

Given the points matrix

$$[\mathbf{P}] = \begin{bmatrix} 1.2 & 1.2 & 0 \\ 4.2 & 1.2 & 0 \\ 1.2 & 5.2 & 0 \end{bmatrix}$$

which describes the right triangle shown, determine the effect of the following two transformation matrices:

$$[\mathbf{T}_1] = \begin{bmatrix} 0.5 & 0 & 0 \\ 0 & 0.5 & 0 \\ 0 & 0 & 0 \end{bmatrix}$$

$$[\mathbf{T}_2] = \begin{bmatrix} 0.5 & 0.5 & 0 \\ 0 & 0.5 & 0 \\ 0 & 0 & 0 \end{bmatrix}$$

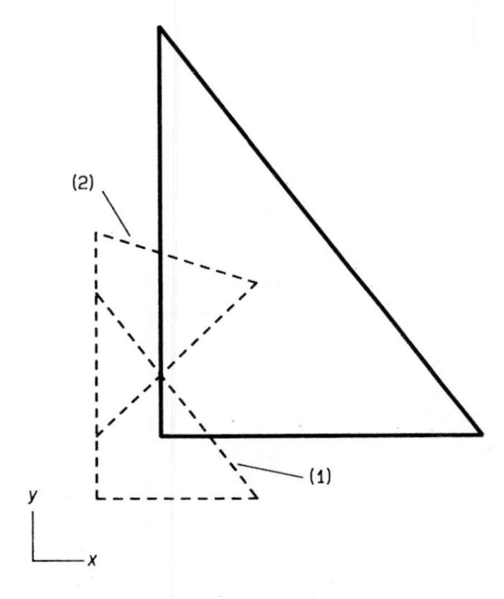

Solution. The matrix multiplication $[\mathbf{P}][\mathbf{T}_1]$ yields

$$[\mathbf{P}_1^*] = \begin{bmatrix} 0.6 & 0.6 & 0 \\ 2.1 & 0.6 & 0 \\ 0.6 & 2.6 & 0 \end{bmatrix}$$

which results in the uniform scaling represented by the dashed triangle labeled (1).

The multiplication $[\mathbf{P}][\mathbf{T}_2]$ yields

$$[\mathbf{P}_2^*] = \begin{bmatrix} 0.6 & 1.2 & 0 \\ 2.1 & 2.7 & 0 \\ 0.6 & 3.2 & 0 \end{bmatrix}$$

which results in the nonuniform scaling shown by the dashed triangle labeled (2). Because the matrix $[\mathbf{T}_2]$ has a nonzero off-diagonal term, triangle (2) is sheared as well as changed in size. Shearing as a modeling transformation does not find a great amount of use in actual applications. As a viewing transformation, shearing may be used to produce three-dimensional effects.

3.2 TRANSLATION

Translation of an object implies that every point of the object experiences the same displacement. A points matrix for a single point $[\mathbf{P}] = [x \quad y \quad 0]$ when translated thus becomes

$$[\mathbf{P}^*] = [x + \Delta x \quad y + \Delta y \quad 0] \tag{3.7}$$

where Δx and Δy are the amounts of translation in the x and y directions. Although it is obvious how to write Eq. (3.7) as the sum of two matrices, it is not so obvious how to write it as a product,

$$[\mathbf{P}^*] = [\mathbf{P}][\mathbf{T}_t] \tag{3.8}$$

where $[\mathbf{T}_t]$ is the transformation matrix for translation.

To provide for such a transformation, an augmented coordinate system, the *homogeneous coordinate* system, is needed. In this system, transformation matrices have an extra row and column to provide for all the needed capabilities. It follows that conformability in matrix multiplication dictates that the points matrix $[\mathbf{P}]$ must have an extra (fourth) column,

$$[\mathbf{P}] = [x \quad y \quad 0 \quad h] \tag{3.9}$$

where it is convenient to assign a value of 1 to the fourth coordinate h.

The homogeneous coordinate system maintains square transformation matrices to allow (1) inverse transformations (only a square matrix may be inverted) and (2) *concatenation* or combination of any number of individual transformations into one single transformation matrix. With this structure, it may be verified that translation is described by

$$[\mathbf{P}^*] = [x \quad y \quad 0 \quad 1]\begin{bmatrix} 1 & 0 & 0 & 0 \\ 0 & 1 & 0 & 0 \\ 0 & 0 & 1 & 0 \\ \Delta x & \Delta y & 0 & 1 \end{bmatrix}$$

$$= [x + \Delta x \quad y + \Delta y \quad 0 \quad 1] \tag{3.10}$$

In the homogeneous coordinate system, the transformation matrix for scaling, Eq. (3.4), must be changed from a 3×3 to a 4×4 matrix. This is done by simply putting 0's in the fourth row and fourth column except on the diagonal, where a 1 is required. Generalized for nonuniform scaling in the x and y (and z)

directions, the resulting transformation matrix is

$$[\mathbf{T}_s] = \begin{bmatrix} s_x & 0 & 0 & 0 \\ 0 & s_y & 0 & 0 \\ 0 & 0 & s_z & 0 \\ 0 & 0 & 0 & 1 \end{bmatrix} \tag{3.11}$$

If only 2-D work is being done, the third column of points matrices and the third row and third column of transformation matrices are redundant. Reduction of the dimensionality from four to three produces significant savings in computer time and storage.

EXAMPLE 3.2

Translate the rectangle with the homogeneous points matrix

$$[\mathbf{P}] = \begin{bmatrix} 2 & 2 & 0 & 1 \\ 2 & 8 & 0 & 1 \\ 10 & 8 & 0 & 1 \\ 10 & 2 & 0 & 1 \end{bmatrix}$$

-0.3 in the x direction and $+1.5$ in the y direction.

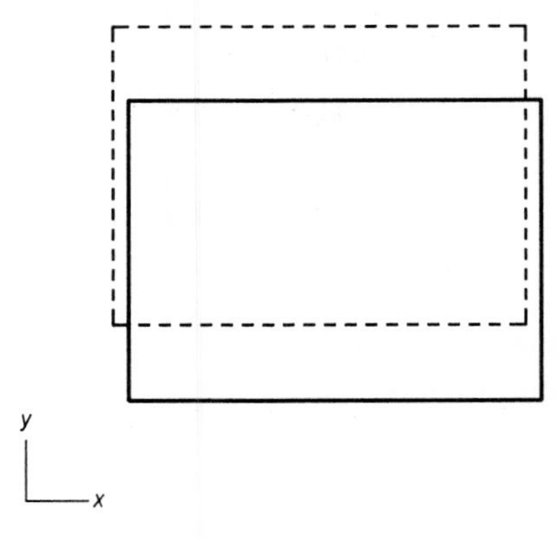

Solution. The defined values of Δx and Δy are substituted into the homogeneous translation matrix, Eq. (3.10),

$$[\mathbf{P}] = \begin{bmatrix} 2 & 2 & 0 & 1 \\ 2 & 8 & 0 & 1 \\ 10 & 8 & 0 & 1 \\ 10 & 2 & 0 & 1 \end{bmatrix} \begin{bmatrix} 1 & 0 & 0 & 0 \\ 0 & 1 & 0 & 0 \\ 0 & 0 & 1 & 0 \\ -0.3 & 1.5 & 0 & 1 \end{bmatrix} = \begin{bmatrix} 1.7 & 3.5 & 0 & 1 \\ 1.7 & 9.5 & 0 & 1 \\ 9.7 & 9.5 & 0 & 1 \\ 9.7 & 3.5 & 0 & 1 \end{bmatrix}$$

It should be noted that, instead of actually using matrix multiplication for this transformation, the result may be computed faster by simply adding Δx to each element in the first column of the points matrix and Δy to each element in the second column. Additional savings in computer time and storage can be achieved by eliminating extra columns and rows since 3-D generalization is not indicated.

3.3 TWO-DIMENSIONAL ROTATION

In plane rotation, points form concentric circular arcs with the center at the axis of rotation. For application to both 2-D and 3-D transformation, the axis of rotation should be visualized as a line, not as a point. In two dimensions, the line that represents the axis of rotation is perpendicular to the x-y plane; the center of the concentric circular arcs is the point formed by the intersection of the rotation axis and the x-y plane.

Formulation of the transformation matrix for plane rotation is based on a trigonometric identity. As shown in Figure 3.2, x^* and y^* are to be expressed in terms of x, y, α, and θ. The point x, y in polar coordinates is

$$\begin{aligned} x &= r\cos\alpha \\ y &= r\sin\alpha \end{aligned} \tag{3.12}$$

and the point x^*, y^* is given by

$$\begin{aligned} x^* &= r\cos(\alpha+\theta) \\ y^* &= r\sin(\alpha+\theta) \end{aligned} \tag{3.13}$$

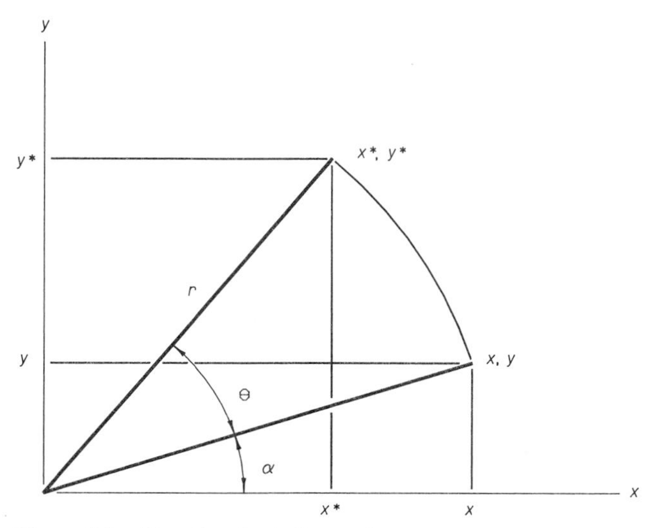

Figure 3.2 Notation for 2-D rotation.

Use of the double-angle formula with Eq. (3.13) yields

$$\begin{aligned} x^* &= r(\cos\alpha\cos\theta - \sin\alpha\sin\theta) \\ y^* &= r(\cos\alpha\sin\theta + \sin\alpha\cos\theta) \end{aligned} \tag{3.14}$$

Substitution of Eqs. (3.12) into Eqs. (3.14) gives

$$\begin{aligned} x^* &= x\cos\theta - y\sin\theta \\ y^* &= x\sin\theta + y\cos\theta \end{aligned} \tag{3.15}$$

In matrix notation, Eq. (3.15) can be written as

$$[x^* \quad y^*] = [x \quad y]\begin{bmatrix} \cos\theta & \sin\theta \\ -\sin\theta & \cos\theta \end{bmatrix} \tag{3.16}$$

Consistent with the scheme for homogeneous transformation matrices that are usable with 2-D and 3-D points matrices, the rotation transformation is

$$[\mathbf{T}_r] = \begin{bmatrix} \cos\theta & \sin\theta & 0 & 0 \\ -\sin\theta & \cos\theta & 0 & 0 \\ 0 & 0 & 1 & 0 \\ 0 & 0 & 0 & 1 \end{bmatrix} \tag{3.17}$$

Here, the axis of rotation is at the origin, and the sign convention is the usual one of counterclockwise positive.

EXAMPLE 3.3

Rotate the line with end points (10, 0) and (0, 8) 90° counterclockwise around the origin.

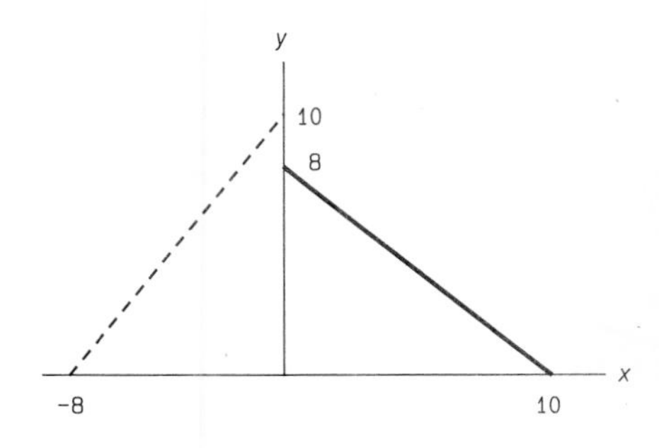

Solution. The homogeneous points matrix and Eq. (3.17) with $\theta = 90°$ can be multiplied to give

$$\begin{bmatrix} 10 & 0 & 0 & 1 \\ 0 & 8 & 0 & 1 \end{bmatrix} \begin{bmatrix} 0 & 1 & 0 & 0 \\ -1 & 0 & 0 & 0 \\ 0 & 0 & 1 & 0 \\ 0 & 0 & 0 & 1 \end{bmatrix} = \begin{bmatrix} 0 & 10 & 0 & 1 \\ -8 & 0 & 0 & 1 \end{bmatrix}$$

The new line is shown dashed in the figure.

3.4 MIRROR REFLECTION

A mirror reflection transformation can be visualized as the image inversion that results from placing a mirror normal to the *x-y* plane. The intersection of the mirror and the *x-y* plane is called the *mirror axis*. Treatment of the mirror reflection transformation as a special case of scaling is correct for both 2-D and 3-D viewing. Although another way to formulate the 2-D mirror transformation is 180° rotation around the mirror axis, it is not correct for 3-D images.

The 2-D mirror reflection transformation is most easily made across the x or y coordinate axis (or, more correctly, across the *x-z* or *y-z* coordinate planes). A point is reflected across the x axis with a sign change on the y term, or

$$[x \quad y \quad 0 \quad 1] \rightarrow [x \quad -y \quad 0 \quad 1] \tag{3.18}$$

and across the y axis with

$$[x \quad y \quad 0 \quad 1] \rightarrow [-x \quad y \quad 0 \quad 1] \tag{3.19}$$

From these relations, it can be deduced that the transformation matrices for mirror reflection are

$$[\mathbf{T}_{mx}] = \begin{bmatrix} 1 & 0 & 0 & 0 \\ 0 & -1 & 0 & 0 \\ 0 & 0 & 1 & 0 \\ 0 & 0 & 0 & 1 \end{bmatrix} \tag{3.20}$$

$$[\mathbf{T}_{my}] = \begin{bmatrix} -1 & 0 & 0 & 0 \\ 0 & 1 & 0 & 0 \\ 0 & 0 & 1 & 0 \\ 0 & 0 & 0 & 1 \end{bmatrix} \tag{3.21}$$

where the first matrix is for reflection around the x axis and the second for reflection around the y axis. These equations are identical to the equation for nonuniform scaling, Eq. (3.11), with $s_y = -1$ in the case of Eq. (3.20) and $s_x = -1$ in the case of Eq. (3.21).

If reflection about some plane other than one defined by a coordinate axis is desired, translation and/or rotation can be used to align the reflection axis with one of the coordinate axes. After the reflection process, the translation and/or rotation must be undone in reverse order to restore the original geometry.

3.5 COMBINING TRANSFORMATIONS

Most applications will require the use of more than one of the basic transformations to achieve desired results. This combination can occur in two ways: (1) by sequential application of transformations or (2) by concatenation of two or more basic transformations into a single transformation.

Sequential application requires that an intermediate result be stored between transformations. Although doing the transformations one at a time is not always efficient, the versatility lends itself to interactive applications.

For combinations of transformations frequently used, it is convenient to use a concatenated transformation matrix obtained by matrix multiplication. Developed below are two very useful combinations: (1) scaling with a given point remaining fixed and (2) rotation around a given axis.

Scaling with an Arbitrary Fixed Point

As can be seen in Example 3.1, all points of an image move under the scaling transformation. The origin itself—the point (0, 0)—does not move. To maintain the position of an arbitrary point, or *pivot point,* under a scaling transformation, a sequence of matrix operations is required. Concatenation of this sequence into a single transformation matrix is done by multiplying the individual transformations in order. The sequence is

$$[\mathbf{P}^*] = [\mathbf{P}][\mathbf{T}_{t1}][\mathbf{T}_s][\mathbf{T}_{t2}] \tag{3.22}$$

where $[\mathbf{T}_{t1}]$ translates $[\mathbf{P}]$ so that the pivot point is moved to the origin, $[\mathbf{T}_s]$ scales the image, and $[\mathbf{T}_{t2}]$ translates the pivot point back to its original position. It is important to note that the order of matrix multiplication progresses from left to right, since altering the order changes the result.

The three transformation matrices in Eq. (3.22) can be concatenated by matrix multiplication to produce a single transformation matrix that uniformly scales an object while keeping the pivot point fixed. Substitution of the homogeneous transformation matrices of Eqs. (3.10) and (3.11) for translation and uniform scaling into the right-hand side of Eq. (3.22) gives

$$[\mathbf{T}_s] = \begin{bmatrix} 1 & 0 & 0 & 0 \\ 0 & 1 & 0 & 0 \\ 0 & 0 & 1 & 0 \\ -x_0 & -y_0 & 0 & 1 \end{bmatrix} \begin{bmatrix} s & 0 & 0 & 0 \\ 0 & s & 0 & 0 \\ 0 & 0 & s & 0 \\ 0 & 0 & 0 & 1 \end{bmatrix} \begin{bmatrix} 1 & 0 & 0 & 0 \\ 0 & 1 & 0 & 0 \\ 0 & 0 & 1 & 0 \\ x_0 & y_0 & 0 & 1 \end{bmatrix} \tag{3.23}$$

where the point (x_0, y_0) is the pivot point. The first matrix of Eq. (3.23) translates the points a distance of $-x_0$, $-y_0$ to the origin, the second does uniform scaling in the x and y directions, and the third translates the origin back to x_0, y_0. The matrix product from Eq. (3.23) is

$$[\mathbf{T}_s] = \begin{bmatrix} s & 0 & 0 & 0 \\ 0 & s & 0 & 0 \\ 0 & 0 & s & 0 \\ x_0 - sx_0 & y_0 - sy_0 & 0 & 1 \end{bmatrix} \tag{3.24}$$

Use of Eq. (3.24) for uniform scaling about a specified pivot point reduces computational effort. This equation degenerates to Eq. (3.11) when the pivot point is the origin.

Rotation around an Arbitrary Point

To accomplish rotation around an axis other than the one through the origin, the rotation matrix, Eq. (3.17), is pre- and postmultiplied by the appropriate translation matrices. The transformation matrix is the product $[\mathbf{T}_{t1}][\mathbf{T}_r][\mathbf{T}_{t2}]$, or

$$\begin{bmatrix} 1 & 0 & 0 & 0 \\ 0 & 1 & 0 & 0 \\ 0 & 0 & 1 & 0 \\ -x_0 & -y_0 & 0 & 1 \end{bmatrix} \begin{bmatrix} \cos\theta & \sin\theta & 0 & 0 \\ -\sin\theta & \cos\theta & 0 & 0 \\ 0 & 0 & 1 & 0 \\ 0 & 0 & 0 & 1 \end{bmatrix} \begin{bmatrix} 1 & 0 & 0 & 0 \\ 0 & 1 & 0 & 0 \\ 0 & 0 & 1 & 0 \\ x_0 & y_0 & 0 & 1 \end{bmatrix}$$

$$= \begin{bmatrix} \cos\theta & \sin\theta & 0 & 0 \\ -\sin\theta & \cos\theta & 0 & 0 \\ 0 & 0 & 1 & 0 \\ -x_0(\cos\theta - 1) + y_0 \sin\theta & -x_0 \sin\theta - y_0(\cos\theta - 1) & 0 & 1 \end{bmatrix} \tag{3.25}$$

where (x_0, y_0) is the intersection of the x-y plane and the axis of rotation, which is in the z direction. In the foregoing sequences of operations, the order of transformation affects the result.

Usage in Graphics Tools Libraries

The basic transformations of scaling, translation, and rotation are usually found in graphics tools subroutine libraries. Sometimes the library transformation routines will be in a concatenated form, such as Eq. (3.24). If the needed concatenated form is not available, it is easier to apply two or more routines sequentially than to derive a new concatenated transformation matrix.

Transformations of 2-D points matrices can be defined with 3×3 homogeneous transformation matrices. However, for compatibility with three dimensions, 4×4 matrices are necessary. For scaling and translation, the 2-D and 3-D transformations are identical. Rotation of points around axes other than the z axis is treated in Chapter 4.

Some graphics tools libraries use transformation matrices other than the 4×4 matrices developed here. For example, Graphical Kernel System (GKS), the 2-D applications programming standard (Section 2.10), uses a 2×3 transformation matrix. Here, the third column contains the two translation factors, the order of computation is changed to [T][P], and the transformation must be applied to a segment.

Although matrix multiplication may not always be the most computationally efficient way to perform transformations, its use is convenient and consistent. In applying the 4×4 homogeneous transformation matrices, it is necessary that the points matrix have the dimension $n \times 4$. Storage can be saved by not storing the fourth column of 1's. In some graphics tools libraries, it is required that the points be stored in three separate column matrices, as in Eq. (2.3). Routines typical of those in a graphics tools library appear in Appendix B.

EXAMPLE 3.4

Given the triangle described by the homogeneous points matrix

$$[\mathbf{P}] = \begin{bmatrix} 2 & 2 & 0 & 1 \\ 2 & 5 & 0 & 1 \\ 5 & 5 & 0 & 1 \end{bmatrix}$$

scale it to three-fourths size keeping the centroid in the same location. Use (a) separate operations and (b) Eq. (3.24).

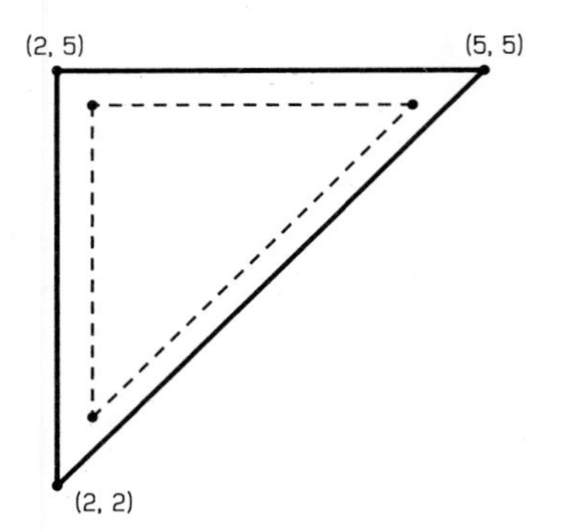

Solution. (**a**) The centroid of the triangle is at $x = (2 + 2 + 5)/3$ and $y = (2 + 5 + 5)/3$ or the point (3, 4). Thus, Δx and Δy are set equal to -3 and -4, respectively, in Eq. (3.10) to first translate the centroid to the origin,

$$\begin{bmatrix} 2 & 2 & 0 & 1 \\ 2 & 5 & 0 & 1 \\ 5 & 5 & 0 & 1 \end{bmatrix} \begin{bmatrix} 1 & 0 & 0 & 0 \\ 0 & 1 & 0 & 0 \\ 0 & 0 & 1 & 0 \\ -3 & -4 & 0 & 1 \end{bmatrix} = \begin{bmatrix} -1 & -2 & 0 & 1 \\ -1 & 1 & 0 & 1 \\ 2 & 1 & 0 & 1 \end{bmatrix}$$

Scaling with Eq. (3.11) (z-direction scaling is irrelevant) yields

$$\begin{bmatrix} -1 & -2 & 0 & 1 \\ -1 & 1 & 0 & 1 \\ 2 & 1 & 0 & 1 \end{bmatrix} \begin{bmatrix} 0.75 & 0 & 0 & 0 \\ 0 & 0.75 & 0 & 0 \\ 0 & 0 & 1 & 0 \\ 0 & 0 & 0 & 1 \end{bmatrix} = \begin{bmatrix} -0.75 & -1.50 & 0 & 1 \\ -0.75 & 0.75 & 0 & 1 \\ 1.50 & 0.75 & 0 & 1 \end{bmatrix}$$

Finally, the image is translated so that the centroid returns to its original position

$$\begin{bmatrix} -0.75 & -1.50 & 0 & 1 \\ -0.75 & 0.75 & 0 & 1 \\ 1.50 & 0.75 & 0 & 1 \end{bmatrix} \begin{bmatrix} 1 & 0 & 0 & 0 \\ 0 & 1 & 0 & 0 \\ 0 & 0 & 1 & 0 \\ 3 & 4 & 0 & 1 \end{bmatrix} = \begin{bmatrix} 2.25 & 2.50 & 0 & 1 \\ 2.25 & 4.75 & 0 & 1 \\ 4.50 & 4.75 & 0 & 1 \end{bmatrix}$$

(**b**) The foregoing set of three operations can be reduced to a single operation using the concatenated matrix of Eq. (3.24) with $x_0 = 3$ and $y_0 = 4$,

$$\begin{bmatrix} 2 & 2 & 0 & 1 \\ 2 & 5 & 0 & 1 \\ 5 & 5 & 0 & 1 \end{bmatrix} \begin{bmatrix} 0.75 & 0 & 0 & 0 \\ 0 & 0.75 & 0 & 0 \\ 0 & 0 & 0.75 & 0 \\ 3 - 0.75(3) & 4 - 0.75(4) & 0 & 1 \end{bmatrix} = \begin{bmatrix} 2.25 & 2.50 & 0 & 1 \\ 2.25 & 4.75 & 0 & 1 \\ 4.50 & 4.75 & 0 & 1 \end{bmatrix}$$

which produces the same result with less computation.

EXAMPLE 3.5

Mirror the given right triangle around axis A–A, which is parallel to the hypotenuse.

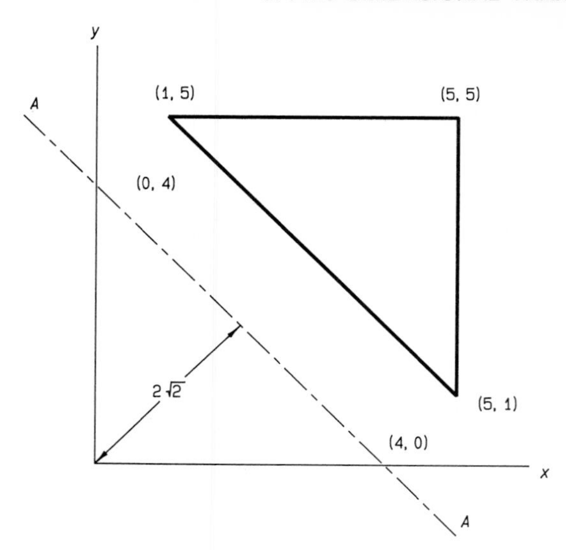

Solution. The figure that consists of the axis and the triangle is (1) rotated 45° clockwise, (2) translated to the left so that the mirror axis passes through the origin, (3) mirrored around the y axis, (4) translated back to the right, and (5) rotated back 45° counterclockwise. The matrices to be multiplied come from Eqs. (3.10), (3.17), and (3.21). The complete transformation process is

$$[\mathbf{P}^*] = [\mathbf{P}][\mathbf{T}_{r1}][\mathbf{T}_{t1}][\mathbf{T}_{my}][\mathbf{T}_{t2}][\mathbf{T}_{r2}]$$

The six matrices on the right-hand side are written out as

$$\begin{bmatrix} 1 & 5 & 0 & 1 \\ 5 & 1 & 0 & 1 \\ 5 & 5 & 0 & 1 \end{bmatrix} \begin{bmatrix} \cos(-45) & \sin(-45) & 0 & 0 \\ -\sin(-45) & \cos(-45) & 0 & 0 \\ 0 & 0 & 1 & 0 \\ 0 & 0 & 0 & 1 \end{bmatrix} \begin{bmatrix} 1 & 0 & 0 & 0 \\ 0 & 1 & 0 & 0 \\ 0 & 0 & 1 & 0 \\ -2\sqrt{2} & 0 & 0 & 1 \end{bmatrix}$$

$$\times \begin{bmatrix} -1 & 0 & 0 & 0 \\ 0 & 1 & 0 & 0 \\ 0 & 0 & 1 & 0 \\ 0 & 0 & 0 & 1 \end{bmatrix} \begin{bmatrix} 1 & 0 & 0 & 0 \\ 0 & 1 & 0 & 0 \\ 0 & 0 & 1 & 0 \\ 2\sqrt{2} & 0 & 0 & 1 \end{bmatrix} \begin{bmatrix} \cos 45 & \sin 45 & 0 & 0 \\ -\sin 45 & \cos 45 & 0 & 0 \\ 0 & 0 & 1 & 0 \\ 0 & 0 & 0 & 1 \end{bmatrix}$$

Note that it is not necessary to include the two points defining the mirror axis in [**P**]. Upon completion of all the indicated multiplications,

$$[\mathbf{P}^*] = \begin{bmatrix} -1 & 3 & 0 & 1 \\ 3 & -1 & 0 & 1 \\ -1 & -1 & 0 & 1 \end{bmatrix}$$

The order of operations and the result may be verified from the figure below.

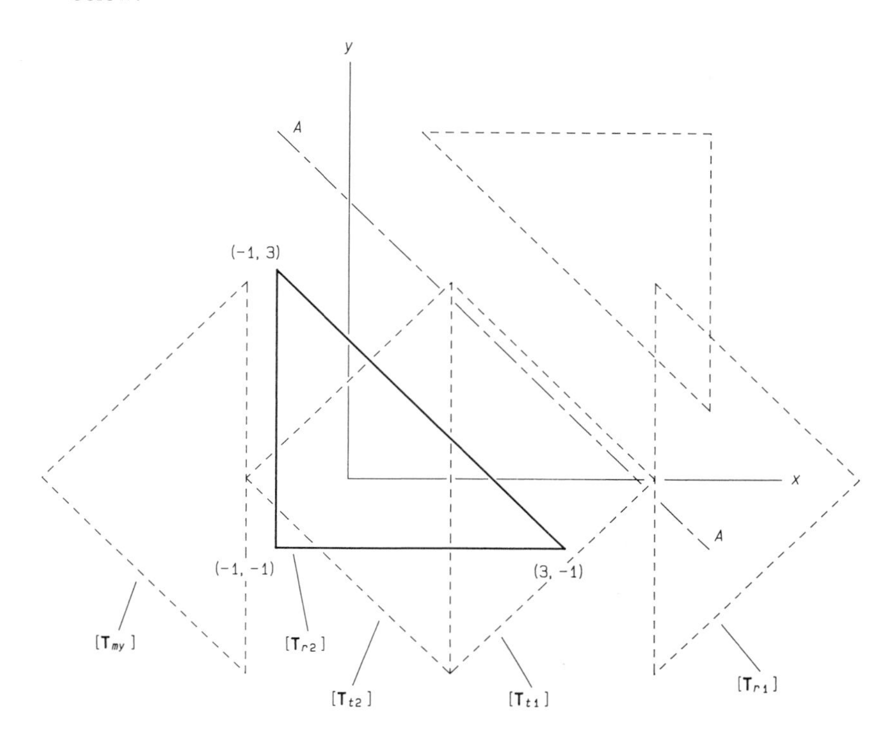

3.6 SIX-POINT TRANSFORMATION

The transformation matrices derived in the preceding sections can be concatenated to produce any arbitrary combination of scaling, shearing, translation, mirror reflection, and rotation. An alternative to this approach is the *six-point transformation,*[29] which combines several transformations and is useful as a copying routine in interactive 2-D CAD programs. Here, one, two, or three pairs of *key points* define a correspondence between the original image and the new image.

A hierarchy is adopted for the pairs of key points where (1) one pair defines the x and y components of translation, (2) two pairs define the x and y components of translation, the angle of rotation, and the factor for uniform scaling, and (3) three pairs define all the operations of the two-pair transformation plus shearing and nonuniform scaling. These three types of transformation are illustrated in Figure 3.3.

As indicated in Section 3.2, the third row and the third column of a homogeneous transformation matrix are redundant for 2-D transformations. Elimination of this row and column reduces the homogeneous 2-D transformation matrix to the dimension 3×3. Conformability in matrix multiplication is

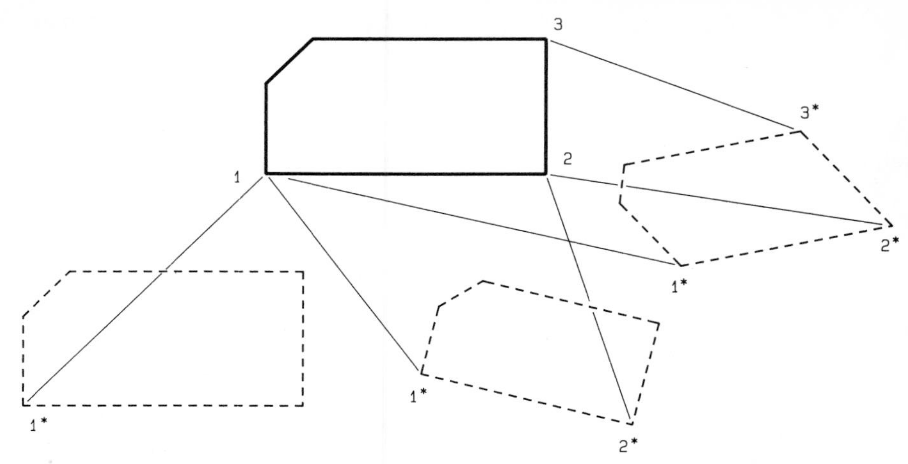

Figure 3.3 Hierarchy of one-, two-, and three-point pairs as used in the six-point transformation. All operations are in two dimensions.

maintained by deleting the z column from the points matrix, making its dimension $n \times 3$.

With the six-point transformation, one, two, or three pairs of key points can be specified in "before"—[**P**]—and "after"—[**P***]—positions. These points can be used to derive the transformation matrix $[\mathbf{T}_{6P}]$, which is then used to postmultiply all the points in the image. To determine $[\mathbf{T}_{6P}]$, the equation used is

$$[\mathbf{P}][\mathbf{T}_{6P}] = [\mathbf{P}^*] \tag{3.26}$$

which defines a set of three linear equations involving three pairs of key points contained in [**P**] and [**P***]. Expansion in 2-D homogeneous coordinates yields

$$\begin{bmatrix} x_1 & y_1 & 1 \\ x_2 & y_2 & 1 \\ x_3 & y_3 & 1 \end{bmatrix} \begin{bmatrix} T_{11} & T_{12} & T_{13} \\ T_{21} & T_{22} & T_{23} \\ T_{31} & T_{32} & T_{33} \end{bmatrix} = \begin{bmatrix} x_1^* & y_1^* & 1 \\ x_2^* & y_2^* & 1 \\ x_3^* & y_3^* & 1 \end{bmatrix} \tag{3.26a}$$

which contains three sets of three equations each with three unknowns. For solution, $[\mathbf{T}_{6P}]$ and $[\mathbf{P}^*]$ are partitioned into three column submatrices such that $[\mathbf{T}_{6P}] = [[\mathbf{T}_x][\mathbf{T}_y][\mathbf{T}_h]]$ and $[\mathbf{P}^*] = [[\mathbf{P}_x^*][\mathbf{P}_y^*][\mathbf{P}_h^*]]$, where the subscripts indicate the columns. Thus, Eq. (3.26a) may be rewritten as three separate systems,

$$\begin{aligned} [\mathbf{P}][\mathbf{T}_x] &= [\mathbf{P}_x^*] \\ [\mathbf{P}][\mathbf{T}_y] &= [\mathbf{P}_y^*] \\ [\mathbf{P}][\mathbf{T}_h] &= [\mathbf{P}_h^*] \end{aligned} \tag{3.27}$$

The third of Eqs. (3.28) yields $T_{13} = 0$, $T_{23} = 0$, and $T_{33} = 1$. The six remaining unknowns, determined numerically, correspond to the x and y

coordinates of the three specified points. If need be, the 3×3 $[\mathbf{T}_{6P}]$ matrix may be expanded to 4×4 format for consistency with the 3-D homogeneous coordinate system as

$$[\mathbf{T}_{6P}]_3 = \begin{bmatrix} T_{11} & T_{12} & 0 & 0 \\ T_{21} & T_{22} & 0 & 0 \\ 0 & 0 & 1 & 0 \\ T_{31} & T_{32} & 0 & 1 \end{bmatrix} \tag{3.28}$$

The special case with only one pair of points is simply translation. Direct use of Eq. (3.10) gives

$$[\mathbf{T}_{6P}]_1 = \begin{bmatrix} 1 & 0 & 0 & 0 \\ 0 & 1 & 0 & 0 \\ 0 & 0 & 1 & 0 \\ x_1^* - x_1 & y_1^* - y_1 & 0 & 1 \end{bmatrix} \tag{3.29}$$

The special case with two pairs of points is a combination of translation, rotation, and scaling. The first translation moves key point 1 to the origin. After rotation and scaling, key point 1 is moved to its new position. Concatenation of the transformation matrices in Eqs. (3.10), (3.17), and (3.11) (set for uniform scaling) in the order $[\mathbf{T}_{t1}][\mathbf{T}_r][\mathbf{T}_s][\mathbf{T}_{t2}]$ yields

$$[\mathbf{T}_{6P}]_2 = \begin{bmatrix} s\cos\theta & s\sin\theta & 0 & 0 \\ -s\sin\theta & s\cos\theta & 0 & 0 \\ 0 & 0 & 1 & 0 \\ x_1^* - sx_1\cos\theta + sy_1\sin\theta & y_1^* - sx_1\sin\theta - sy_1\cos\theta & 0 & 1 \end{bmatrix} \tag{3.30}$$

where the scaling factor s is the ratio

$$s = \sqrt{\frac{(x_2^* - x_1^*)^2 + (y_2^* - y_1^*)^2}{(x_2 - x_1)^2 + (y_2 - y_1)^2}} \tag{3.30a}$$

and the angle θ is defined by

$$\tan\theta = \frac{(y_2^* - y_1^*)(x_2 - x_1) - (x_2^* - x_1^*)(y_2 - y_1)}{(x_2^* - x_1^*)(x_2 - x_1) + (y_2^* - y_1^*)(y_2 - y_1)} \tag{3.30b}$$

EXAMPLE 3.6

Given the six-point transformation defined by the three pairs of key points

$$(1, 1) \rightarrow (2, 2)$$

$$(4, 1) \rightarrow (5, 3)$$

$$(1, 5) \rightarrow (3, 4)$$

determine the corresponding homogeneous transformation matrix $[\mathbf{T}_{6P}]$.

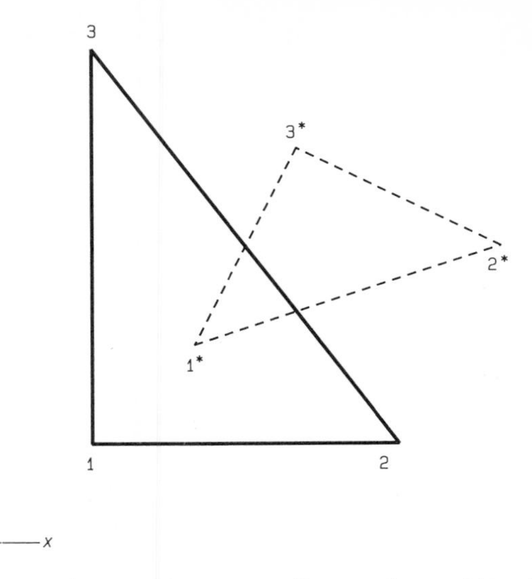

Solution. In 2-D homogeneous coordinates the problem is stated as

$$\begin{bmatrix} 1 & 1 & 1 \\ 4 & 1 & 1 \\ 1 & 5 & 1 \end{bmatrix} \begin{bmatrix} T_{11} & T_{12} & T_{13} \\ T_{21} & T_{22} & T_{23} \\ T_{31} & T_{32} & T_{33} \end{bmatrix} = \begin{bmatrix} 2 & 2 & 1 \\ 5 & 3 & 1 \\ 3 & 4 & 1 \end{bmatrix}$$

Numerical solution for $[\mathbf{T}_{6P}]$ yields

$$[\mathbf{T}_{6P}] = \begin{bmatrix} 1 & 1/3 & 0 \\ 1/4 & 1/2 & 0 \\ 3/4 & 7/6 & 1 \end{bmatrix}$$

which is expanded into the 3-D homogeneous coordinate system as

$$\begin{bmatrix} 1 & 1/3 & 0 & 0 \\ 1/4 & 1/2 & 0 & 0 \\ 0 & 0 & 1 & 0 \\ 3/4 & 7/6 & 0 & 1 \end{bmatrix}$$

Multiplication of the given points matrix by this result provides a check of the solution.

3.7 FURTHER PROPERTIES OF HOMOGENEOUS COORDINATES

The vector for the homogeneous representation of a point in 2-D space has three components, while that of a point in 3-D space has four components. Transformation matrices have a corresponding additional row and column that facilitate the translation transformation. This scheme accommodates the concatenation of transformation matrices in a consistent way. It also happens that the extra column in the points and transformation matrices may convey certain additional information.

Specialized for uniform scaling ($s = s_x = s_y = s_z$), Eq. (3.11) can be rewritten as

$$[\mathbf{T}_s] = s\begin{bmatrix} 1 & 0 & 0 & 0 \\ 0 & 1 & 0 & 0 \\ 0 & 0 & 1 & 0 \\ 0 & 0 & 0 & 1/s \end{bmatrix} \tag{3.31}$$

When multiplied by the points matrix $[\mathbf{P}] = [x \quad y \quad 0 \quad 1]$, Eq. (3.31) yields

$$[\mathbf{P}^*] = s[x \quad y \quad 0 \quad 1/s] = [sx \quad sy \quad 0 \quad 1] \tag{3.32}$$

Actually, the scalar s, as written in front of the intermediate matrix, is made redundant by the conventions of the homogeneous coordinate system. Here, when there are elements in the fourth column of the points matrix that are not unity, the points matrix is to be "normalized" by dividing all other elements in the row by the fourth element, as

$$[x \quad y \quad 0 \quad h] \equiv [x/h \quad y/h \quad 0 \quad h/h] = [x^* \quad y^* \quad 0 \quad 1] \tag{3.33}$$

With this convention, the matrices $[9 \quad 6 \quad 0 \quad 3]$, $[6 \quad 4 \quad 0 \quad 2]$, and $[3 \quad 2 \quad 0 \quad 1]$ all represent the same point, the last one being the preferred form.

Infinite values result when the homogeneous coordinate representation of a point contains 0 in the fourth column. For example, the matrix $[1 \quad 0 \quad 0 \quad 0]$ represents a point at infinity on the x axis, since the indeterminate form $1/0 \equiv \infty$.

The fourth column of a transformation matrix can be modified to produce a different type of scaling. If the point $[x \quad y \quad 0 \quad 1]$ is multiplied by an identity matrix modified by having the zeros in the fourth column replaced by arbitrary constants, there results

$$[x \quad y \quad 0 \quad 1]\begin{bmatrix} 1 & 0 & 0 & a \\ 0 & 1 & 0 & b \\ 0 & 0 & 1 & c \\ 0 & 0 & 0 & 1 \end{bmatrix} = [x \quad y \quad 0 \quad ax + by + 1] \tag{3.34}$$

Normalization yields

$$\begin{aligned} [\mathbf{P}^*] &= \left[\frac{x}{ax + by + 1} \quad \frac{y}{ax + by + 1} \quad 0 \quad 1\right] \\ &= [x^* \quad y^* \quad 0 \quad 1] \end{aligned} \tag{3.35}$$

Examination of this equation shows that the scaling of the points x and y is affected by their values. For example, if a, b, x, and y are all positive, it can be seen that x^* and y^* are inversely proportional to the distance from the origin. This makes points farther from the origin closer together. In other words, objects farther from the origin appear to be smaller. This type of scaling is applied in the perspective transformation, discussed in Section 4.6.

3.8 CLOSURE

Four-by-four transformation matrices in the homogeneous coordinate system provide elements in 3-D space for the viewing and modeling operations shown in submatrices as

$$\begin{array}{cc} 3\times 3 & 3\times 1 \\ \multicolumn{2}{c}{\left[\begin{array}{c:c} \begin{array}{c}\text{Rotation,}\\ \text{nonuniform scaling,}\\ \text{and shearing}\end{array} & \text{Perspective} \\ \hdashline \text{Translation} & \text{Scaling} \end{array}\right]} \\ 1\times 3 & 1\times 1 \end{array} \tag{3.36}$$

The third row and the third column of the transformation matrix can be omitted if only 2-D entities are treated. Square transformation matrices permit concatenation of two or more transformations into a single square matrix. Furthermore, the inverse exists for such matrices† so that any transformation can be undone.

Points matrices in the 3-D homogeneous coordinate system provide three columns for the x-y-z points coordinates and an extra column. The fourth column gives conformability for matrix multiplication with transformation matrices and supports uniform scaling through the normalization transformation.

PROBLEMS

3.1. The lower left corner of a square is A:(2, 2) and the upper right corner is B:(6, 6). If these points undergo the nonuniform scaling transformation given in Eq. (3.11) such that the new coordinates of A are (1.5, 2.5), determine the new coordinates of B.

3.2. The points matrix for a triangle is

$$[\mathbf{P}] = \begin{bmatrix} 0 & 0 & 0 & 1 \\ 3 & 1 & 0 & 1 \\ 1 & 2 & 0 & 1 \end{bmatrix}$$

†The exception is a singular matrix that results from one or more zero scaling factors.

Use nonuniform scaling to transform [**P**] so that the screen transformation *appears* that the triangle has been rotated 30° around the y axis.

3.3. A circle with a radius of 10 in. and its center at the origin undergoes nonuniform scaling transformation with $s_x = 1.2$ and $s_y = 0.8$. Does this transformation make an ellipse? Support your answer.

3.4. An algorithm for computation of the translation transformation is obviously faster if Δx and Δy are simply added to each term of the points matrix. Compare the number of arithmetic operations for this scheme with the number of arithmetic operations for the matrix multiplication method, Eq. (3.10).

3.5. A 2-in. diameter circle has its center at C:(2, 3) (inches). A radius is drawn vertically up from the center. Rotate this image $-30°$ around the origin. Determine the new coordinates of the center and sketch the transformed image.

3.6. Derive a concatenated transformation matrix in 2-D for scaling followed by rotation. Under what circumstances does the order make a difference?

3.7. Find the mirror reflection across the line $x = 0.5$ of AB where A:(2, 1) and B:(2, 3). Sketch the mirror axis, the original line, and the transformed line.

3.8. Mirror the line segment going from point (2, 2) to point (4, -2) around the axis described by $y = -1$. Sketch the line before and after.

3.9. Mirror the line segment that connects the points A:(-1, 3) and B:(2, -1) around the line CD where C:(2, 2) and D:(0, 3). Make a sketch showing the result.

3.10. Rederive Eq. (3.24) for the case of *nonuniform* scaling.

3.11. Derive a concatenated transformation matrix to create a mirror image of a planar points matrix around an arbitrary axis in the plane. (See Example 3.5.)

3.12. A two-point pair application of the six-point transformation maps [**P**] to [**P***] where

$$[\mathbf{P}] = \begin{bmatrix} 5 & 2 & 0 & 1 \\ 4 & 1 & 0 & 1 \end{bmatrix} \quad \text{and} \quad [\mathbf{P}^*] = \begin{bmatrix} 3 & 1 & 0 & 1 \\ 1 & 0 & 0 & 1 \end{bmatrix}$$

List the combination of basic transformations performed and determine the new coordinates of the point A:(1, 1) under this transformation. The key points are in the order given.

3.13. The three vertices of triangle ABC have the coordinates A:(0, 0), B:(4, 0), and C:(0, 4). Given that A^*:(2, 2) and B^*:(4, 1), use the six-point transformation as defined for two point pairs to determine the new coordinates of ABC if (a) A is the first key point and B is the second, and (b) B is the first key point and A is the second. Make a sketch showing ABC and $A^*B^*C^*$ for both cases (a) and (b). Based on the result of this problem, what can you say about the effect of the order in which the two point pairs are used?

3.14. Triangle ABC has vertices at A:(0, 0), B:(3, 0), and C:(0, 3). A six-point transformation is applied in the order given where A^*:(1, 1), B^*:(5, 1), and C^*:(1, 4). Determine the transformation matrix in the form of Eq. (3.28).

3.15. Discuss the advantages and disadvantages of using homogeneous coordinates for 2-D transformations.

3.16. In the transformation of Eq. (3.34), set $a = b = -0.5$ and $c = 0$. Thus, transform the point $[1 \quad 1 \quad 0 \quad 1]$. Explain the physical significance of your result.

PROJECTS

3.1. Prepare input data for a simple 2-D view of a vehicle. Write a program to display, on the same screen, the image as input and a transformed image that is translated to another position, scaled to three-fourths size, and rotated 30°.

3.2. Prepare 2-D input data for one-half span of a four-span arch bridge. Write a program that completes the image with mirror reflections and displays the result.

3.3. Develop data for a four-panel Howe truss. Show the members as lines. Write a program that uses the nonuniform scaling transformation to make a scale drawing of the truss for any given span and depth.

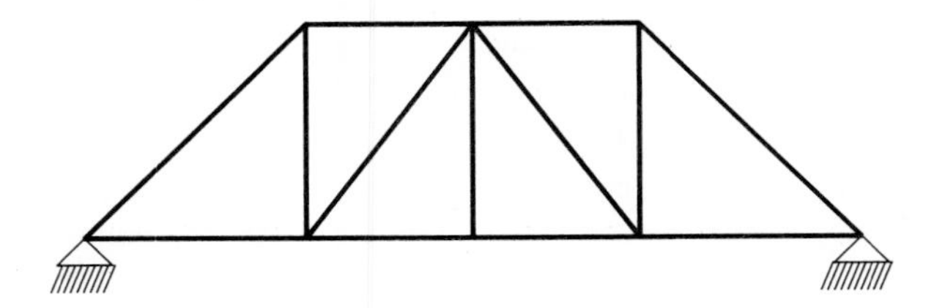

3.4. Prepare input data for a power transmission tower. Write a program that creates the illusion of a transmission line that vanishes into the distance by translating several copies of the tower with decreasing sizes.

3.5. Write a program that demonstrates the six-point transformation. Your program should

(a) Draw a figure from a data base of points and topology, with numbers displayed for each of the vertices.

(b) Ask for user input to specify whether one, two, or three point pairs will be used for a six-point transformation.

(c) Ask for the number of any vertex to be arbitrarily selected as the first key point.

(d) Provide for locating the new position of this key point with the screen cursor.

(e) Repeat (c) and (d) as necessary to enter the second and third key-point pairs.

(f) Draw the resulting image with a line style different from the original figure.

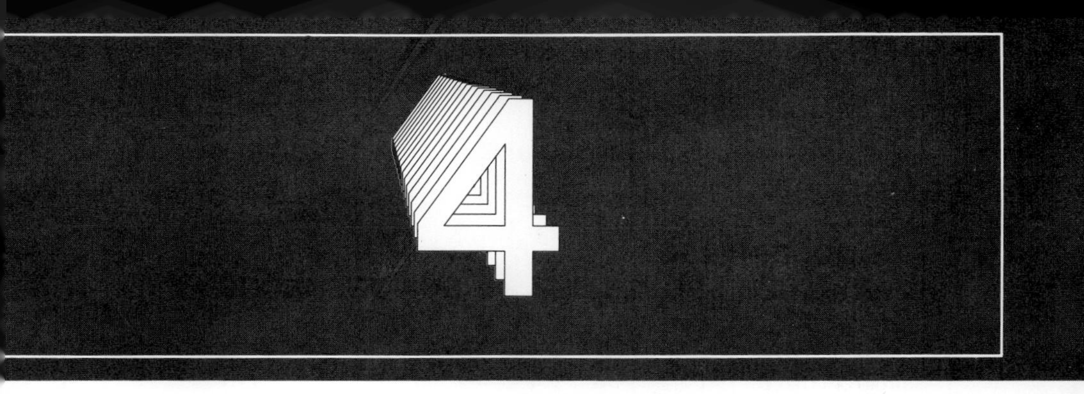

Three-Dimensional Transformations

A distinct advantage of computer graphics is the ability to deal with three-dimensional images. Although graphics display surfaces show only a 2-D representation of the image, the data base can contain 3-D information. Manipulation by geometric transformations provides views from different directions, from different distances, under different lighting, and so forth without having to make an actual physical model.

The objective of this chapter is to develop 3-D modeling or viewing transformations that deal with scaling, translation, rotation, and projection. The construction of pictorial projections is included.

As seen in the previous chapter, transformations on points have the form

$$[\mathbf{P}^*] = [\mathbf{P}][\mathbf{T}] \tag{4.1}$$

where $[\mathbf{P}^*]$ is the new points matrix, $[\mathbf{P}]$ contains the old points matrix, and $[\mathbf{T}]$ is the transformation matrix. As before, the points matrix has four columns, the first three containing the x, y, and z coordinates and the fourth containing the scale factor (usually 1) for homogeneous coordinates.

Display of 3-D images is accomplished with the appropriate viewing or modeling transformations followed by a transformation to make a *screen projection*. The display process, shown in Figure 4.1, uses the conventions that

> The x and y axes define the picture plane (screen), with the z axis outwardly normal from this plane. The x-y-z coordinates obey the right-hand rule.
>
> The viewer is located along the positive z axis. Thus, viewing transformations manipulate the object so that the *line of sight* corresponds to the z axis.

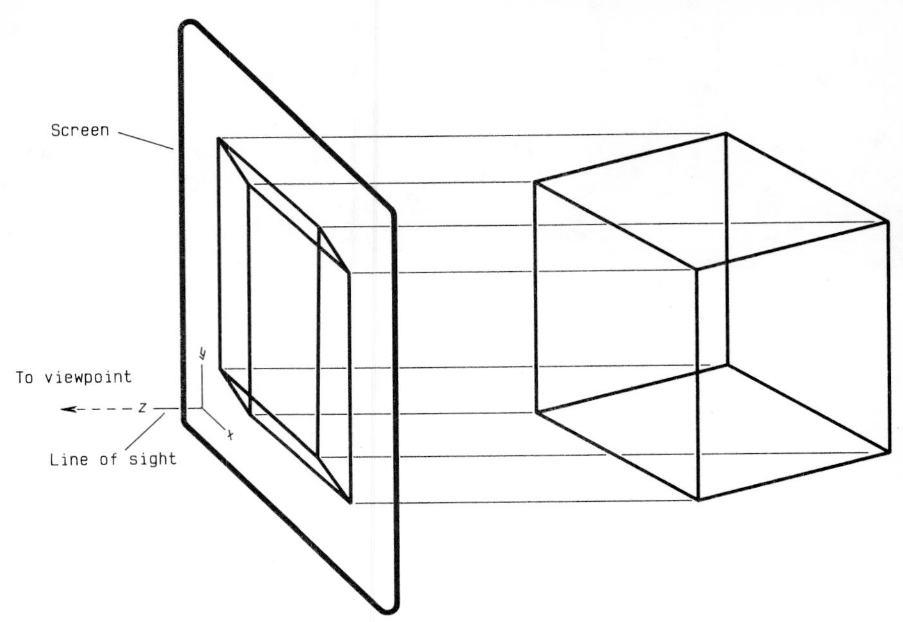

Figure 4.1 Illustration of parallel projection. Here, the intersection of the projection lines with the picture plane (screen) defines a 2-D representation of a 3-D object.

> The representation of the image in the *x-y* plane, the *screen projection,* is produced by using only the *x* and *y* components of the points matrix to draw the lines, panels, and so forth. (The *z* components of [**P**] do affect the *x* and *y* components in any transformation done before or after screen projection.)

Shown in Figure 4.1 is a *parallel* projection. In contrast, a *perspective* projection results when the projection lines converge at one point. Parallel projections may be further classified as *axonometric* or *oblique,* depending on whether or not the projection lines are perpendicular to the picture plane.

The term *affine* is applied to transformations that do not change the [**P**] matrix in such a way that further transformations will be incorrect—only affine transformations should be used for modeling. All the transformations in the following sections are affine with the exception of the perspective and oblique projections.

4.1 SCALING, TRANSLATION, AND REFLECTION

The background for 3-D scaling and translation was developed in Chapter 3. With reference to Eqs. (3.10) and (3.11), an additional term in the 4×4 transformation matrix controls the translation or scaling in the *z* direction.

Thus,

$$[\mathbf{T}_t] = \begin{bmatrix} 1 & 0 & 0 & 0 \\ 0 & 1 & 0 & 0 \\ 0 & 0 & 1 & 0 \\ \Delta x & \Delta y & \Delta z & 1 \end{bmatrix} \tag{4.2}$$

and

$$[\mathbf{T}_s] = \begin{bmatrix} s_x & 0 & 0 & 0 \\ 0 & s_y & 0 & 0 \\ 0 & 0 & s_z & 0 \\ 0 & 0 & 0 & 1 \end{bmatrix} \tag{4.3}$$

are for translation and nonuniform scaling. Uniform scaling can be accomplished with

$$[\mathbf{T}_s] = \begin{bmatrix} s & 0 & 0 & 0 \\ 0 & s & 0 & 0 \\ 0 & 0 & s & 0 \\ 0 & 0 & 0 & 1 \end{bmatrix} \tag{4.4}$$

If the results are normalized using Eq. (3.33), an alternative form of Eq. (4.4) is

$$[\mathbf{T}_s] = \begin{bmatrix} 1 & 0 & 0 & 0 \\ 0 & 1 & 0 & 0 \\ 0 & 0 & 1 & 0 \\ 0 & 0 & 0 & 1/s \end{bmatrix} \tag{4.5}$$

The scaling factor s is greater than one to enlarge the image and less than one to shrink the image. As before, negative scaling factors cause a mirror reflection transformation of the image. Hence, the three resulting mirror reflection transformations for the planes formed by the coordinate axes are for reflection with respect to the x-y plane:

$$[\mathbf{T}_{mxy}] = \begin{bmatrix} 1 & 0 & 0 & 0 \\ 0 & 1 & 0 & 0 \\ 0 & 0 & -1 & 0 \\ 0 & 0 & 0 & 1 \end{bmatrix} \tag{4.6}$$

for reflection with respect to the y-z plane:

$$[\mathbf{T}_{myz}] = \begin{bmatrix} -1 & 0 & 0 & 0 \\ 0 & 1 & 0 & 0 \\ 0 & 0 & 1 & 0 \\ 0 & 0 & 0 & 1 \end{bmatrix} \tag{4.7}$$

and for reflection with respect to the x-z plane:

$$[\mathbf{T}_{mxz}] = \begin{bmatrix} 1 & 0 & 0 & 0 \\ 0 & -1 & 0 & 0 \\ 0 & 0 & 1 & 0 \\ 0 & 0 & 0 & 1 \end{bmatrix} \tag{4.8}$$

All three transformations simply change the signs of one column of the points matrix. These results are the same as Eqs. (3.20) and (3.21). However, the x axis is more correctly visualized as the x-z plane and the y axis as the y-z plane.

Shearing in 3-D is handled similarly to 2-D shearing. The off-diagonal terms in the upper 3×3 submatrix of the 4×4 transformation matrix control shear,

$$\left[\begin{array}{ccc|c} 1 & T_{xy} & T_{xz} & 0 \\ T_{yx} & 1 & T_{yz} & 0 \\ T_{zx} & T_{zy} & 1 & 0 \\ \hline 0 & 0 & 0 & 1 \end{array}\right] \tag{4.9}$$

where the subscripts on the terms T_{xy}, T_{xz}, and so forth correspond to coordinate directions.

4.2 ROTATION WITH RESPECT TO COORDINATE AXES

The 2-D rotation transformation, Eq. (3.17), represents rotation around the z axis, which, by convention, is always perpendicular to the viewing plane. For completeness, that equation with more descriptive notation is given by

$$[\mathbf{T}_{rz}] = \begin{bmatrix} \cos\theta_z & \sin\theta_z & 0 & 0 \\ -\sin\theta_z & \cos\theta_z & 0 & 0 \\ 0 & 0 & 1 & 0 \\ 0 & 0 & 0 & 1 \end{bmatrix} \tag{4.10}$$

The familiar right-hand rule applies to the sign of θ_z: when the thumb points along the positive z axis, the fingers of the right hand show the positive direction of rotation.

To rotate an object around the x or the y coordinate axis, one simply interchanges the rows and columns of Eq. (4.10). For rotation around the x axis, the existing or "old" terms for z (the third row and column) are put into the first row and column. Next in order, the old x terms are put into the second row and column (new y), and the old y terms are put into the third row and column. Diagrammatically, one may write $z \rightarrow x$, $x \rightarrow y$, $y \rightarrow z$, (this means that old T_{xz} equals new T_{yx}, and so on) or

$$[\mathbf{T}_{rx}] = \begin{bmatrix} 1 & 0 & 0 & 0 \\ 0 & \cos\theta_x & \sin\theta_x & 0 \\ 0 & -\sin\theta_x & \cos\theta_x & 0 \\ 0 & 0 & 0 & 1 \end{bmatrix} \tag{4.11}$$

Similarly, a matrix for rotation around the y axis is formed from Eq. (4.11) with the shift $x \rightarrow y$, $y \rightarrow z$, and $z \rightarrow x$, or

$$[\mathbf{T}_{ry}] = \begin{bmatrix} \cos\theta_y & 0 & -\sin\theta_y & 0 \\ 0 & 1 & 0 & 0 \\ \sin\theta_y & 0 & \cos\theta_y & 0 \\ 0 & 0 & 0 & 1 \end{bmatrix} \tag{4.12}$$

Rotation in three dimensions is accomplished by either the concatenation or the sequential application of the transformation matrices in Eqs. (4.10), (4.11), and (4.12). The order of these transformations affects the final position since $[\mathbf{T}_{ry}][\mathbf{T}_{rz}] \neq [\mathbf{T}_{rz}][\mathbf{T}_{ry}]$ as demonstrated in Figure 4.2.

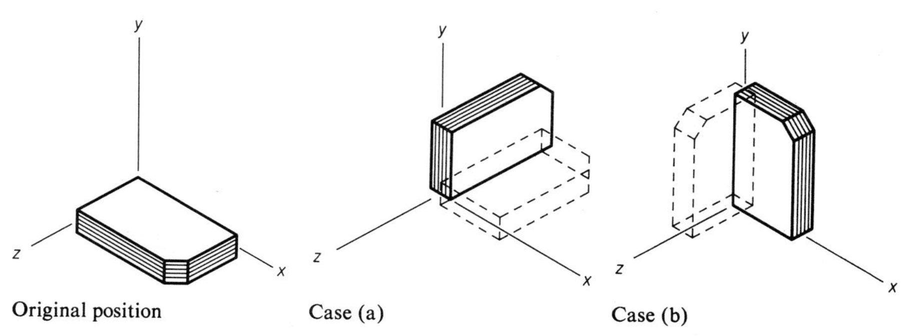

Original position Case (a) Case (b)

Figure 4.2 Demonstration of rotations around the y axis and z axis performed in opposite order. Case (a): rotation of 90° around the y axis followed by rotation of 90° around the z axis. Case (b): rotation of 90° around the z axis followed by rotation of 90° around the y axis.

EXAMPLE 4.1

The points matrix for a wedge is

$$[\mathbf{P}] = \begin{bmatrix} 0 & 0 & 0 & 1 \\ 0 & 0 & 2 & 1 \\ 4 & 0 & 0 & 1 \\ 4 & 0 & 2 & 1 \\ 0 & 3 & 0 & 1 \\ 0 & 3 & 2 & 1 \end{bmatrix}$$

Rotate the object 30° around the x axis and $-45°$ around the y axis. Show the screen (x-y) projection of the result.

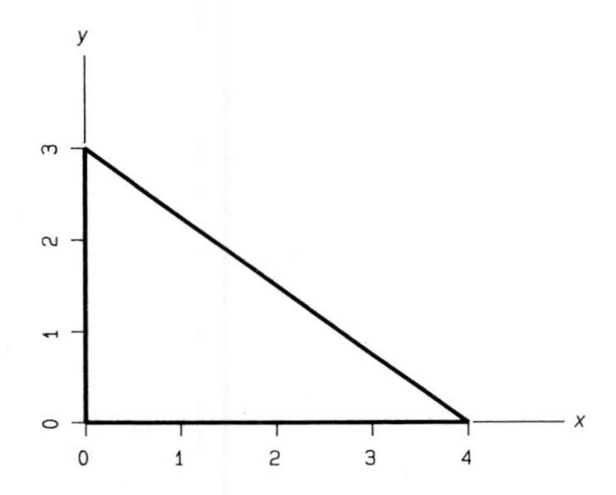

Solution. The matrices to be multiplied are

$$\begin{bmatrix} 0 & 0 & 0 & 1 \\ 0 & 0 & 2 & 1 \\ 4 & 0 & 0 & 1 \\ 4 & 0 & 2 & 1 \\ 0 & 3 & 0 & 1 \\ 0 & 3 & 2 & 1 \end{bmatrix} \begin{bmatrix} 1 & 0 & 0 & 0 \\ 0 & \cos 30° & \sin 30° & 0 \\ 0 & -\sin 30° & \cos 30° & 0 \\ 0 & 0 & 0 & 1 \end{bmatrix} \begin{bmatrix} \cos(-45°) & 0 & -\sin(-45°) & 0 \\ 0 & 1 & 0 & 0 \\ \sin(-45°) & 0 & \cos(-45°) & 0 \\ 0 & 0 & 0 & 1 \end{bmatrix}$$

which yields the new points matrix

$$[\mathbf{P}^*] = \begin{bmatrix} 0 & 0 & 0 & 1 \\ -1.2247 & -1.0 & 1.2247 & 1 \\ 2.8284 & 0 & 2.8284 & 1 \\ 1.6036 & -1.0 & 4.0532 & 1 \\ -1.0605 & 2.5981 & 1.0605 & 1 \\ -2.2853 & 1.5980 & 2.2853 & 1 \end{bmatrix}$$

The screen projection of the result is shown below.

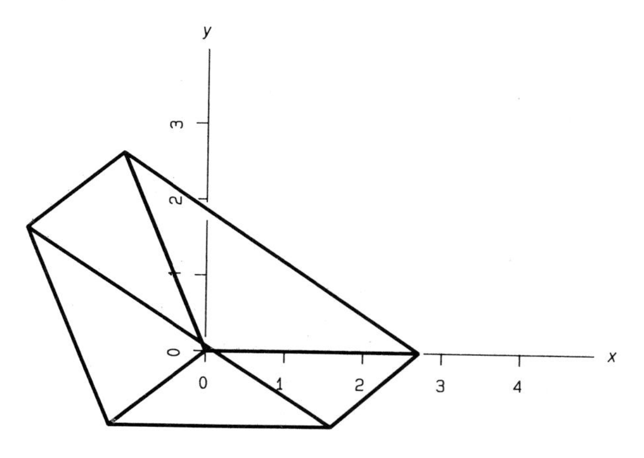

4.3 ROTATION WITH RESPECT TO AN OBLIQUE AXIS

An object can be revolved into any position by combining rotations around the coordinate axes. Another way to accomplish the same result is to use an arbitrary oblique axis of rotation. For convenience, this axis will be defined as an axis through the origin.

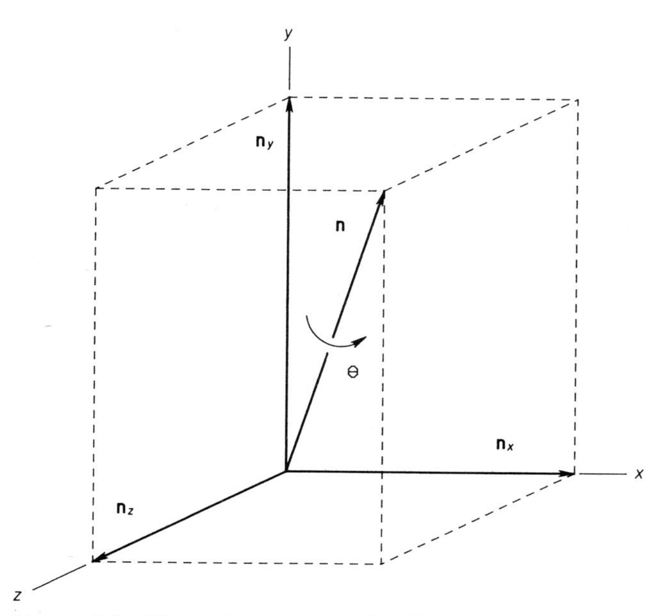

Figure 4.3 The unit vector **n** describes an oblique axis through the origin. The components of **n** (n_x, n_y, n_z) are the direction cosines.

The oblique axis is described with the unit vector

$$\mathbf{n} = n_x\mathbf{i} + n_y\mathbf{j} + n_z\mathbf{k} \tag{4.13}$$

where $\mathbf{n}$ is directed from the origin along the axis of rotation with the positive sense describing rotation according to the right-hand rule; $\mathbf{i}$, $\mathbf{j}$, and $\mathbf{k}$ are unit vectors in the directions of the coordinate axes; n_x, n_y, and n_z are direction cosines. The cosines of the three angles formed by the axis of rotation and the three respective coordinate axes obey the relationship

$$n_x^2 + n_y^2 + n_z^2 = 1 \tag{4.14}$$

Although only two of the three direction cosines are necessary to describe the axis of rotation, the signs on all three are necessary to determine the correct *direction* of rotation. The direction cosines are shown in Figure 4.3 as components of $\mathbf{n}$.

The rotation of point P around axis $\mathbf{n}$ is shown in Figure 4.4. The perpendicular distance from $\mathbf{n}$ to P is the line segment QP, which is denoted as $r\mathbf{v}$, where r is the radius and $\mathbf{v}$ is a unit vector normal to $\mathbf{n}$. The third unit vector $\mathbf{u}$ is normal to $\mathbf{n}$ and $\mathbf{v}$. The plane defined by $\mathbf{u}$ and $\mathbf{v}$ is shown in Figure 4.5.

As the vector r rotates through the angle θ, P moves to P^*. For convenience in notation, vectors from the origin O to points Q, P, and P^*, respectively, are denoted at $\mathbf{Q}$, $\mathbf{P}$, and $\mathbf{P^*}$. Because of the orthogonality of the unit vectors,

$$\mathbf{u} = \mathbf{n} \times \mathbf{v} \tag{4.15}$$

Addition of the vectors seen in Figures 4.4 and 4.5 yields

$$\mathbf{P^*} = \mathbf{Q} + r\cos\theta\mathbf{v} + r\sin\theta\mathbf{u} \tag{4.16}$$

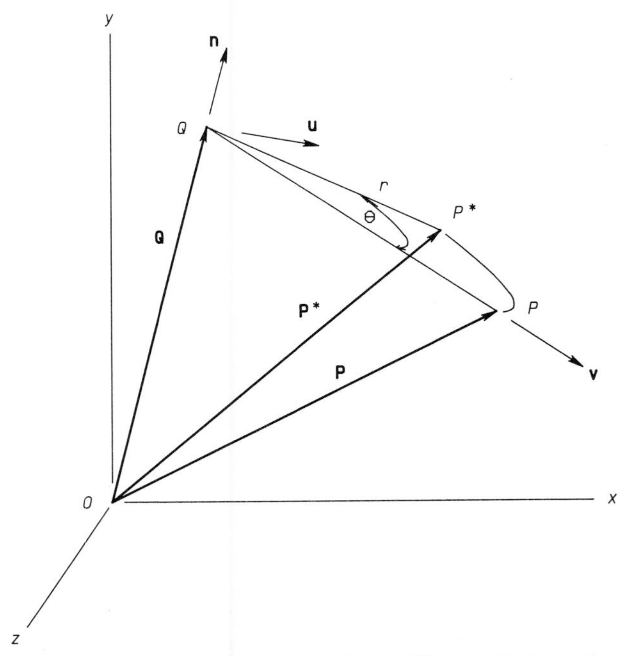

Figure 4.4 Rotation of P around an oblique axis through the origin.

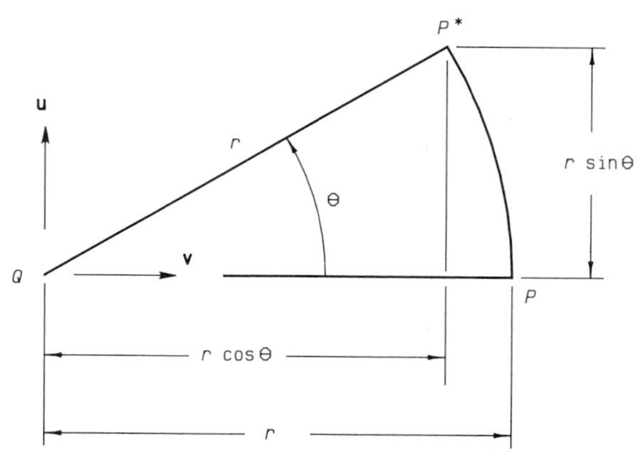

Figure 4.5 Plane of **u** and **v** with circular arc described by rotation of P to P^*.

Furthermore, vector subtraction (see Figure 4.4) yields

$$r\mathbf{v} = \mathbf{P} - \mathbf{Q} \tag{4.17}$$

From Eqs. (4.15) and (4.17) it follows that

$$\mathbf{u} = \frac{\mathbf{n} \times (\mathbf{P} - \mathbf{Q})}{r} \tag{4.18}$$

Next, Eqs. (4.17) and (4.18) are substituted into Eq. (4.16) to give

$$\mathbf{P}^* = \mathbf{Q} + (\mathbf{P} - \mathbf{Q}) \cos\theta + r \sin\theta \, \frac{\mathbf{n} \times (\mathbf{P} - \mathbf{Q})}{r} \tag{4.19}$$

Because of collinearity, $\mathbf{n} \times \mathbf{Q} = 0$. Expansion of Eq. (4.19) leads to

$$\mathbf{P}^* = \mathbf{Q}(1 - \cos\theta) + \mathbf{P}\cos\theta + (\mathbf{n} \times \mathbf{P})\sin\theta \tag{4.20}$$

The projection of **P** on the axis of rotation is **Q**, where

$$\mathbf{Q} = (\mathbf{P} \cdot \mathbf{n})\mathbf{n} \tag{4.21}$$

which when substituted into Eq. (4.20) produces

$$\mathbf{P}^* = (\mathbf{P} \cdot \mathbf{n})\mathbf{n}(1 - \cos\theta) + \mathbf{P}\cos\theta + (\mathbf{n} \times \mathbf{P})\sin\theta \tag{4.22}$$

In matrix notation, with reference to Eqs. (A.17) and (A.20) of Appendix A, Eq. (4.22) may be written as

$$\mathbf{P}^* = [\mathbf{P}]\begin{bmatrix} n_x \\ n_y \\ n_z \end{bmatrix}[n_x \quad n_y \quad n_z]\begin{bmatrix} \mathbf{i} \\ \mathbf{j} \\ \mathbf{k} \end{bmatrix}(1 - \cos\theta) + [\mathbf{P}]\begin{bmatrix} \mathbf{i} \\ \mathbf{j} \\ \mathbf{k} \end{bmatrix}(\cos\theta)$$

$$+ [\mathbf{P}]\begin{bmatrix} 0 & n_z & -n_y \\ -n_z & 0 & n_x \\ n_y & -n_x & 0 \end{bmatrix}\begin{bmatrix} \mathbf{i} \\ \mathbf{j} \\ \mathbf{k} \end{bmatrix}(\sin\theta) \tag{4.23}$$

where [**P**] is the nonhomogeneous row matrix for point **P**. (The reversal of signs in the cross-product term reflects the reversal of order of **n** and **P**.)

The vector **P*** is the product of the Cartesian components and the unit vectors,

$$\mathbf{P}^* = [\mathbf{P}^*]\begin{bmatrix}\mathbf{i}\\ \mathbf{j}\\ \mathbf{k}\end{bmatrix} = x^*\mathbf{i} + y^*\mathbf{j} + z^*\mathbf{k} \tag{4.24}$$

where [**P***] is the nonhomogeneous row matrix for the point **P***. It follows that the scalar portion of Eq. (4.23) is

$$[\mathbf{P}^*] = [\mathbf{P}]\begin{bmatrix} n_x^2 & n_xn_y & n_xn_z \\ n_xn_y & n_y^2 & n_yn_z \\ n_xn_z & n_yn_z & n_z^2 \end{bmatrix}(1 - \cos\theta) + [\mathbf{P}](\cos\theta)$$

$$+ [\mathbf{P}]\begin{bmatrix} 0 & n_z & -n_y \\ -n_z & 0 & n_x \\ n_y & -n_x & 0 \end{bmatrix}(\sin\theta) \tag{4.25}$$

By matrix addition, Eq. (4.25) becomes

$$[\mathbf{P}^*] = [\mathbf{P}]\begin{bmatrix} n_x^2 + (1 - n_x^2)\cos\theta & n_xn_y(1 - \cos\theta) + n_z\sin\theta & n_xn_z(1 - \cos\theta) - n_y\sin\theta \\ n_xn_y(1 - \cos\theta) - n_z\sin\theta & n_y^2 + (1 - n_y^2)\cos\theta & n_yn_z(1 - \cos\theta) + n_x\sin\theta \\ n_xn_z(1 - \cos\theta) + n_y\sin\theta & n_yn_z(1 - \cos\theta) - n_x\sin\theta & n_z^2 + (1 - n_z^2)\cos\theta \end{bmatrix} \tag{4.26}$$

At this juncture, it is noted that [**P**] and [**P***] could just as well be an n-row matrix for n points as a single-row matrix for one point.

Finally, in homogeneous coordinates, the transformation matrix for rotation around an arbitrary axis through the origin is

$$[\mathbf{T}_{rg}] = \begin{bmatrix} n_x^2 + (1 - n_x^2)\cos\theta & n_xn_y(1 - \cos\theta) + n_z\sin\theta & n_xn_z(1 - \cos\theta) - n_y\sin\theta & 0 \\ n_xn_y(1 - \cos\theta) - n_z\sin\theta & n_y^2 + (1 - n_y^2)\cos\theta & n_yn_z(1 - \cos\theta) + n_x\sin\theta & 0 \\ n_xn_z(1 - \cos\theta) + n_y\sin\theta & n_yn_z(1 - \cos\theta) - n_x\sin\theta & n_z^2 + (1 - n_z^2)\cos\theta & 0 \\ 0 & 0 & 0 & 1 \end{bmatrix} \tag{4.27}$$

A limitation of Eq. (4.27) is that it is applicable only when the axis of rotation passes through the origin of the coordinate system. In the case that the axis of rotation does not pass through the origin, it will be necessary to pre- and postmultiply the rotation transformation matrix by the appropriate translation transformation matrices. It is easily verified that Eq. (4.27) degenerates to the special cases of rotation with respect to the coordinate axes, Eqs. (4.10), (4.11), and (4.12).

EXAMPLE 4.2

Rotate the cube given by the points

$$[\mathbf{P}] = \begin{bmatrix} 1 & 1 & 0 & 1 \\ 2 & 1 & 0 & 1 \\ 2 & 2 & 0 & 1 \\ 1 & 2 & 0 & 1 \\ 1 & 1 & 1 & 1 \\ 2 & 1 & 1 & 1 \\ 2 & 2 & 1 & 1 \\ 1 & 2 & 1 & 1 \end{bmatrix}$$

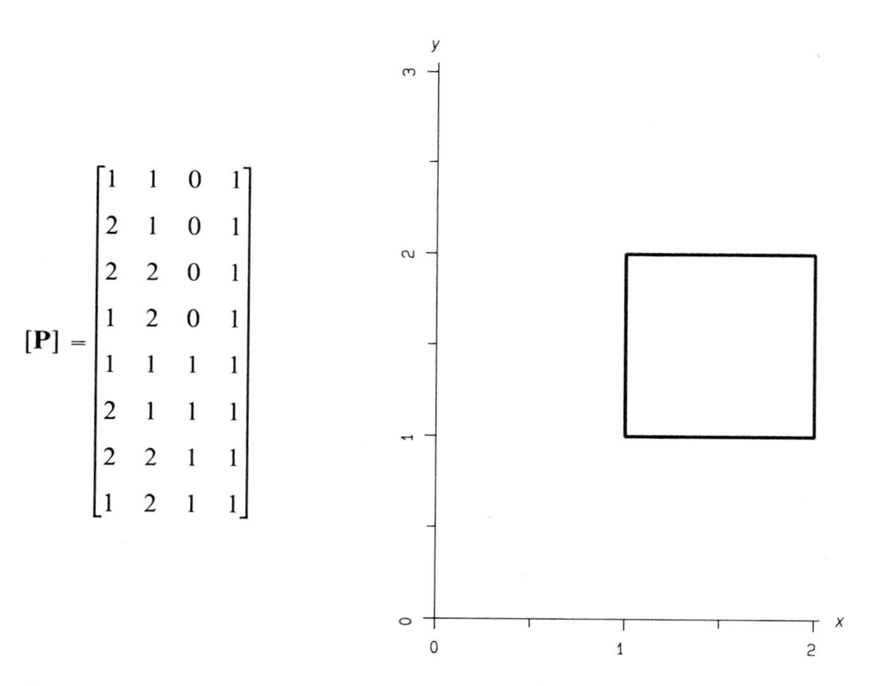

through an angle of 60° around an axis from the origin to the point (0, 3, 4).

Solution. First, the direction cosines for the axis of rotation are computed. The distance from the origin to (0, 3, 4) is

$$\sqrt{0^2 + 3^2 + 4^2} = 5$$

The unit vector is then

$$\mathbf{n} = 0\mathbf{i} + \frac{3}{5}\mathbf{j} + \frac{4}{5}\mathbf{k} = 0\mathbf{i} + 0.6\mathbf{j} + 0.8\mathbf{k}$$

The matrix multiplication for the transformation is given by Eq. (4.27),

$$\begin{bmatrix} 1 & 1 & 0 & 1 \\ 2 & 1 & 0 & 1 \\ 2 & 2 & 0 & 1 \\ 1 & 2 & 0 & 1 \\ 1 & 1 & 1 & 1 \\ 2 & 1 & 1 & 1 \\ 2 & 2 & 1 & 1 \\ 1 & 2 & 1 & 1 \end{bmatrix} \begin{bmatrix} 0 + (1-0)\cos 60 & 0 + 0.8 \sin 60 & 0 - 0.6 \sin 60 & 0 \\ 0 - 0.8 \sin 60 & 0.6^2 + (1 - 0.6^2)\cos 60 & 0.6(0.8)(1 - \cos 60) + 0 & 0 \\ 0 + 0.6 \sin 60 & 0.6(0.8)(1 - \cos 60) - 0 & 0.8^2 + (1 - 0.8^2)\cos 60 & 0 \\ 0 & 0 & 0 & 1 \end{bmatrix}$$

which results in a points matrix of

$$[\mathbf{P}^*] = \begin{bmatrix} -0.1928 & 1.3728 & -0.2796 & 1 \\ 0.3072 & 2.0656 & -0.7992 & 1 \\ -0.3856 & 2.7456 & -0.5592 & 1 \\ -0.8856 & 2.0528 & -0.0396 & 1 \\ 0.3268 & 1.6128 & 0.5404 & 1 \\ 0.8268 & 2.3056 & 0.0208 & 1 \\ 0.1340 & 2.9856 & 0.2608 & 1 \\ -0.3660 & 2.2928 & 0.7804 & 1 \end{bmatrix}$$

The screen projection of the result is shown below.

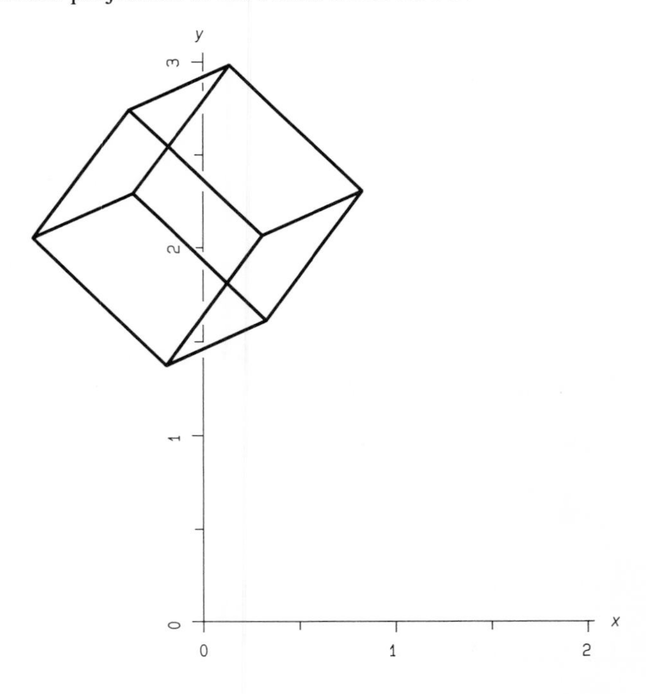

The general rotation matrix in homogeneous coordinates may be concatenated with any of the other transformation matrices. Often, the result of the concatenation of these transformation matrices is not very tractable. Consequently, it is simpler to do separate operations (computer subroutines) in the required order. This is illustrated in the next example.

EXAMPLE 4.3

Rotate the point (2, 0, 2) through an angle of $-90°$ around the axis described by the line segment from point (0, 1, 0) to the point (4, 4, 0).

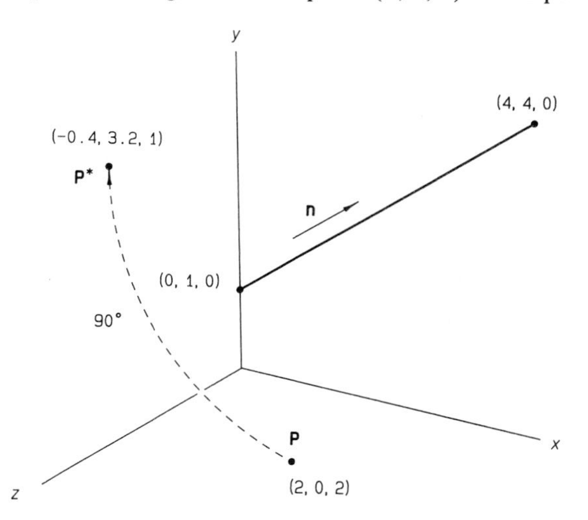

Solution. The system, which consists of three points, is first translated so that the axis of rotation passes through the origin. This is done by Eq. (4.2) with $\Delta x = 0$, $\Delta y = -1$, and $\Delta z = 0$,

$$\begin{bmatrix} 2 & 0 & 2 & 1 \\ 0 & 1 & 0 & 1 \\ 4 & 4 & 0 & 1 \end{bmatrix} [\mathbf{T}_{t0}] = \begin{bmatrix} 2 & -1 & 2 & 1 \\ 0 & 0 & 0 & 1 \\ 4 & 3 & 0 & 1 \end{bmatrix}$$

Or, more simply, Δx, Δy, and Δz may be added to the respective columns.

The unit vector along the axis of rotation is

$$\mathbf{n} = \frac{4\mathbf{i} + 3\mathbf{j}}{\sqrt{25}} = 0.8\mathbf{i} + 0.6\mathbf{j}$$

Substitution of direction cosines in Eq. (4.27) gives

$$[\mathbf{P}^*] = [2 \quad -1 \quad 2 \quad 1] \begin{bmatrix} 0.64 & 0.48 & 0.60 & 0 \\ 0.48 & 0.36 & -0.80 & 0 \\ -0.60 & 0.80 & 0 & 0 \\ 0 & 0 & 0 & 1 \end{bmatrix}$$

$$= [-0.4 \quad 2.2 \quad 2.0 \quad 1]$$

Note that only the original point, but not the end points of the axis of rotation, must be transformed.

Finally, the point is translated back to the original position with $\Delta x = 0$, $\Delta y = +1$, and $\Delta z = 0$, or

$$[\mathbf{P}^*] = [-0.4 \quad 2.2 \quad 2.0 \quad 1][\mathbf{T}_{t1}]$$
$$= [-0.4 \quad 3.2 \quad 2.0 \quad 1]$$

The new location of the point is $(-0.4, 3.2, 2.0)$.

4.4 VIEWPOINT TRANSFORMATION

A convenient method for generating a pictorial view of a 3-D object, the *viewpoint* transformation, is based on a vector defined by the line of sight. This vector extends from the *viewsite* to the *viewpoint*. The new view is produced by transforming the line-of-sight vector so that it is outwardly normal to the viewing surface. In other words, if the line-of-sight vector were part of the object, it would appear as a point after the transformation.

The basic procedure in the viewpoint transformation is (1) translation of the viewsite to the origin, (2) rotation of the line-of-sight vector to correspond to the screen z axis, and (3) translation of the viewsite back to its original position. These three operations may be coded into a single routine.

In Figure 4.6, the viewsite has been translated to the origin. When the image is rotated appropriately, the vector **R** will be normal to the screen (the x-y plane). In other words, R and O will appear to be the same point. The viewpoint

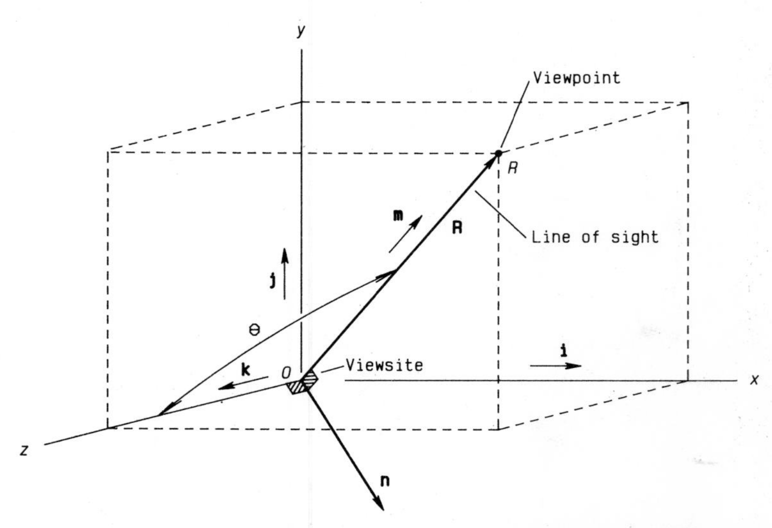

Figure 4.6 Notation for the viewpoint transformation; **R** defines the line-of-sight vector.

transformation, therefore, must rotate **R** until it is collinear with the z axis, which, by convention, is perpendicular to the screen. A unit vector along **R** is

$$\mathbf{m} = \frac{R_x\mathbf{i} + R_y\mathbf{j} + R_z\mathbf{k}}{\sqrt{R_x^2 + R_y^2 + R_z^2}}$$

$$= m_x\mathbf{i} + m_y\mathbf{j} + m_z\mathbf{k} \qquad (4.28)$$

The third direction cosine m_z defines the angle between **m** and the z axis. Thus, the angle of rotation is given by

$$\theta = \cos^{-1}(m_z) \qquad (4.29)$$

Next, the axis of rotation is defined. The unit vector **n** defining the axis of rotation, shown in Figure 4.6, is mutually perpendicular to **m** and the z axis, which is described by the unit vector **k**. Hence, use of the cross product yields

$$\mathbf{n} = \frac{\mathbf{m} \times \mathbf{k}}{\sin\theta} \qquad (4.30)$$

where the angle θ between **m** and **k** is given in Eq. (4.29). Combining Eqs. (4.28) and (4.30) yields

$$\mathbf{n} = \frac{m_y\mathbf{i} - m_x\mathbf{j}}{\sin\theta} \qquad (4.31)$$

The direction cosines describing the axis of rotation can be separated from Eq. (4.31) as

$$n_x = \frac{m_y}{\sin\theta} \qquad n_y = \frac{-m_x}{\sin\theta} \qquad n_z = 0 \qquad (4.32)$$

The viewpoint transformation uses the general axis rotation algorithm, Eq. (4.27), with θ given by Eq. (4.29) and the direction cosines by Eq. (4.32).

The magnitude of the distance between the viewsite and the viewpoint does not enter into the transformation, unless a perspective transformation is used.

EXAMPLE 4.4

Determine the angle and axis of rotation for performing the viewpoint transformation on the impeller shown with a viewsite of (0, 15, 0) and a viewpoint of (25, 10, 15). The outer diameter of the impeller is 42 cm, the inner diameter is 28 cm, and the thickness in the z direction is 7 cm.

Solution. Translation of the viewsite to the origin yields the unit vector along the line of sight

$$\mathbf{m} = \frac{25\mathbf{i} - 5\mathbf{j} + 15\mathbf{k}}{\sqrt{25^2 + (-5)^2 + 15^2}}$$

$$= 0.8452\mathbf{i} - 0.1690\mathbf{j} + 0.5071\mathbf{k}$$

Hence, from Eq. (4.29) the angle of rotation is

$$\theta = \cos^{-1}(0.5071) = 59.5°$$

and the unit vector along the axis of rotation is, from Eq. (4.31),

$$\mathbf{n} = \frac{-0.1690\mathbf{i} - 0.8452\mathbf{j}}{\sin 59.5} = -0.196\mathbf{i} - 0.981\mathbf{j}$$

Using these values in Eq. (4.27), followed by translating the viewsite (and image) back to (0, 15, 0), yields the result shown below. Hidden surfaces are removed in the third view to aid in visualization.

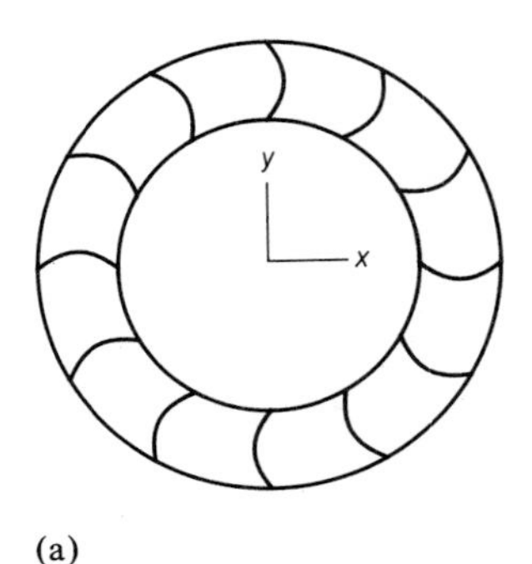

(a)

Before:
impeller in x-y plane

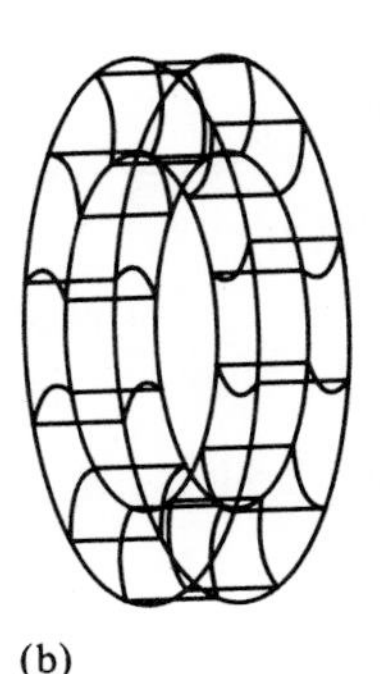

(b)

After:
wire frame

(c)

After:
hidden surfaces removed

4.5 AXONOMETRIC PROJECTIONS

The rotation transformations in the foregoing sections produce *axonometric* projections, which (1) obey the familiar rules of Euclidean geometry such as parallel lines remaining parallel and (2) are affine—any number of these transformations can be combined without distortion.

According to the angles made between the coordinate axes in the screen projection, axonometric projections can be classified into the three standard types known as *isometric, dimetric,* and *trimetric.* The distinctions between these classifications are summarized in Table 4.1 and illustrated in Figure 4.7.

The isometric projection, familiar to engineers, gives equal emphasis to all three directions of viewing. The popularity of isometric drawings is due to a

TABLE 4.1 CLASSIFICATION OF AXONOMETRIC PROJECTIONS

Type	Angle between axis images	Example viewpoint[a]	Example of corresponding y rotation	Example of corresponding x rotation
Isometric	All equal (120°)	(1, 1, 1)	−45°	35.264°
Dimetric	Any two equal	1, 1, any z	−45°	Any
Trimetric	None equal	Any x, y, z	Any	Any

[a]The correspondence between viewpoints and rotation angles above will be developed in this section.

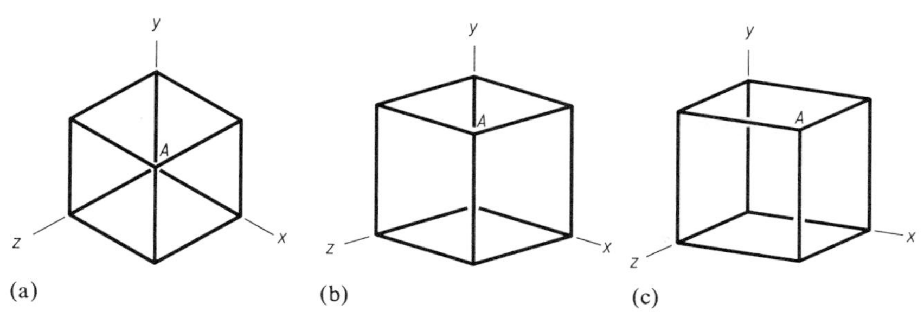

Figure 4.7 Example axonometric projections for a 1 × 1 × 1 cube with the x-y-z axes and origin indicated. The coordinates of point A are (1, 1, 1). The object is most clearly seen in the trimetric projection due to the nonoverlapping lines. (a) Isometric with y axis vertical; (b) dimetric with $\theta_y = -45°$, $\theta_x = 15°$; (c) trimetric with $\theta_y = -30°$, $\theta_x = 15°$.

systematic technique that has evolved for manual construction where the 120° angles between the x, y, and z coordinate axes are handled with a 30°-60°-90° triangle. Manually constructed isometrics should have the dimensions in the x, y, and z directions foreshortened to 82% of the original, a convention which is often disregarded. (The reason for 82% is shown below.) A circle can be drawn in isometric projection with an ellipse template or approximated with a construction that uses four circular arcs.

Hand construction of isometric drawings is, however, time-consuming. Computer-aided design systems with 3-D capability greatly facilitate the construction of isometric drawings as well as other axonometric projections.

Because of the widespread use of isometric projection, it is worthwhile to develop a transformation matrix for this special case. Two approaches will be demonstrated that produce the same results.

Isometric Projection by Rotations around Two Coordinate Axes

In order to keep the y axis vertical, rotation is first around the y axis and then around the x axis. The two rotational matrices, Eqs. (4.12) and (4.11), are multiplied as

$$[\mathbf{T}_{ry}][\mathbf{T}_{rx}] = \begin{bmatrix} \cos\theta_y & 0 & -\sin\theta_y & 0 \\ 0 & 1 & 0 & 0 \\ \sin\theta_y & 0 & \cos\theta_y & 0 \\ 0 & 0 & 0 & 1 \end{bmatrix} \begin{bmatrix} 1 & 0 & 0 & 0 \\ 0 & \cos\theta_x & \sin\theta_x & 0 \\ 0 & -\sin\theta_x & \cos\theta_x & 0 \\ 0 & 0 & 0 & 1 \end{bmatrix}$$

$$= \begin{bmatrix} \cos\theta_y & \sin\theta_y \sin\theta_x & -\sin\theta_y \cos\theta_x & 0 \\ 0 & \cos\theta_x & \sin\theta_x & 0 \\ \sin\theta_y & -\cos\theta_y \sin\theta_x & \cos\theta_y \cos\theta_x & 0 \\ 0 & 0 & 0 & 1 \end{bmatrix} \quad (4.33)$$

With reference to Figure 4.7(a), symmetry dictates that the first rotation is $\theta_y = -45°$. The second angle of rotation, that around the x axis, is determined from the transformation of point A, which is originally at (1, 1, 1) and is transformed to $(0, 0, z_p)$. When these values are substituted into $[\mathbf{P}^*] = [\mathbf{P}][\mathbf{T}_i]$, using Eq. (4.33) gives

$$[0 \quad 0 \quad z_p \quad 1] = [1 \quad 1 \quad 1 \quad 1]\begin{bmatrix} \cos 45 & -\sin 45 \sin\theta_x & \sin 45 \cos\theta_x & 0 \\ 0 & \cos\theta_x & \sin\theta_x & 0 \\ -\sin 45 & -\cos 45 \sin\theta_x & \cos 45 \cos\theta_x & 0 \\ 0 & 0 & 0 & 1 \end{bmatrix} \tag{4.34}$$

Expansion of the y term (which is 0) of $[\mathbf{P}^*]$ gives

$$0 = -\sin 45 \sin\theta_x + \cos\theta_x - \cos 45 \sin\theta_x \tag{4.35}$$

which may be solved for θ_x

$$\theta_x = \tan^{-1}\left(\frac{1}{\sqrt{2}}\right) = 35.264°$$

The magnitude of z_p does not affect the result.

Substitution of $\theta_x = 35.264°$ in Eq. (4.34) makes the transformation matrix for the isometric projection

$$[\mathbf{T}_i] = \begin{bmatrix} 0.7071 & -0.4082 & 0.5774 & 0 \\ 0 & 0.8165 & 0.5774 & 0 \\ -0.7071 & -0.4082 & 0.5774 & 0 \\ 0 & 0 & 0 & 1 \end{bmatrix} \tag{4.36}$$

Isometric Projection by Viewpoint Transformation

For an isometric projection, the viewpoint is (1, 1, 1) or any other three equal values. A unit vector along the viewing axis is given by

$$\mathbf{m} = \left(\frac{1}{\sqrt{3}}\right)\mathbf{i} + \left(\frac{1}{\sqrt{3}}\right)\mathbf{j} + \left(\frac{1}{\sqrt{3}}\right)\mathbf{k} \tag{4.37}$$

From Eq. (4.29) the angle of rotation is

$$\theta = \cos^{-1}\left(\frac{1}{\sqrt{3}}\right) = 54.74° \tag{4.38}$$

The direction cosines, from Eq. (4.32), are

$$n_x = \frac{(1/\sqrt{3})}{\sin 54.74} = 0.7071$$

$$n_y = -\frac{(1/\sqrt{3})}{\sin 54.74} = -0.7071 \qquad (4.39)$$

$$n_z = 0$$

Substituting these values into the general rotation matrix, Eq. (4.27), produces

$$[\mathbf{T}_i] = \begin{bmatrix} 0.7885 & -0.2115 & 0.5774 & 0 \\ -0.2115 & 0.7885 & 0.5774 & 0 \\ -0.5774 & -0.5774 & 0.5774 & 0 \\ 0 & 0 & 0 & 1 \end{bmatrix} \qquad (4.40)$$

It is seen that Eq. (4.40) is not the same as Eq. (4.36). Although Eq. (4.40) does produce a correct isometric transformation, the y axis of the transformed image is not vertical. The tilt of the y axis is determined from the transformation of a point lying on the y axis, such as $[\mathbf{P}] = [0 \quad 1 \quad 0 \quad 1]$. Multiplication of this $[\mathbf{P}]$ by the transformation matrix of Eq. (4.40) yields

$$[\mathbf{P}^*] = [-0.2115 \quad 0.7885 \quad 0.5774 \quad 1] \qquad (4.41)$$

The angle between the y axis of the transformed image and the vertical is

$$\theta_z = \tan^{-1}\left(\frac{\Delta x}{\Delta y}\right) = \tan^{-1}\left(\frac{-0.2115}{0.7885}\right) = -15° \qquad (4.42)$$

Postmultiplication of Eq. (4.40) by Eq. (4.10) with $\theta_z = -15°$ yields Eq. (4.36), the isometric projection transformation with the y axis vertical.

The amount of foreshortening of lines in an isometric projection can be determined from the transformation of a vector originally in the x-y plane. For example, if the unit vector along the y axis $[\mathbf{P}] = [0 \quad 1 \quad 0 \quad 1]$ is multiplied by Eq. (4.36) the product is $[\mathbf{P}^*] = [0 \quad 0.8165 \quad 0.5774 \quad 1]$. The length of this unit vector, displayed on the screen (x-y plane), is 0.8165. Thus, the unit vector is foreshortened to approximately 82% of its original length. From symmetry, it is clear that the same foreshortening occurs in the x and z directions for isometric projections. The zero x component in the result for $[\mathbf{P}^*]$ shows that Eq. (4.36) keeps the y axis vertical.

In contrast, the foreshortening in dimetric and trimetric projection is not equal in all directions. In these cases, the foreshortening factors can be determined from the transformation of the three unit vectors corresponding to the coordinate axes.

4.6 OBLIQUE PROJECTIONS

As has been demonstrated, one type of pictorial representation of 3-D objects for computer generation is the axonometric projection. Parallel projection lines from the object are perpendicular to the picture plane, as shown in Figure 4.1. Axonometric projection finds wide use in 3-D interactive graphics, where a pictorial view from any viewpoint may be produced routinely.

With 2-D CAD systems and manual drafting, however, the simplest method for making a pictorial view is the *oblique* drawing. Such drawings are best used for showing boxlike objects with most of the detail on one surface. In the oblique drawings of a cube with a hole shown in Figure 4.8, the parallel diagonal edges defining the top and side of the cube are drawn at any convenient angle. Common choices for this angle are 30°, 45°, or 60°—corresponding to standard drafting triangles. These edges are drawn at true length for a *cavalier* oblique drawing and are foreshortened to half of true length for a *cabinet* oblique drawing.

Oblique projection is a parallel projection where the projectors are *not* perpendicular to the picture plane. The transformation matrix[24] used to produce oblique projection is

$$[\mathbf{T}_o] = \begin{bmatrix} 1 & 0 & 0 & 0 \\ 0 & 1 & 0 & 0 \\ k\cos\theta & k\sin\theta & 1 & 0 \\ 0 & 0 & 0 & 1 \end{bmatrix} \tag{4.43}$$

where k is the foreshortening factor and θ is the arbitrary angle made with the x direction. When $k = 1$ a cavalier projection results; when $k = 0.5$ a cabinet projection results. Any other positive value of k, particularly in the range 0.5 to 1, may also be used. After application of Eq. (4.43), the screen projection is made in the usual way.

In computer graphics, oblique projections find little use for at least two reasons: (1) they are unnatural and tend not to look "right," and (2) they involve

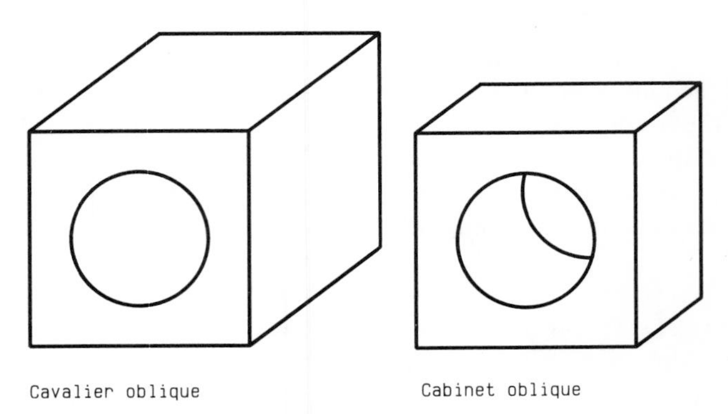

Figure 4.8 Oblique drawings.

about the same amount of computation as axonometric projections but are not as versatile. It turns out that the most realistic pictorial views are produced with perspective projection, explained in the next section.

4.7 PERSPECTIVE PROJECTIONS

More realistic renditions of images can be made with perspective projections, because size-distance relationships are observed. Unlike the parallel projection lines shown in Figure 4.1, perspective projection lines converge at one point, the *viewpoint,* as shown in Figure 4.9. Perspective projection is fairly cumbersome with manual drafting methods.

The geometry of perspective projection is shown in Figure 4.10, where the distances that control the apparent size of objects are shown. The focal length of the lens depicted in Figure 4.10(a) is analogous to the distance d_z in Figure 4.10(b). The most realistic photographs are made with a "normal" lens, one that subtends approximately the same angle—45° to 55°—as human vision. A "wide-angle" lens has a shorter than normal focal length, which makes objects appear smaller when the distance from the camera to the object is unchanged. However, if the camera is moved closer to increase image size, the object will look unrealistically large because of the distortion of perspective. The opposite is true for a "telephoto" lens, which has a longer than normal focal length. This apparent distortion of perspective arises from the "unnatural" angle of view subtended by wide-angle and telephoto lenses.

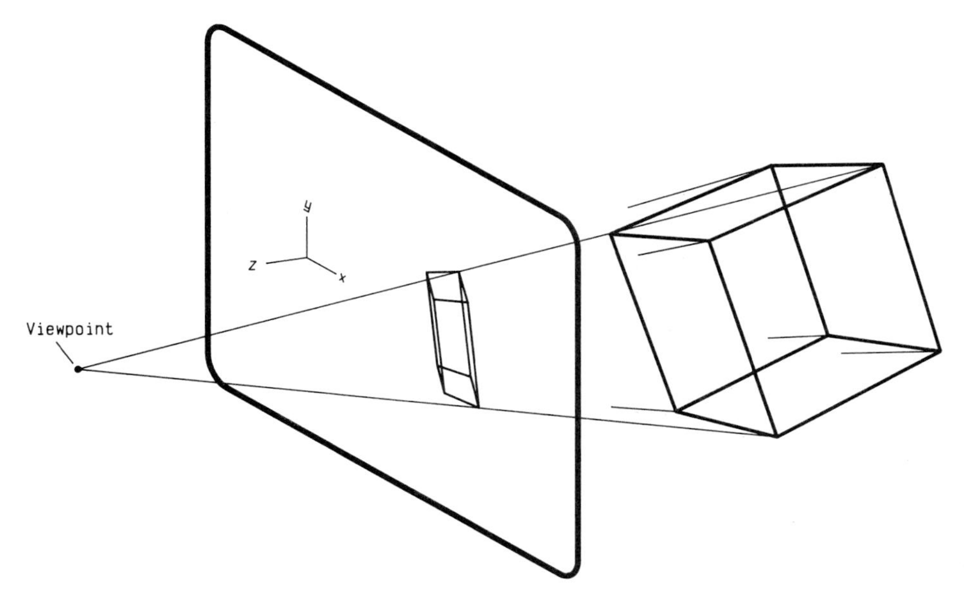

Figure 4.9 Perspective projection. The projectors converge at a viewpoint in the positive z direction.

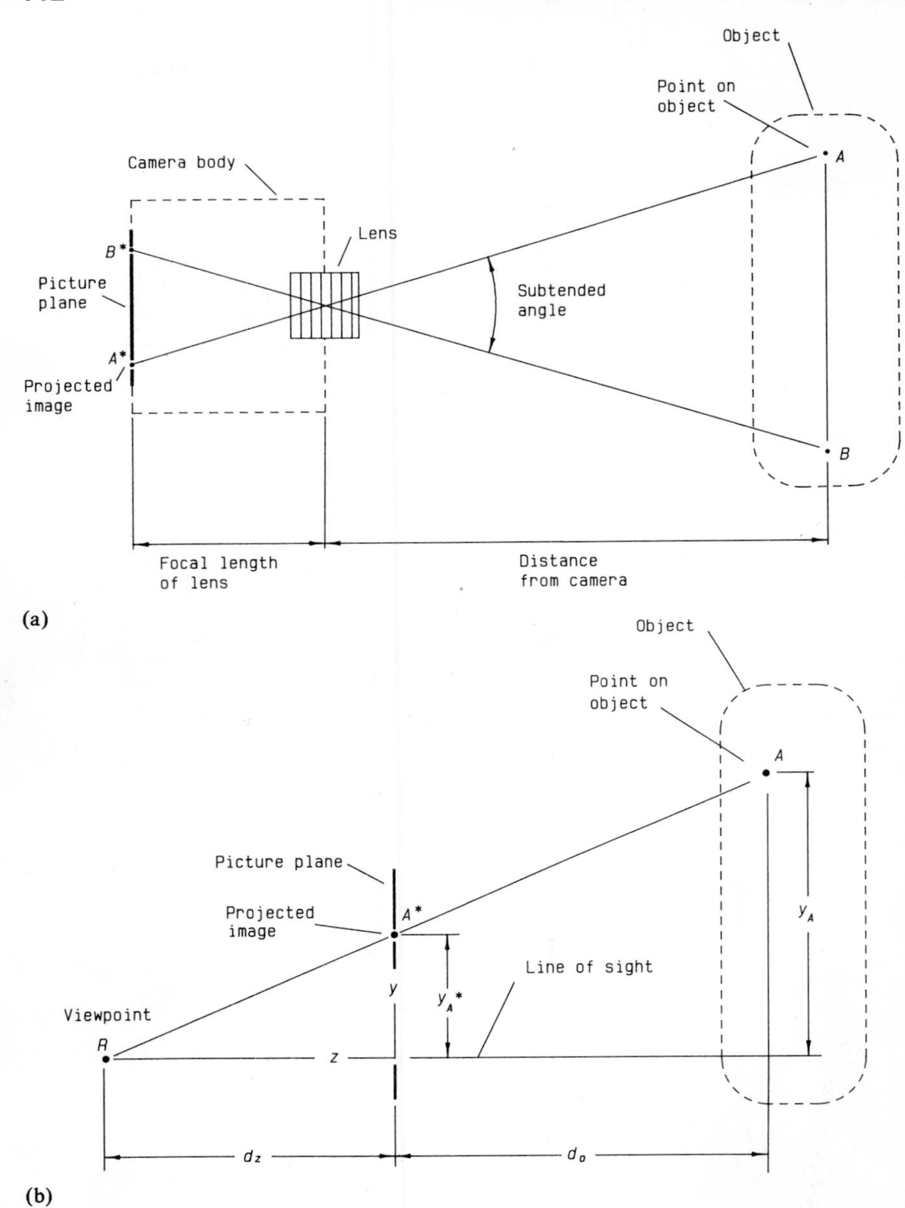

Figure 4.10 Geometry of perspective projection. (a) In photography, as the focal length of the lens is increased, the subtended angle of view decreases. The image of an object is scaled according to the ratio of the focal length and the distance from the camera. (b) In the perspective transformation, the scaling of the image is related to the distance d_z and the location of points on the object. Note that the z coordinates are negative for all points in the object shown.

The geometry in Figure 4.10(b) is used to develop the mathematical transformation for perspective projection. From similar triangles,

$$y_A^* = \frac{d_z}{d_z + d_o} y_A = \frac{d_z}{d_z - z_A} y_A \tag{4.44}$$

where $d_o = -z_A$. This equation shows that the farther a point is in the negative z direction, the smaller its projection.

Creation of realistic-looking perspective drawings is basically a three-step process:

1. Necessary transformations, translation and rotation, are performed on the points matrix to position the object so that the line of sight is aligned with the z axis.
2. The points matrix resulting from step 1 is translated or clipped as required in the z direction. Simple perspectives are made by translating (or clipping) the points matrix to $z \leq 0$. Alternatively, translating (clipping) to $z < d_z$ may be acceptable if the window or viewport is changed.
3. The points matrix resulting from step 2 is postmultiplied by the homogeneous perspective transformation matrix

$$[\mathbf{T}_p] = \begin{bmatrix} 1 & 0 & 0 & -1/d_x \\ 0 & 1 & 0 & -1/d_y \\ 0 & 0 & 1 & -1/d_z \\ 0 & 0 & 0 & 1 \end{bmatrix} \tag{4.45}$$

which, applied to a points matrix where $y = y_A$ and $z = -z_A$, will produce Eq. (4.44) when $d_x = d_y = \infty$.

Positioning the z axis along the line-of-sight vector in step 1 is required because only this axis remains perpendicular to the picture plane after the perspective transformation. In contrast, *axonometric* projections are not affected by the position of the z axis because lines parallel to the z axis remain parallel. To illustrate the effect of the z axis placement in perspective projections, Figure 4.11 illustrates how a 35-story building looks from the level of the 35th story, from the level of the 17th story, and from the ground.

The z position of the origin in the points matrix which is postmultiplied by Eq. (4.45) affects the size of the projected image. The length of a line connecting two points in the $z = 0$ plane is unchanged by the perspective transformation. If the two points are behind the $z = 0$ plane (negative z) the length is decreased, which is desirable. However, if one or both of the points are in front of the $z = 0$ plane, the resulting enlargement of the image may make it necessary to change the window or viewpoint. The perspective projection of any point with $z = d_z$ becomes infinite, and that of a point with $z > d_z$ becomes inverted. Hence, it is simplest if we make all z values in the points matrix *negative* before using Eq. (4.45).

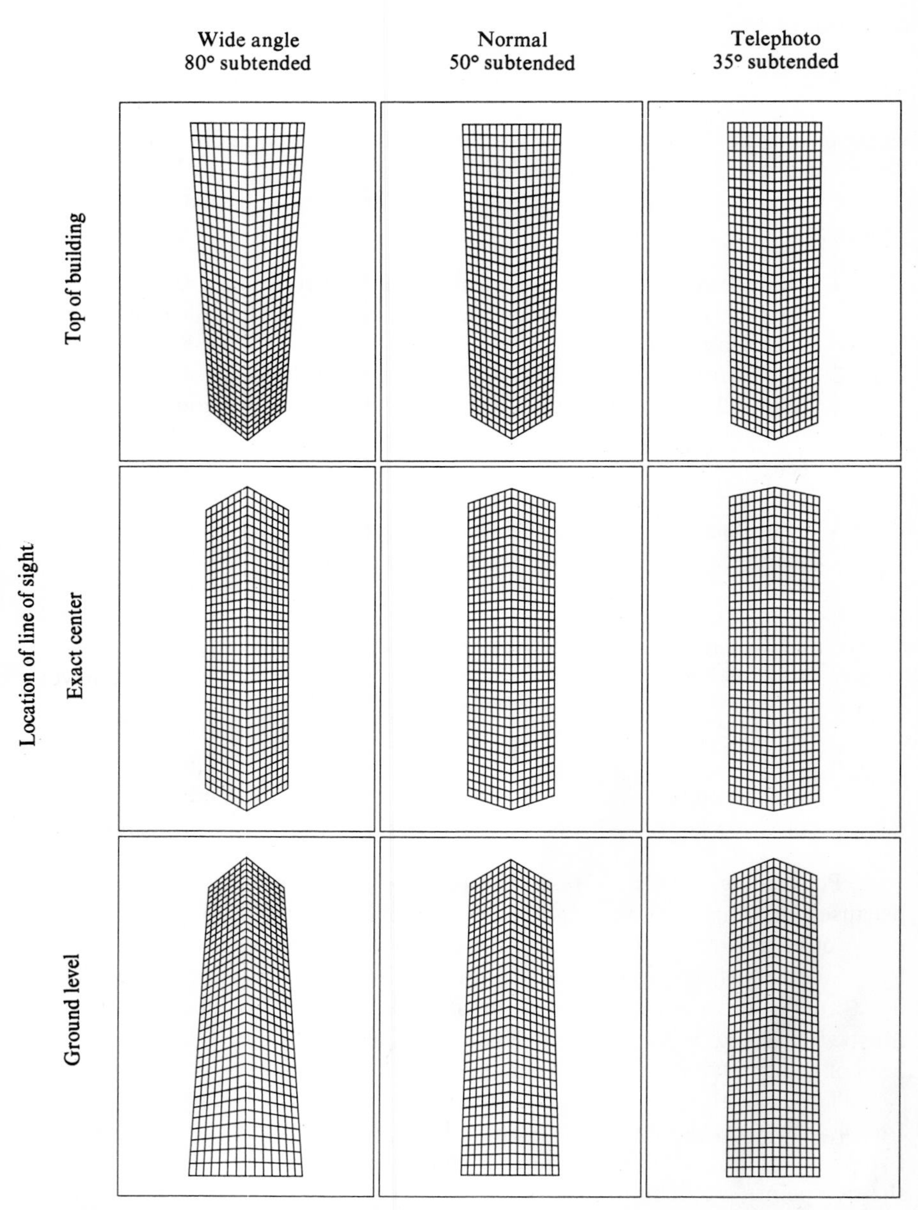

Figure 4.11 Demonstration of the effect of varying the angle subtended (viewing distance d_z) and vertical location of z axis on appearance of a perspective projection of a 35-story building. The height of the building is five times the width (or depth). The three-point perspectives of the top and bottom rows result from the use of finite values of d_x and d_y in Eq. (4.45).

The factors d_x and d_y take into account the horizontal and vertical components of distance between points on the object and the viewpoint. These effects can be neglected for simple projections. Hence, very satisfactory results are obtained with $d_x = d_y = \infty$.

A basic guide for computing the viewing distance in Eq. (4.45) can be based on Figure 4.12. The distance w^* is approximately equal to the desired screen projection of the longest dimension of the object. An angle of 50°, which is roughly the angle subtended by human vision, is used to produce an undistorted perspective. Thus the viewing distance d_z is estimated by

$$d_z = \frac{w/2}{\tan 25} \approx w^* \tag{4.46}$$

The columns of Figure 4.11 illustrate the effect of the subtended angle of view on the appearance of a perspective projection.

Perspective transformation with a 4 × 4 matrix is consistent with the homogeneous coordinate system. However, when the perspective transformation is used, the resultant points matrix must be normalized to maintain unity in the fourth column.

The perspective ponts matrix cannot be rotated in three dimensions without distortion unless it is first multiplied by the inverse of the perspective transformation matrix. Actually, it is simpler to store an unmodified copy of the points matrix and discard the perspective-transformed points matrix after display. Thus, the perspective projection is always a *viewing* transformation and never a *modeling* transformation.

One-, Two-, and Three-Point Perspective

Engineers are familiar with the classification of perspective drawings by the number of *vanishing points,* locations where parallel lines in an object appear to

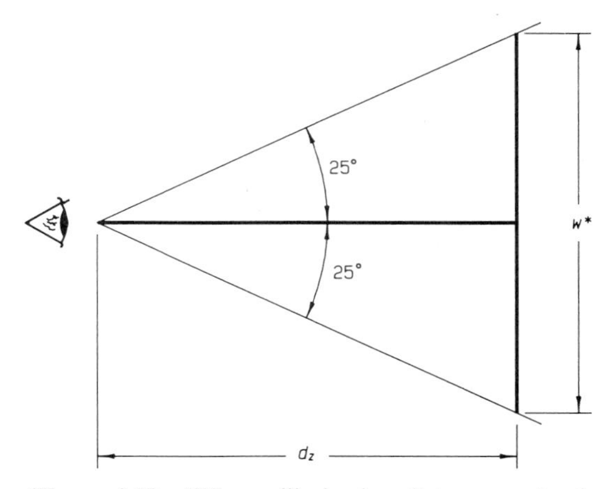

Figure 4.12 "Normal" viewing distance and subtended angle.

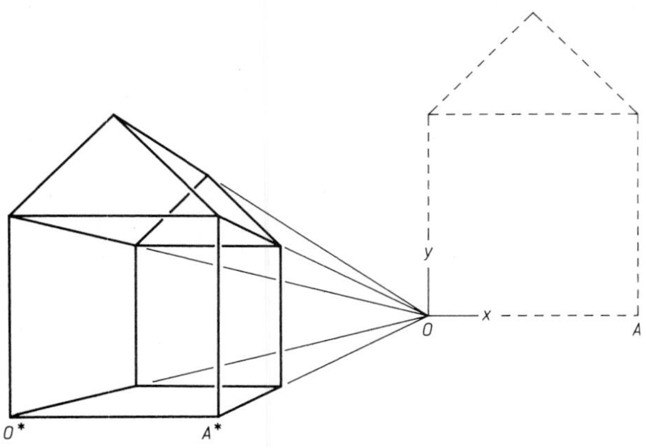

Figure 4.13 One-point perspective of a 2 × 2 × 3 high "house," with the front face lying on the picture plane where $z = 0$. The image is produced by (1) translation with $\Delta x = -4$ and $\Delta y = -1$, which positions the line-of-sight vector above the ground on the right side of the house, and (2) application of the perspective transformation $[\mathbf{T}_p]$ with $d_x = d_y = \infty$ and $d_z = 5$. The transformation moves points O and A to O^* and A^*.

converge. If in a perspective projection parallel lines remain parallel, the associated vanishing point is at infinity.

One-point perspectives result when one plane of the object is parallel to the screen. The object is translated so that the z axis is along the desired line of sight. The viewing distance d_z for Eq. (4.45) may be estimated with Eq. (4.46). A typical one-point perspective projection is shown in Figure 4.13.

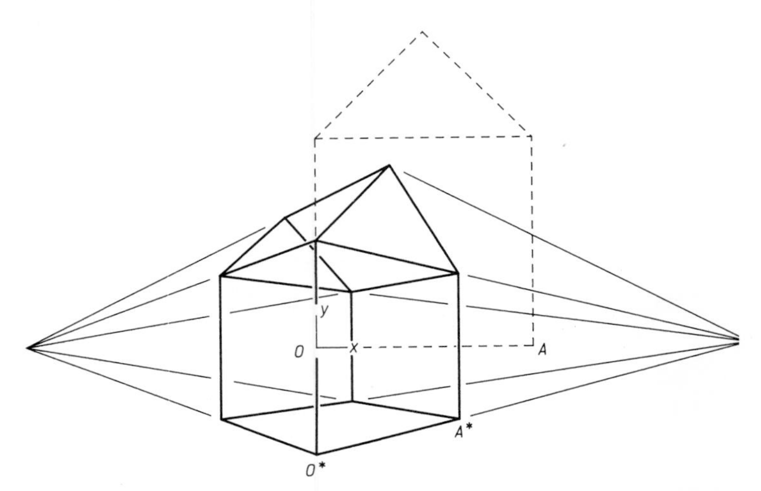

Figure 4.14 Two-point perspective of the same "house" as in Figure 4.13. This is produced by (1) translation with $\Delta y = -1$, (2) 35° rotation around the y axis, and (3) application of $[\mathbf{T}_p]$ with $d_x = d_y = \infty$ and $d_z = 3$.

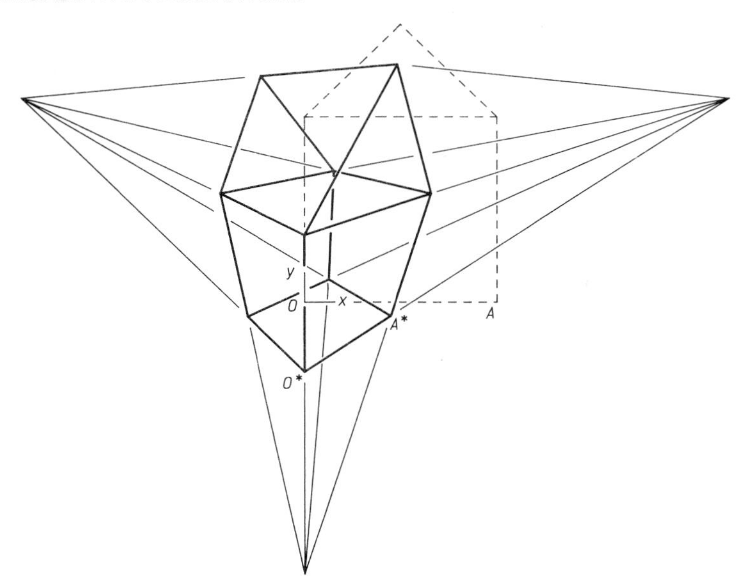

Figure 4.15 Three-point perspective of the same "house" as in Figure 4.13. This is produced by (1) translation with $\Delta y = -1$, (2) 35° rotation around the y axis, (3) 40° rotation around the x axis, and (4) application of $[\mathbf{T}_p]$ with $d_x = d_y = \infty$ and $d_z = 2.5$.

Two-point perspectives, illustrated in the middle row of Figure 4.11, are popular for manual drafting because there is a reasonable balance between the amount of work involved and the realism of the result. This projection is usually seen with vertical lines parallel to the screen y axis and horizontal lines converging to a right and to a left vanishing point. Although unusual, a two-point projection with vanishing points above and below the object may also be produced. To construct a two-point projection, the user performs the necessary translations and a rotation around the y axis (or x axis for the case of vanishing points above or below). An example of this projection is shown in Figure 4.14.

Three-point perspectives, shown in the top and bottom rows of Figure 4.11 and in Figure 4.15, are more complex to produce manually because of the lack of parallel lines in the finished drawing. In this projection none of the original axis directions is parallel to a screen axis.

In the following example, an isometric drawing is converted to a three-point perspective.

EXAMPLE 4.5

Produce a perspective transformation of the unit cube shown in Figure 4.7(a). Place the viewpoint on the same axis as the isometric viewpoint, 2 units from the nearest vertex of the cube.

Solution. The line of sight is the diagonal from the origin through point A. Application of the isometric transformation, Eq. (4.36), yields

$$\begin{bmatrix} 0 & 0 & 0 & 1 \\ 0 & 0 & 1 & 1 \\ 0 & 1 & 0 & 1 \\ 0 & 1 & 1 & 1 \\ 1 & 0 & 0 & 1 \\ 1 & 0 & 1 & 1 \\ 1 & 1 & 0 & 1 \\ 1 & 1 & 1 & 1 \end{bmatrix} \begin{bmatrix} 0.7071 & -0.4082 & 0.5774 & 0 \\ 0 & 0.8165 & 0.5774 & 0 \\ -0.7071 & -0.4082 & 0.5774 & 0 \\ 0 & 0 & 0 & 1 \end{bmatrix} = \begin{bmatrix} 0 & 0 & 0 & 1 \\ -0.707 & -0.408 & 0.577 & 1 \\ 0 & 0.816 & 0.577 & 1 \\ -0.707 & 0.408 & 1.155 & 1 \\ 0.707 & -0.408 & 0.577 & 1 \\ 0 & -0.816 & 1.155 & 1 \\ 0.707 & 0.408 & 1.155 & 1 \\ 0 & 0 & 1.732 & 1 \end{bmatrix}$$

which is illustrated in Figure 4.7(a).

The largest z value in the transformed points matrix is 1.732 units. For the proper viewing distance, a translation of -1.732 units in the $-z$ direction is necessary. Finally, postmultiplication by Eq. (4.45) with $d_x = d_y = \infty$ and $d_z = 2$ gives

$$\begin{bmatrix} 0 & 0 & -1.732 & 1 \\ -0.707 & -0.408 & -1.155 & 1 \\ 0 & 0.816 & -1.155 & 1 \\ -0.707 & 0.408 & -0.577 & 1 \\ 0.707 & -0.408 & -1.155 & 1 \\ 0 & -0.816 & -0.577 & 1 \\ 0.707 & 0.408 & -0.577 & 1 \\ 0 & 0 & 0 & 1 \end{bmatrix} \begin{bmatrix} 1 & 0 & 0 & 0 \\ 0 & 1 & 0 & 0 \\ 0 & 0 & 1 & -0.5 \\ 0 & 0 & 0 & 1 \end{bmatrix} = \begin{bmatrix} 0 & 0 & -1.732 & 1.866 \\ -0.707 & -0.408 & -1.155 & 1.578 \\ 0 & 0.816 & -1.155 & 1.578 \\ -0.707 & 0.408 & -0.577 & 1.289 \\ 0.707 & -0.408 & -1.155 & 1.578 \\ 0 & -0.816 & -0.577 & 1.289 \\ 0.707 & 0.408 & -0.577 & 1.289 \\ 0 & 0 & 0 & 1 \end{bmatrix}$$

Normalization of the result yields

$$[\mathbf{P^*}] = \begin{bmatrix} 0 & 0 & -0.928 & 1 \\ -0.448 & -0.259 & -0.732 & 1 \\ 0 & 0.517 & -0.732 & 1 \\ -0.548 & 0.317 & -0.448 & 1 \\ 0.448 & -0.259 & -0.732 & 1 \\ 0 & -0.663 & -0.448 & 1 \\ 0.548 & 0.317 & -0.448 & 1 \\ 0 & 0 & 0 & 1 \end{bmatrix}$$

which produces the three-point perspective projection shown.

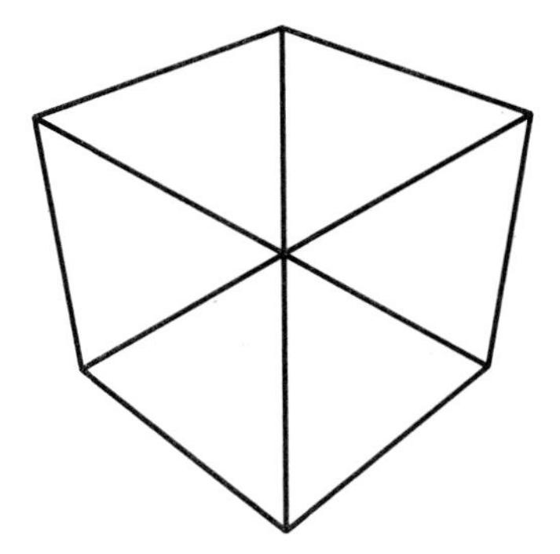

A two-point perspective results from rotation around the y axis only [Eq. (4.36) is a rotation around both the y and x axes], the required z translation, and the application of Eq. (4.45) with $d_x = d_y = \infty$.

4.8 IMAGE RECONSTRUCTION

One of the properties of pictorial projections is that the resulting 2-D images contain depth information. As a consequence, the inverse of the pictorial projection process can be used to deduce 3-D points matrices from 2-D images. One of the uses of image reconstruction is in creating data bases from digitized images for modeled animation, discussed in Section 2.9.

In general, any transformed image can be represented by

$$h[\mathbf{P}^*] = [\mathbf{P}][\mathbf{T}_c] \tag{4.47}$$

where h is the normalization factor in Eq. (3.33), $[\mathbf{P}^*]$ is the transformed points matrix, $[\mathbf{P}]$ is the original points matrix, and $[\mathbf{T}_c]$ is the transformation matrix that may include perspective information. Because the *screen projection* $[\mathbf{P}^*]$ has only x and y components, the third column of $[\mathbf{T}_c]$ must be zero, or

$$[\mathbf{T}_c] = \begin{bmatrix} T_{11} & T_{12} & 0 & T_{14} \\ T_{21} & T_{22} & 0 & T_{24} \\ T_{31} & T_{32} & 0 & T_{34} \\ T_{41} & T_{42} & 0 & T_{44} \end{bmatrix} \tag{4.48}$$

A points matrix with one point,

$$[\mathbf{P}] = [x \quad y \quad z \quad 1] \tag{4.49}$$

will be considered in the following discussion.[50] Combination of Eqs. (4.47) through (4.49) yields

$$\begin{aligned} T_{11}x + T_{21}y + T_{31}z + T_{41} &= hx^* \\ T_{12}x + T_{22}y + T_{32}z + T_{42} &= hy^* \\ T_{14}x + T_{24}y + T_{34}z + T_{44} &= h \end{aligned} \tag{4.50}$$

Elimination of h from Eqs. (4.50) yields

$$
\begin{aligned}
&(T_{11} - T_{14}x^*)x + (T_{21} - T_{24}x^*)y + (T_{31} - T_{34}x^*)z + (T_{41} - T_{44}x^*) = 0 \\
&(T_{12} - T_{14}y^*)x + (T_{22} - T_{24}y^*)y + (T_{32} - T_{34}y^*)z + (T_{42} - T_{44}y^*) = 0
\end{aligned}
\tag{4.51}
$$

The x^* and y^* are known values. Since the number of unknowns in Eq. (4.51) is 15, more information is needed. Two cases will be considered. In the simpler one, the transformation is known in advance; two screen projections are needed to completely determine the coordinates of any point. In the other case, the transformation is not known in advance; one screen projection of several points with known coordinates is needed to compute the transformation.

Transformation Known in Advance

When the 12 remaining elements of $[\mathbf{T}_c]$ are known, there are three unknowns (x, y, and z) for each point under consideration. The reconstruction requires *two* sets of x^* and y^* information. Therefore, two different screen projections of the unknown point are necessary. If the superscripts 1 and 2 are used to denote the point in the two screen projections made with the two known transformations, Eq. (4.51) yields

$$
\begin{aligned}
&(\overset{1}{T}_{11} - \overset{1}{T}_{14}\overset{1}{x}{}^*)x + (\overset{1}{T}_{21} - \overset{1}{T}_{24}\overset{1}{x}{}^*)y + (\overset{1}{T}_{31} - \overset{1}{T}_{34}\overset{1}{x}{}^*)z + (\overset{1}{T}_{41} - \overset{1}{T}_{44}\overset{1}{x}{}^*) = 0 \\
&(\overset{1}{T}_{12} - \overset{1}{T}_{14}\overset{1}{y}{}^*)x + (\overset{1}{T}_{22} - \overset{1}{T}_{24}\overset{1}{y}{}^*)y + (\overset{1}{T}_{32} - \overset{1}{T}_{34}\overset{1}{y}{}^*)z + (\overset{1}{T}_{42} - \overset{1}{T}_{44}\overset{1}{y}{}^*) = 0 \\
&(\overset{2}{T}_{11} - \overset{2}{T}_{14}\overset{2}{x}{}^*)x + (\overset{2}{T}_{21} - \overset{2}{T}_{24}\overset{2}{x}{}^*)y + (\overset{2}{T}_{31} - \overset{2}{T}_{34}\overset{2}{x}{}^*)z + (\overset{2}{T}_{41} - \overset{2}{T}_{44}\overset{2}{x}{}^*) = 0 \\
&(\overset{2}{T}_{12} - \overset{2}{T}_{14}\overset{1}{y}{}^*)x + (\overset{2}{T}_{22} - \overset{2}{T}_{24}\overset{2}{y}{}^*)y + (\overset{2}{T}_{32} - \overset{2}{T}_{34}\overset{2}{y}{}^*)z + (\overset{2}{T}_{42} - \overset{2}{T}_{44}\overset{2}{y}{}^*) = 0
\end{aligned}
\tag{4.52}
$$

In matrix form, Eq. (4.52) can be rewritten as

$$
\begin{bmatrix}
\overset{1}{T}_{11} - \overset{1}{T}_{14}\overset{1}{x}{}^* & \overset{1}{T}_{21} - \overset{1}{T}_{24}\overset{1}{x}{}^* & \overset{1}{T}_{31} - \overset{1}{T}_{34}\overset{1}{x}{}^* \\
\overset{1}{T}_{12} - \overset{1}{T}_{14}\overset{1}{y}{}^* & \overset{1}{T}_{22} - \overset{1}{T}_{24}\overset{1}{y}{}^* & \overset{1}{T}_{32} - \overset{1}{T}_{34}\overset{1}{y}{}^* \\
\overset{2}{T}_{11} - \overset{2}{T}_{14}\overset{2}{x}{}^* & \overset{2}{T}_{21} - \overset{2}{T}_{24}\overset{2}{x}{}^* & \overset{2}{T}_{31} - \overset{2}{T}_{34}\overset{2}{x}{}^* \\
\overset{2}{T}_{12} - \overset{2}{T}_{14}\overset{2}{y}{}^* & \overset{2}{T}_{22} - \overset{2}{T}_{24}\overset{2}{y}{}^* & \overset{2}{T}_{32} - \overset{2}{T}_{34}\overset{2}{y}{}^*
\end{bmatrix}
\begin{bmatrix} x \\ y \\ z \end{bmatrix}
=
\begin{bmatrix}
\overset{1}{T}_{44}\overset{1}{x}{}^* - \overset{1}{T}_{41} \\
\overset{1}{T}_{44}\overset{1}{y}{}^* - \overset{1}{T}_{42} \\
\overset{2}{T}_{44}\overset{2}{x}{}^* - \overset{2}{T}_{41} \\
\overset{2}{T}_{44}\overset{2}{y}{}^* - \overset{2}{T}_{42}
\end{bmatrix}
\tag{4.53}
$$

Because any three of the four equations can be used to determine the three unknowns, Eq. (4.53) is overspecified. Solution in the mean sense[59] can be used advantageously to reduce errors that might occur in digitizing the images. Representing the form of Eq. (4.53) by $[\mathbf{A}][\mathbf{X}] = [\mathbf{B}]$, a mean squares solution results from solving the system

$$
[\mathbf{A}]^T[\mathbf{A}][\mathbf{X}] = [\mathbf{A}]^T[\mathbf{B}]
\tag{4.54}
$$

which now contains three equations and three unknowns.

Transformation Unknown in Advance

In the more common case, the transformation is not known in advance. Here the problem is to determine the transformation $[\mathbf{T}_c]$ by knowing the world coordinates of enough points that appear in the 2-D image. These key points typically would be marked and numbered in a real-world situation.

Expansion of Eq. (4.51) yields

$$
\begin{aligned}
T_{11}x - T_{14}xx^* + T_{21}y - T_{24}yx^* + T_{31}z - T_{34}zx^* + T_{41} - T_{44}x^* &= 0 \\
T_{12}x - T_{14}xy^* + T_{22}y - T_{24}yy^* + T_{32}z - T_{34}zy^* + T_{42} - T_{44}y^* &= 0
\end{aligned}
\tag{4.55}
$$

Because there are 12 unknown components of $[\mathbf{T}_c]$, six discrete points with known locations are required for reconstruction. With subscripts denoting the individual points in the usual way, Eq. (4.55) may be written as

$$
\begin{bmatrix}
x_1 & 0 & -x_1x_1^* & y_1 & 0 & -y_1x_1^* & z_1 & 0 & -z_1x_1^* & 1 & 0 & -x_1^* \\
0 & x_1 & -x_1y_1^* & 0 & y_1 & -y_1y_1^* & 0 & z_1 & -z_1y_1^* & 0 & 1 & -y_1^* \\
x_2 & 0 & -x_2x_2^* & y_2 & 0 & -y_2x_2^* & z_2 & 0 & -z_2x_2^* & 1 & 0 & -x_2^* \\
0 & x_2 & -x_2y_2^* & 0 & y_2 & -y_2y_2^* & 0 & z_2 & -z_2y_2^* & 0 & 1 & -y_2^* \\
x_3 & 0 & -x_3x_3^* & y_3 & 0 & -y_3x_3^* & z_3 & 0 & -z_3x_3^* & 1 & 0 & -x_3^* \\
0 & x_3 & -x_3y_3^* & 0 & y_3 & -y_3y_3^* & 0 & z_3 & -z_3y_3^* & 0 & 1 & -y_3^* \\
x_4 & 0 & -x_4x_4^* & y_4 & 0 & -y_4x_4^* & z_4 & 0 & -z_4x_4^* & 1 & 0 & -x_4^* \\
0 & x_4 & -x_4y_4^* & 0 & y_4 & -y_4y_4^* & 0 & z_4 & -z_4y_4^* & 0 & 1 & -y_4^* \\
x_5 & 0 & -x_5x_5^* & y_5 & 0 & -y_5x_5^* & z_5 & 0 & -z_5x_5^* & 1 & 0 & -x_5^* \\
0 & x_5 & -x_5y_5^* & 0 & y_5 & -y_5y_5^* & 0 & z_5 & -z_5y_5^* & 0 & 1 & -y_5^* \\
x_6 & 0 & -x_6x_6^* & y_6 & 0 & -y_6x_6^* & z_6 & 0 & -z_6x_6^* & 1 & 0 & -x_6^* \\
0 & x_6 & -x_6y_6^* & 0 & y_6 & -y_6y_6^* & 0 & z_6 & -z_6y_6^* & 0 & 1 & -y_6^*
\end{bmatrix}
\begin{bmatrix} T_{11} \\ T_{12} \\ T_{14} \\ T_{21} \\ T_{22} \\ T_{24} \\ T_{31} \\ T_{32} \\ T_{34} \\ T_{41} \\ T_{42} \\ T_{44} \end{bmatrix}
= 0 \tag{4.56}
$$

To obtain a nontrivial solution of these equations, one of the unknowns must be specified. From Eq. (3.31), T_{44} is used as a scaling factor. Therefore, $T_{44} \equiv 1$, which reduces Eq. (4.56) to a system of 11 unknowns,

$$
\begin{bmatrix}
x_1 & 0 & -x_1x_1^* & y_1 & 0 & -y_1x_1^* & z_1 & 0 & -z_1x_1^* & 1 & 0 \\
0 & x_1 & -x_1y_1^* & 0 & y_1 & -y_1y_1^* & 0 & z_1 & -z_1y_1^* & 0 & 1 \\
x_2 & 0 & -x_2x_2^* & y_2 & 0 & -y_2x_2^* & z_2 & 0 & -z_2x_2^* & 1 & 0 \\
0 & x_2 & -x_2y_2^* & 0 & y_2 & -y_2y_2^* & 0 & z_2 & -z_2y_2^* & 0 & 1 \\
x_3 & 0 & -x_3x_3^* & y_3 & 0 & -y_3x_3^* & z_3 & 0 & -z_3x_3^* & 1 & 0 \\
0 & x_3 & -x_3y_3^* & 0 & y_3 & -y_3y_3^* & 0 & z_3 & -z_3y_3^* & 0 & 1 \\
x_4 & 0 & -x_4x_4^* & y_4 & 0 & -y_4x_4^* & z_4 & 0 & -z_4x_4^* & 1 & 0 \\
0 & x_4 & -x_4y_4^* & 0 & y_4 & -y_4y_4^* & 0 & z_4 & -z_4y_4^* & 0 & 1 \\
x_5 & 0 & -x_5x_5^* & y_5 & 0 & -y_5x_5^* & z_5 & 0 & -z_5x_5^* & 1 & 0 \\
0 & x_5 & -x_5y_5^* & 0 & y_5 & -y_5y_5^* & 0 & z_5 & -z_5y_5^* & 0 & 1 \\
x_6 & 0 & -x_6x_6^* & y_6 & 0 & -y_6x_6^* & z_6 & 0 & -z_6x_6^* & 1 & 0
\end{bmatrix}
\begin{bmatrix} T_{11} \\ T_{12} \\ T_{14} \\ T_{21} \\ T_{22} \\ T_{24} \\ T_{31} \\ T_{32} \\ T_{34} \\ T_{41} \\ T_{42} \end{bmatrix}
=
\begin{bmatrix} x_1^* \\ y_1^* \\ x_2^* \\ y_2^* \\ x_3^* \\ y_3^* \\ x_4^* \\ y_4^* \\ x_5^* \\ y_5^* \\ x_6^* \end{bmatrix}
\tag{4.57}
$$

Because there are zeros on the diagonals of Eq. (4.57), the routine used for solution of the system of linear equations must provide for pivoting.[32,59] The resulting solution for the transformation matrix $[\mathbf{T}_c]$ is used to determine the world coordinates of points other than the original six that defined the transformation. If the ith point has the coordinates (x_i^*, y_i^*) on the screen projection, the world coordinates can be found from

$$[x_i \quad y_i \quad z_i]\begin{bmatrix} T_{11} & T_{12} & T_{14} \\ T_{21} & T_{22} & T_{24} \\ T_{31} & T_{32} & T_{34} \end{bmatrix} = [x_i^* - T_{41} \quad y_i^* - T_{42} \quad 0] \tag{4.58}$$

The choice of the original six points is not entirely arbitrary. For example, the solution of Eq. (4.57) will not be unique if all six points are coplanar. Because this and certain other combinations of points can produce numerical difficulties, the solution should always be checked.

EXAMPLE 4.6

Two different projections of point **P** and the transformed original coordinate axes (marked by *) are shown in the accompanying figure.

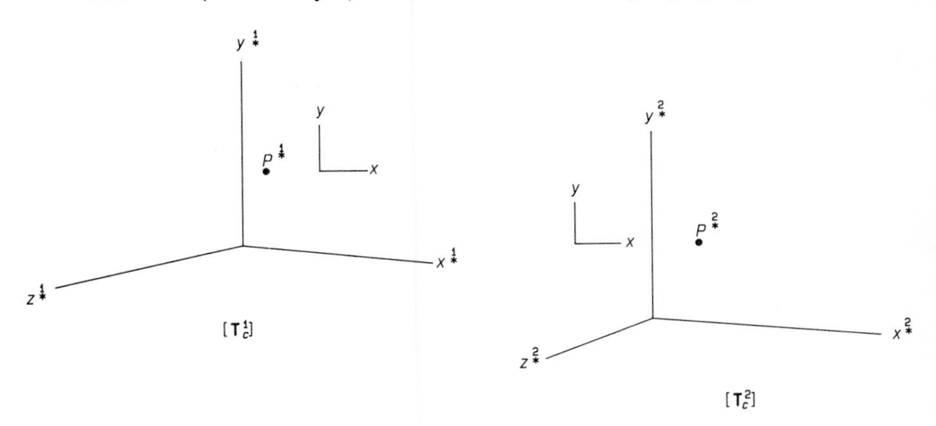

The projections have been prepared with the transformations

$$[\mathbf{T}_c^1] = \begin{bmatrix} 0.346 & 0 & 0 & -0.0200 \\ 0 & 0.400 & 0 & 0 \\ -0.200 & 0 & 0 & -0.0346 \\ -1.60 & -1.60 & 0 & 1 \end{bmatrix}$$

$$[\mathbf{T}_c^2] = \begin{bmatrix} 0.346 & 0 & 0 & -0.0200 \\ 0 & 0.400 & 0 & 0 \\ -0.200 & 0 & 0 & -0.0346 \\ 1.60 & -1.60 & 0 & 1 \end{bmatrix}$$

The screen projections of **P** are located at

$$\overset{1}{\mathbf{P}}{}^{*}:(-1.10, 0) \quad \text{and} \quad \overset{2}{\mathbf{P}}{}^{*}:(2.57, 0)$$

Determine the location of **P** in world coordinates.

Solution. Substitution of the given information into Eq. (4.53) yields

$$\begin{bmatrix} 0.3240 & 0 & -0.2381 \\ 0 & 0.4000 & 0 \\ 0.3974 & 0 & -0.1111 \\ 0 & 0.4000 & 0 \end{bmatrix} \begin{bmatrix} x \\ y \\ z \end{bmatrix} = \begin{bmatrix} 0.5000 \\ 1.600 \\ 0.9700 \\ 1.600 \end{bmatrix}$$

Application of Eq. (4.54) reduces the system to

$$\begin{bmatrix} 0.2629 & 0 & -0.1213 \\ 0 & 0.3200 & 0 \\ -0.1213 & 0 & 0.06903 \end{bmatrix} \begin{bmatrix} x \\ y \\ z \end{bmatrix} = \begin{bmatrix} 0.5475 \\ 1.280 \\ -0.2268 \end{bmatrix}$$

which has the solution

$$[\mathbf{P}] = [x \quad y \quad z \quad 1] = [2.99 \quad 4.00 \quad 1.97 \quad 1]$$

4.9 CLOSURE

Various pictorial representations of 3-D images can be made using the projections described in this chapter. The projection process consists of applying one or more transformations to the points matrix followed by projecting the image to the screen. Pictorial projections may be classified as shown in Figure 4.16.

Axonometric projection is easy to use, since only rotation transformations are needed to obtain the desired view. In some cases the projection may require adjustment of the window, since rotations can move the image to a new position on the picture plane. Because the rotation transformation is affine, subsequent transformations to the points matrix may be performed.

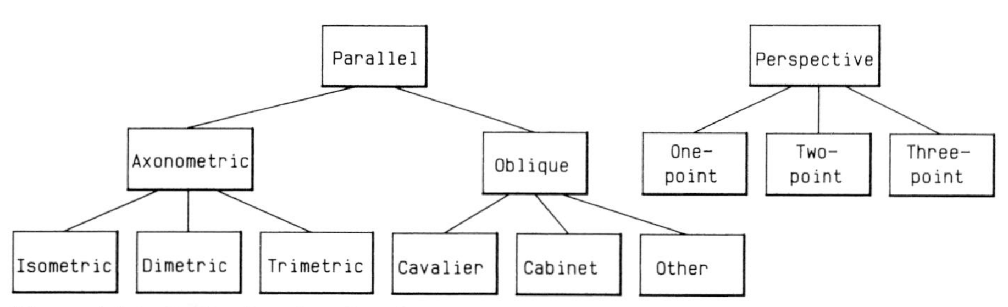

Figure 4.16 Projections for pictorial representations.

Axonometric and oblique projections are both parallel projections; they differ in whether or not the picture plane is perpendicular to the projectors. Oblique projections do not produce natural-appearing images, and there is little reason to use them in 3-D interactive graphics systems.

Perspective transformations give the most realistic pictorials. Before the perspective transformation is applied, translation and rotation transformations are required to align the line-of-sight vector properly and to place the object behind the picture plane. The factors d_x and d_y are usually set to infinity; the factor d_z must be consistent with the size of the given object to make a realistic-appearing perspective projection. Should subsequent operations be needed, provision in the program must be made for recovery of the undistorted points matrix since the perspective transformation is not affine.

Depth information contained in perspective projections can be used to reconstruct the 3-D points matrix. Image reconstruction is useful for applications such as modeled animation.

PROBLEMS

4.1. For the line AB where A:(1, 1, 1) and B:(2, 3, −1), determine the mirror reflection with respect to the plane $y = 0.5$. Sketch, approximately to scale, the screen (x-y) projection of the original and the copied line.

4.2. Generalize Eq. (3.24) for 3-D nonuniform scaling around a point.

4.3. Mirror point A:(2, 3, 1) with respect to the plane defined by the three points B:(0, 2, 0), C:(1, 2, 0), and D:(0, 2, 2).

4.4. Determine the new coordinates of the point B:(1, 2, 1) (a) if it is first rotated −30° around the x axis and then 90° around the y axis and (b) if the order of rotation is reversed.

4.5. A 10-cm (as measured on the screen) line that originally lies in the x-y plane parallel to the x axis is rotated 20° around the y axis and 10° around the x axis. Determine the length and direction of the screen projection of this line.

4.6. A vector **P** that initially goes from (0, 0, 0) to (0, 0, 2) is rotated 30° around the x axis. Determine the subsequent rotation around the y axis if the screen projection of **P** forms an angle of −120° with the screen x axis.

4.7. A line segment extends from (3, 4, 0) to the origin. If the point A:(2, 0, −1) is rotated −60° around the given line segment, determine the new coordinates of A.

4.8 Repeat Problem 4.3, except that B: (0, 0, 0). *Hint:* Rotate the points around BD.

4.9. A line segment connects from the point A:(−1, −2, −1) to the point B:(2, 0, 3). Determine the new coordinates of the point C:(0, 3, 0) if it is rotated 120° around AB.

4.10. Determine the transformation matrix to show the image of an object from a viewpoint of (10, −5, 10) to a viewsite of (0, 5, 0).

4.11. Determine the amount of foreshortening seen in the screen projection of the three unit vectors in the x, y, and z coordinate directions in the case where a parallel projection is made from a viewpoint of (2, 1, 0).

4.12. Determine the direction cosines and the angle of rotation for use in Eq. (4.27) that give the same axonometric projection as a rotation of 30° around the y axis followed by a rotation of $-36.8°$ around the x axis.

4.13. For a perspective transformation, determine the length of the screen projection of a line with the original end points

$$\begin{bmatrix} 2 & 2 & 3 & 1 \\ 5 & 6 & 3.5 & 1 \end{bmatrix}$$

if $d_x = d_y = \infty$ and $d_z = 4$, with the condition that the z translation is *not* done prior to application of the perspective transformation.

4.14. For the transformation in Problem 4.10, determine the foreshortening of the three unit vectors **i**, **j**, and **k** if (a) an axonometric projection is made and (b) a perspective projection is made with a z translation such that all z values are 0 or less, $d_x = d_y = \infty$, and d_z is equal to the viewsite-viewpoint distance.

4.15. Determine and sketch the screen projection of the perspective transformation of a unit cube with one vertex at the origin and one vertex at $(1, 1, -1)$ if $d_x = d_y = \infty$, and $d_z = 3$. Do not do any other transformations. Is this result a one-, two-, or three-point perspective?

4.16. Repeat Problem 4.15, except that $d_x = 4$.

4.17. Discuss the use of the factors that modify the x and y distances in the perspective transformation.

4.18. Show, with a numerical example, that the perspective can "blow up" if an attempt is made to transform a point "inside" the model. (That is, illustrate the importance of handling the distance z properly.)

4.19. Translate the origin of the coordinate system in Figure 4.10(b) a distance d_p to the right, such that the z coordinate of A is $+z_A$. Show, that as an alternative to Eq. (4.45), the perspective transformation matrix can be written as

$$[\mathbf{T}_p] = \begin{bmatrix} 1 & 0 & 0 & 0 \\ 0 & 1 & 0 & 0 \\ 0 & 0 & 1 & -1/d_z \\ 0 & 0 & 0 & (d_z + d_p)/d_z \end{bmatrix}$$

where $d_x = d_y = \infty$ and the distance d_o is redundant. (This form eliminates the need to translate the points matrix in the z direction before application of the perspective transformation.)

4.20. In the use of Eq. (4.55), discuss the consequences of using a screen projection that is axonometric rather than perspective.

PROJECTS

4.1. Prepare a points matrix for a simple wireframe drawing of a table. The edges of the table and the legs should be parallel to coordinate axes. If two lines are drawn for the legs, the entire image can be drawn with one call to polyline. Use rotation transformations in a program to produce an isometric, a dimetric, and a trimetric projection. Label the projections.

4.2. Develop a program that generates a wireframe of a conical surface. Start with a line (generatrix) that lies in the *x-y* plane. Rotate the points, say, 11 times around the *y* axis, giving a total of 12 lines describing the elements. Put all 24 points that describe the 12 lines in one [**P**], the odd-numbered points all being the same end of the line. Write a loop that does MOVE to the odd points followed by DRAW to the even points. A 10° rotation around the *y* axis followed by any rotation around the *x* axis makes a suitable axonometric projection. The image can be improved by generating another points matrix [**PC**] from just the odd-numbered points in [**P**]. A polyline connecting the points of [**PC**] and copies of [**PC**] made from scaling and translating along the *y* axis gives form to the image.

4.3. Write a program that draws a wireframe of a sphere. First, generate the points matrix for a circle in the *x-y* (centered on the origin) plane by making, say, 72 points at 5° increments using a rotation routine. A good-looking wireframe can be produced by making five copies of the first circle by rotation at 30° increments around the *y* axis. The six great circles intersect at the north and south poles of the sphere. An equator can be produced by rotating the first circle around the *x* axis. Circles at a few selected latitudes of 30° and 60° north and south can be generated by translating and scaling the equator circle. The final axonometric projection result will look best with a 10° rotation around the *x* axis.

4.4. Prepare the 3-D data for an image of a floppy disk drive. Write a program to display this image in an isometric projection. If each surface has its own points matrix, a polygon (panel) fill routine can be used to generate each surface.

4.5. Prepare points matrices in the form of flat polygonal panels to describe an image of a computer terminal. Panels for the areas of the screen and keyboard can be superimposed. Develop a program to make an isometric projection with hidden surfaces removed. (Hint: Plan to have only the visible surfaces in the data base.)

4.6. Write an *interactive* program that asks the user to enter three angles around coordinate axes in any order. Show the result of the rotational transformations with a wireframe display of a simple image, such as a wedge or brick.

4.7. Prepare points matrices of panels describing the exterior view of a residence. Unnecessary detail should be avoided. Develop a program to display this image as an axonometric projection and as a perspective projection. If panel filling is available, the quality of the result can be enhanced by using the priority fill hidden-surface technique of Section 2.5.

4.8. Prepare data for one section of a multispan truss bridge. Show the structural members as lines. Write a program that uses mirror and copy routines to generate the whole bridge and displays the result as a projection. (Note: Perspective helps make a wireframe image easier to visualize. Compare an axonometric projection with a perspective projection.)

4.9. Write a program to show one wall with a laboratory bench and cabinets (or kitchen cabinets) in perspective.

4.10. Develop the data for two walls, the floor, and the ceiling of a room. Also develop data for a table, bed, chest, and so forth. Develop a program that will display the furnished room in a pleasing perspective. (Note: The priority fill of raster graphics is extremely useful for hiding the walls and floor behind the furniture.)

4.11. Modify Project 4.6 to use the viewpoint as input instead of the axes of rotation. Give attention to keeping vertical lines upright.

4.12. Enhance Project 4.6 or 4.11 to include the ability to remove hidden surfaces. The simplest way is to use the priority filling algorithm, Section 2.5, which requires sorting on the z coordinates of the centroids of each panel.

4.13. Enhance Project 4.10 to move one or more of the furniture pieces under control of the screen cursor. The program should redraw the entire image after a new location is sent for a piece of furniture. (Note: As seen in Section 4.8, *two* projections are needed to describe x, y, and z of a point. However, since furniture stays on the floor, the other two coordinates can be computed from the location of the cursor.)

4.14. Obtain a large photograph of the exterior of a building or other structure with known dimensions. For six points, write down the points matrix [**P***] in screen coordinates from the photograph. Use the reconstruction algorithm in Section 4.8 to deduce the transformation. Digitize other points in the structure and develop a program that makes projections of the structure.

4.15. A further enhancement of Project 4.11 involves the addition of light source shading. In Section 2.7, an algorithm is given to compute the intensity of light reflected from a surface. Determine the reflected light intensity from each surface after a transformation and change in *lightness* parameter of the color map for each panel accordingly. Use a single light source parallel to the z axis.

3

Geometric Modeling

Curves

The 2-D and 3-D transformations in Chapters 3 and 4 treat arrays of points that are used to define the end points of straight line segments. Clearly, these transformations can deal with curved lines when the curve is approximated with a number of points connected by straight line segments or chords. In fact, curves are drawn on terminals and plotters with this scheme, where the number of chords affects the apparent smoothness of the result.

In solid modeling and other applications, however, it is much less cumbersome to deal with the equations of curves rather than a large points matrix. In the case of a circle (knowing it is a circle) the rectangular coordinates of the center and of one point on the periphery provide a complete description. By contrast, 72 points are required to approximate a circle constructed from chords subtending 5° arcs—a description that may not be an adequate approximation. A mathematical model of such a curve not only saves storage but also enhances accuracy.

The mathematical representation of a curve can be classified as either *nonparametric* or *parametric*. An example of a nonparametric (or natural) curve is the kinematics expression for position as a function of time for rectilinear motion with constant acceleration,

$$x = \frac{1}{2}at^2 + v_0t + x_0 \tag{5.1}$$

where t is the independent variable, x is the dependent variable, and a, v_0, and x_0 are specified constant values. Another example of a nonparametric curve is the cubic polynomial

$$y = c_1 + c_2x + c_3x^2 + c_4x^3 \tag{5.2}$$

where x is the independent variable, y is the dependent variable, and c_1, c_2, c_3, and c_4 are constants. Both of the above are further classified as explicit forms because there is a unique single value of the dependent variable for each value of the independent variable.

Multiple-valued relationships among the variables can be expressed with implicit nonparametric forms. An example is the equation of a circle,

$$(x - x_c)^2 + (y - y_c)^2 = r^2 \tag{5.3}$$

where x_c and y_c are the location of the center of the circle and r is the radius of the circle. No distinction is made between dependent and independent variables.

Often, formulas are simplified by replacing nonparametric curves with parametric curves, where the variables are expressed in terms of some other parameter. For example, the trajectory of a projectile is given by the equation

$$y = (\tan \theta_0)(x - x_0) - \frac{g(x - x_0)^2}{2v_0^2 \cos \theta_0} + y_0 \tag{5.4}$$

which is the explicit nonparametric form relating altitude y to horizontal distance x, initial angle θ_0, initial velocity v_0, initial position x_0 and y_0, and gravitational constant g. Engineers are more familiar with the parametric form

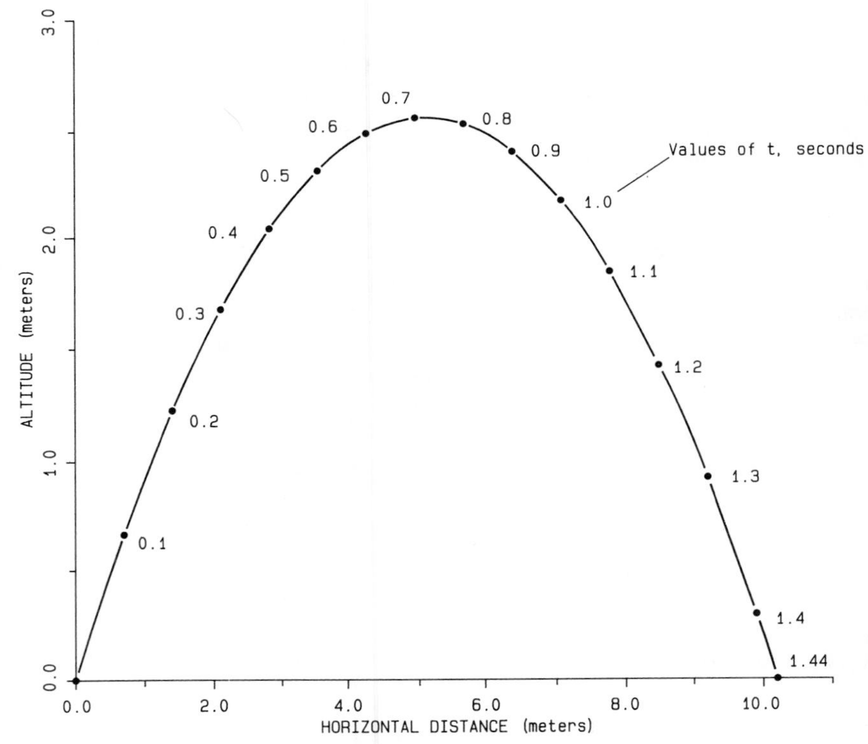

Figure 5.1 Plot of Eq. (5.4) or (5.5) with $x_0 = y_0 = 0$, $\theta_0 = 45°$, $v_0 = 10$ m/s, and $g = 9.8$ m/s^2. Plotted points indicate values of parameter t at 1/10-second intervals.

of the trajectory

$$x = (v_0 \cos \theta_0)t + x_0$$
$$y = (v_0 \sin \theta_0)t - \frac{1}{2}gt^2 + y_0 \tag{5.5}$$

which taken together describe the x and y coordinates of the curve in terms of the parameter time t. It is noted that Eq. (5.4) can be produced by elimination of the parameter t in Eqs. (5.5). These equations are plotted in Figure 5.1.

In this chapter it will be seen that parametric forms allow great versatility in constructing space curves that are multivalued and easily manipulated.

5.1 NONPARAMETRIC EQUATIONS

In some applications, a nonparametric equation that fits certain given points may be needed. Examples include computer-aided drafting (CAD) program algorithms such as blending radii and three-point circles. Fitting nonparametric equations to pass exactly through selected data points requires that the number of unknown constants in the equation equal the number of data. For example, the parabola of Eq. (5.4) can be expressed as

$$y = c_1 + c_2 x + c_3 x^2 \tag{5.6}$$

and numerical values for the constants can be found by the solution of three simultaneous equations that are based on three known values of y as a function of x. If these three known pairs are numbered 1, 2, and 3 the system of equations indicated by Eq. (5.6) is

$$\begin{bmatrix} 1 & x_1 & x_1^2 \\ 1 & x_2 & x_2^2 \\ 1 & x_3 & x_3^2 \end{bmatrix} \begin{bmatrix} c_1 \\ c_2 \\ c_3 \end{bmatrix} = \begin{bmatrix} y_1 \\ y_2 \\ y_3 \end{bmatrix} \tag{5.7}$$

which can be solved for $c_1, c_2, \ldots, c_n$ using standard techniques[32,59] such as Gaussian elimination or Cholesky decomposition.

When there are more data than unknown coefficients in the desired equation, some kind of "best" fit is sought. One of the most popular techniques in data presentation graphics, the "least squares" technique, is treated in Section 8.5. In least squares fitting, a curve with the desired form is positioned to minimize the squares of the differences between the data and the value predicted by the curve.

EXAMPLE 5.1

A system with a single degree of freedom is assumed to have undamped free vibration. If the natural circular frequency is known to be 8.3 rad/s and the displacements x are equal to 1.4 mm at $t = 0.2$ s and -0.2 mm at $t = 2.1$ s, determine the equation of motion.

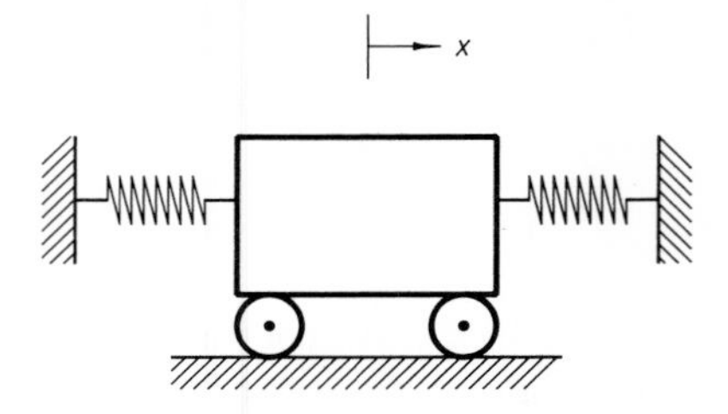

Solution. For simple harmonic motion, the displacement x is given by

$$x = c_1 \cos \omega t + c_2 \sin \omega t \tag{a}$$

Substitution of the given data into Eq. (a) yields

$$\begin{aligned} 1.4 &= c_1 \cos(8.3 \times 0.2) + c_2 \sin(8.3 \times 0.2) \\ -0.2 &= c_1 \cos(8.3 \times 2.1) + c_2 \sin(8.3 \times 2.1) \end{aligned} \tag{b}$$

Solution of these two simultaneous equations yields $c_1 = 19.111$ mm and $c_2 = 3.115$ mm. The equation of motion is thus

$$x = 19.111 \cos(8.3t) + 3.115 \sin(8.3t) \tag{c}$$

If three displacements as a function of time were given and not the circular frequency, the three equations resulting from the substitution into Eq. (a) would be nonlinear. An iterative solution would be required to solve for c_1, c_2, and ω.

EXAMPLE 5.2

Determine the radius and the coordinates of the center of a circle that passes through the points (1, 1), (2, 3), and (3, 1) by solving three simultaneous equations formed from Eq. (5.3).

Solution. Expansion of Eq. (5.3) yields

$$x^2 + y^2 - 2x_c x - 2y_c y + (x_c^2 + y_c^2) - r^2 = 0 \tag{a}$$

which may be simplified to

$$x^2 + y^2 - ax - by + c = 0 \tag{b}$$

Algebra is simplified if the given points matrix is translated so that one point (the first point) is coincident with the origin. This makes a nonhomogeneous two-dimensional points matrix

$$[\mathbf{P}^*] = \begin{bmatrix} 0 & 0 \\ 1 & 2 \\ 2 & 0 \end{bmatrix}$$

Substitution of the points in [**P***] into Eq. (b) yields

$$c = 0$$

$$5 - a - 2b + c = 0$$

$$4 - 2a \qquad + c = 0$$

which give $a = 2$, $b = 1.5$, and $c = 0$. Comparison of the coefficients of the two terms that are linear in x and y in Eqs. (a) and (b) shows that the translated center points are

$$x_c^* = a/2 = 1$$

$$y_c^* = b/2 = 0.75$$

With $c = 0 = x_c^{*2} + y_c^{*2} - r^2$, $r = 1.25$. Translation of x_c^* and y_c^* back to the original positions gives $x_c = 2$ and $y_c = 1.75$. Shown in the accompanying figure are the two points matrices [**P**] and [**P***] and the resulting three-point circle.

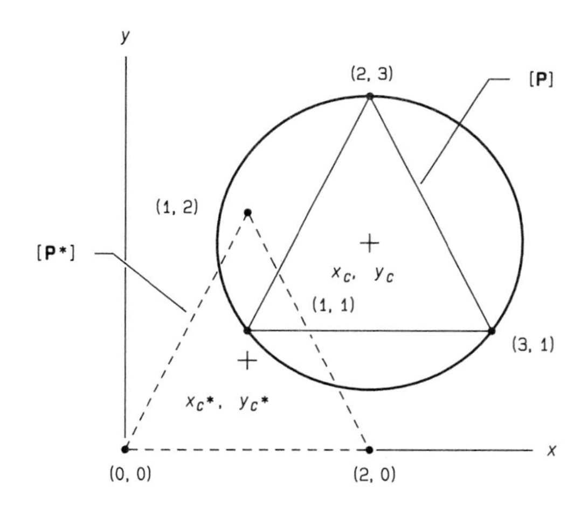

Other techniques for producing a three-point circle can be found in the literature.[12]

5.2 PARAMETRIC CONIC SECTIONS

Conic sections—circles, ellipses, hyperbolas, and parabolas—find common use in engineering applications involving CAD, data presentation, and solid modeling. Parametric techniques that involve recursive formulas for generating representations of conic sections save significant computational time. This section introduces some general features of parametric as opposed to natural or nonparametric formulation.

Circles

The nonparametric form of the equation of a circle is given in Eq. (5.3). This form produces a circle of poor quality when equal increments of x are substituted into this equation solved for y. From the standpoint of quality, it is better to use equal straight line increments or chords to approximate the circle. Thus, it is clearly preferable to describe the circle in polar coordinates, that is, a circle parametric in angle θ and radius r. The equation of a parametric circle is

$$
\begin{aligned}
x &= r\cos\theta + x_c \\
y &= r\sin\theta + y_c
\end{aligned}
\tag{5.8}
$$

where x_c and y_c are the coordinates of the center of the circle.

Small equal increments in θ in Eq. (5.8) produce a good quality circle with the points matrix

$$
[\mathbf{P}] = r\begin{bmatrix} 1 & 0 & 0 & 1 \\ \cos\Delta\theta & \sin\Delta\theta & 0 & 1 \\ \cos 2\Delta\theta & \sin 2\Delta\theta & 0 & 1 \\ \cos 3\Delta\theta & \sin 3\Delta\theta & 0 & 1 \\ \cdot & \cdots & \cdots & \cdot \\ \cos n\Delta\theta & \sin n\Delta\theta & 0 & 1 \end{bmatrix} \tag{5.9}
$$

where for simplicity the center is at $x = 0$, $y = 0$ and the circle lies in the x-y plane. The drawing starts at $x = r$, $y = 0$ and ends at $x = r\cos n\Delta\theta$, $y = r\sin n\Delta\theta$, where the values of n and $\Delta\theta$ are such that $n\Delta\theta = 2\pi$. There are n distinct points and n chords describing the circle. Experience shows that perhaps $n = 12$ produces a good small circle on a display screen, whereas $n = 180$ may be needed to produce a smooth large circle with a pen plotter. Generally, an intermediate value of n will suffice. Circle smoothness is illustrated in Figure 5.2.

The algorithm shown in Eq. (5.9) is more computationally intensive than is necessary. A faster circle-drawing routine uses *recursion,* where the points are

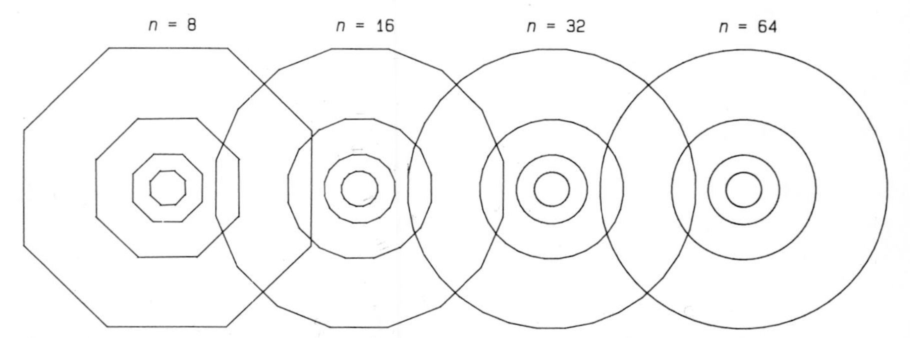

Figure 5.2 Effect of number of segments on circle smoothness. Note that satisfactory smaller circles require fewer segments.

computed from the previous points and the trigonometric functions are computed only once. The circle starts at x_i, y_i, an arbitrary point on the circle with its center at the origin. The radius line from the origin to this point has an angular coordinate θ_i. With an increment of angle $\Delta\theta$, the next $(i + 1)$ point is

$$\begin{aligned} x_{i+1} &= r\cos(\theta_i + \Delta\theta) \\ y_{i+1} &= r\sin(\theta_i + \Delta\theta) \end{aligned} \tag{5.10}$$

Double-angle trigonometric formulas expand the right-hand side to

$$\begin{aligned} x_{i+1} &= r(\cos\theta_i \cos\Delta\theta - \sin\theta_i \sin\Delta\theta) \\ y_{i+1} &= r(\cos\theta_i \sin\Delta\theta + \sin\theta_i \cos\Delta\theta) \end{aligned} \tag{5.11}$$

Since $x_i = r\cos\theta_i$ and $y_i = r\sin\theta_i$, Eq. (5.11) can be rewritten as

$$\begin{aligned} x_{i+1} &= x_i \cos\Delta\theta - y_i \sin\Delta\theta \\ y_{i+1} &= x_i \sin\Delta\theta + y_i \cos\Delta\theta \end{aligned} \tag{5.12}$$

When equal increments of $\Delta\theta$ are used, the constants

$$c = \cos\Delta\theta \quad \text{and} \quad s = \sin\Delta\theta \tag{5.13}$$

need only be computed once. In more compact notation, the recursive relationship of Eq. (5.12) is

$$[x^* \quad y^*] = [x \quad y]\begin{bmatrix} c & s \\ -s & c \end{bmatrix} \tag{5.14}$$

where x^* and y^* are the new values computed from the previous values x and y.

Next, the circle described by Eq. (5.14) is modified so that its center is at an arbitrary point (x_c, y_c) instead of the origin. With use of the translation transformation of Eq. (4.2) and homogeneous coordinates,

$$\begin{aligned} [x^* \quad y^* \quad 0 \quad 1] &= [x \quad y \quad 0 \quad 1]\begin{bmatrix} 1 & 0 & 0 & 0 \\ 0 & 1 & 0 & 0 \\ 0 & 0 & 1 & 0 \\ -x_c & -y_c & 0 & 1 \end{bmatrix} \\ &\times \begin{bmatrix} c & s & 0 & 0 \\ -s & c & 0 & 0 \\ 0 & 0 & 1 & 0 \\ 0 & 0 & 0 & 1 \end{bmatrix}\begin{bmatrix} 1 & 0 & 0 & 0 \\ 0 & 1 & 0 & 0 \\ 0 & 0 & 1 & 0 \\ x_c & y_c & 0 & 1 \end{bmatrix} \end{aligned} \tag{5.15}$$

Expansion yields

$$\begin{aligned} x^* &= x_c + (x - x_c)c - (y - y_c)s \\ y^* &= y_c + (x - x_c)s + (y - y_c)c \end{aligned} \tag{5.16}$$

Circle drawing with the algorithm of Eq. (5.16) can be an order of magnitude faster than with Eq. (5.9).

EXAMPLE 5.3

Implement the recursive algorithm of Eq. (5.16) in a FORTRAN-77 subroutine that draws a circle with the radius and coordinates of the center given as input data.

Solution. The necessary arguments are sent from the calling program to the subroutine listed below.

```
      SUBROUTINE CIRCLE (R,XC,YC,NSEG)
      REAL R, X, Y, XNEW, YNEW, XC, YC, C, S, X, DELT
      INTEGER I, NSEG
C
C   ROUTINE TO DRAW EFFICIENT PARAMETRIC CIRCLE
C   GRAPHICS SUBROUTINES LIBRARY NEEDED FOR MOVE AND DRAW
C   INITIALIZE GRAPHICS ENVIRONMENT BEFORE CALLING THIS ROUTINE
C
C      R - RADIUS OF CIRCLE
C      XC, YC - LOCATION OF CENTER
C      NSEG - NUMBER OF SEGMENTS IN WHOLE CIRCLE
C      X, Y, XNEW, YNEW - COORDINATES OF POINTS ON THE CIRCLE
C      C, S - STORAGE FOR TRIG FUNCTIONS COMPUTED JUST ONCE
C      DELT - INCREMENTAL ANGLE FOR EACH SEGMENT
C
C   DEFINE STARTING POINT OF CIRCLE
C
      X = XC + R
      Y = YC
      CALL MOVE(X,Y)
C
C   DEFINE THE INCREMENTAL ANGLE AND CONSTANTS
C
      DELT = 2.0*3.141596/FLOAT(NSEG)
      C = COS(DELT)
      S = SIN(DELT)
C
C   DRAW CIRCLE CCW
C
      DO 10 I=1,NSEG
         XNEW = XC + (X-XC)*C - (Y-YC)*S
         YNEW = YC + (X-XC)*S + (Y-YC)*C
         CALL DRAW (XNEW,YNEW)
         X = XNEW
         Y = YNEW
 10   CONTINUE
      RETURN
      END
```

Ellipses

The techniques for circle drawing are logically extended to ellipses.[50] The parametric representation† of an ellipse is

$$x = a \cos \theta$$
$$y = b \sin \theta \tag{5.17}$$

where the center of the ellipse is at the origin; the lengths of the semimajor and semiminor axes are a and b. An entire ellipse is swept out with $0 \leq \theta \leq 2\pi$. To generalize the representation of an ellipse so that the semimajor axis makes an angle of ϕ with the x axis and the center of the ellipse is at x_c, y_c, Eq. (5.17) is transformed by use of Eqs. (3.10) and (3.16) (with ϕ as the angle of rotation),

$$x^* = x_c + a \cos \theta \cos \phi - b \sin \theta \sin \phi$$
$$y^* = y_c + a \cos \theta \sin \phi - b \sin \theta \cos \phi \tag{5.18}$$

If the ellipse is drawn at equal increments of $\Delta\theta$, the functions of θ may be computed recursively. From the double-angle formula [cf. Eq. (5.11)] the recursive relations are

$$\cos(\theta + \Delta\theta) = \cos \theta^* = c \cos \theta - s \sin \theta$$
$$\sin(\theta + \Delta\theta) = \sin \theta^* = s \cos \theta + c \sin \theta \tag{5.19}$$

where $c = \cos \Delta\theta$, $s = \sin \Delta\theta$, and the * denotes the "new" value. Thus, from Eqs. (5.18) and (5.19) the formula is

$$x^* = x_c + a_x \cos \theta^* - b_x \sin \theta^*$$
$$y^* = y_c + a_y \cos \theta^* + b_y \sin \theta^* \tag{5.20}$$

where $a_x = a \cos \phi$, $b_x = b \sin \phi$, $a_y = a \sin \phi$, and $b_y = b \cos \phi$ which need only be computed once. Some families of ellipses are illustrated in Figure 5.3.

†Generally, the parameter θ is *not* the polar coordinate location of points on the ellipse.

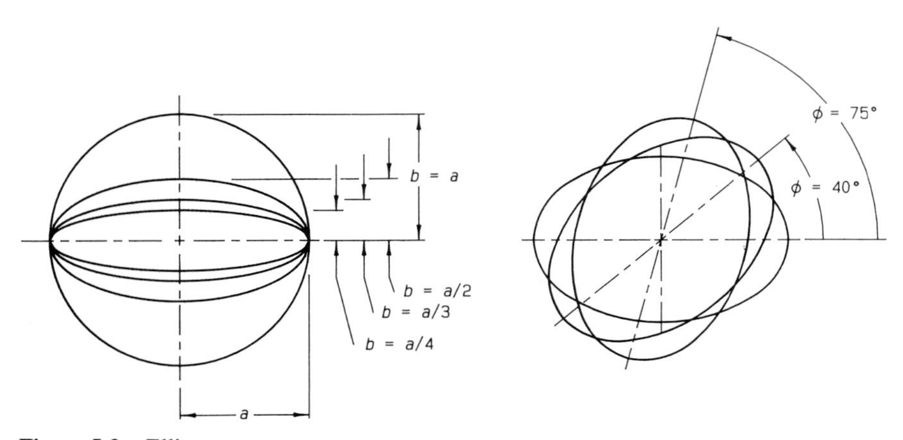

Figure 5.3 Ellipses.

Hyperbolas

Hyperbolic functions can conveniently be used for the parametric representation of a hyperbola.[50] Here, using the coordinate axes for the asymptotes,

$$\begin{aligned} x &= a \cosh u \\ y &= b \sinh u \end{aligned} \tag{5.21}$$

where $\cosh u = (e^u + e^{-u})/2$ and $\sinh u = (e^u - e^{-u})/2$. The entire hyperbola in the first quadrant is generated with $0 \le u < \infty$. A recursion formula for generation of the hyperbola is based on the double-angle formulas,

$$\begin{aligned} \cosh(u + \Delta u) &= \cosh u \cosh \Delta u + \sinh u \sinh \Delta u \\ \sinh(u + \Delta u) &= \sinh u \cosh \Delta u + \cosh u \sinh \Delta u \end{aligned} \tag{5.22}$$

From Eq. (5.21), a point at the parametric location $u + \Delta u$ is

$$\begin{aligned} x^* &= a \cosh(u + \Delta u) \\ y^* &= b \sinh(u + \Delta u) \end{aligned} \tag{5.23}$$

Combination of Eqs. (5.22) and (5.23) yields the recursion formula

$$\begin{aligned} x^* &= x \cosh \Delta u + \frac{a}{b} y \sinh \Delta u \\ y^* &= \frac{b}{a} x \sinh \Delta u + y \cosh \Delta u \end{aligned} \tag{5.24}$$

Note that the hyperbolic functions are computed only once. In the application of Eq. (5.24), the limiting range on x or y determines the range of u for generating the function. The transformation equations of Chapter 3 may be applied to the x and y values to translate, rotate, scale, or mirror the hyperbola.

Parabolas

A parabola symmetrical about a line parallel to the x axis is written parametrically in u as

$$x = au^2 + x_c \quad \text{and} \quad y = 2au + y_c \tag{5.25}$$

where x_c and y_c are the coordinates of the vertex and $-\infty < u < \infty$ defines the entire parabola. The range of u is determined by solving one of Eqs. (5.25) with the required range of x or y. The parabola can be plotted recursively in equal increments of u where

$$u^* = u + \Delta u \tag{5.26}$$

Replacement of u in Eq. (5.25) with u^* from Eq. (5.26) yields

$$\begin{aligned} x^* &= au^{*2} + x_c = au^2 + 2au\,\Delta u + a(\Delta u)^2 + x_c \\ y^* &= 2au^* + y_c = 2au + 2a\,\Delta u + y_c \end{aligned} \tag{5.27}$$

or

$$x^* = x + a(2u + \Delta u)\Delta u$$
$$y^* = y + 2a\,\Delta u \tag{5.28}$$

Sample parabolas generated with Eqs. (5.28) are shown in Figure 5.4. Parabolas with other orientations can be developed by means of the two-dimensional rotation transformation, Eq. (3.17).

Specification of a parabola in terms of the parameter u is awkward. From a user's standpoint, more useful ways to define a parametric parabola include specifying one of the following:

- Three points on the parabola; or
- Vertex, orientation, range of u, value of a; or
- Starting point, vertex, finishing point.

Generation of a parametric parabola with any one of the above specifications requires a lengthy formulation.

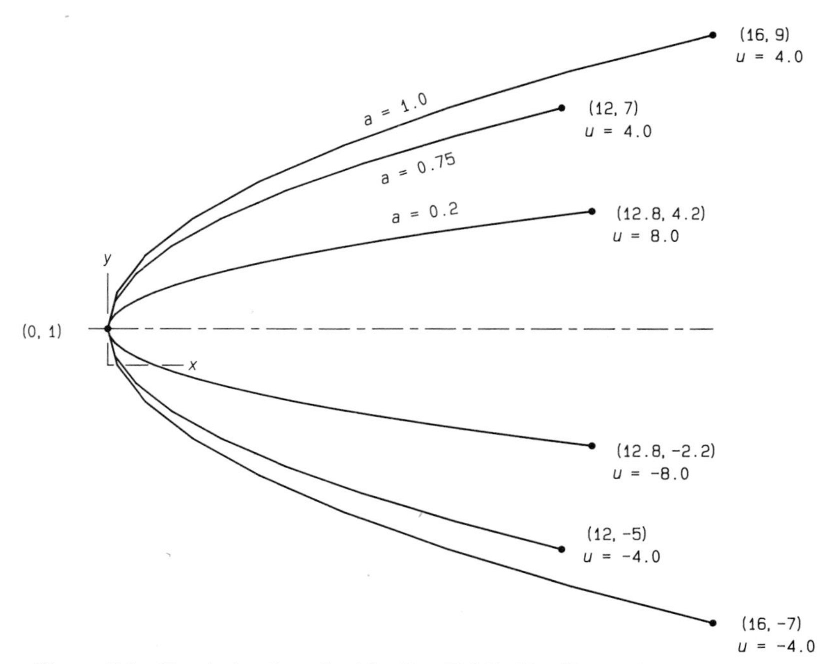

Figure 5.4 Parabolas described by Eq. (5.28). For illustration, $\Delta u = 0.5$, the starting point is at the lower right, and the computed points are connected with straight line segments. Notice that the parabola for $a = 0.2$, generated with twice as many segments as the others, presents a smoother appearance.

A parametric "parabola" that is not necessarily a conic section is the *parametric quadratic curve.* It will be seen that the following formulation overcomes the disadvantages above. The curve is described by the equation

$$[x(u) \quad y(u) \quad z(u)] = a_{1x} + a_{2x}u + a_{3x}u^2 + a_{1y} + a_{2y}u + a_{3y}u^2 + a_{1z} + a_{2z}u + a_{3z}u^2$$

or

$$[\mathbf{P}(u)] = [\mathbf{a}_1] + [\mathbf{a}_2]\mathbf{u} + [\mathbf{a}_3]u^2 \tag{5.29}$$

where the parameter u has the range $0 \leq u \leq 1$, $[\mathbf{P}(u)]$ is the three-dimensional row matrix for a point in space, the matrix $[\mathbf{a}_1]$ is $[a_{1x} \quad a_{1y} \quad a_{1z}]$, and so forth. For brevity, Eq. (5.29) can be written as

$$\mathbf{P}(u) = [1 \quad u \quad u^2][\mathbf{a}_1 \quad \mathbf{a}_2 \quad \mathbf{a}_3]^T = [1 \quad u \quad u^2][\mathbf{A}] \tag{5.30}$$

where the square brackets for the row matrices on $\mathbf{P}(u)$ and $\mathbf{a}$ are dropped. This equation is said to be in *algebraic form.* One way to determine the components of $\mathbf{a}_1$, $\mathbf{a}_2$, and $\mathbf{a}_3$, a total of nine unknowns, is to specify that the curve should pass through three arbitrary points. For the simplicity of having the parametric points evenly spaced, $u = 0, 1/2, 1$. It should be noted that the *order* in which the three points are taken affects the result. The $\mathbf{a}$'s are determined from the system

$$\begin{bmatrix} \mathbf{P}_1 \\ \mathbf{P}_2 \\ \mathbf{P}_3 \end{bmatrix} = \begin{bmatrix} x_1 & y_1 & z_1 \\ x_2 & y_2 & z_2 \\ x_3 & y_3 & z_3 \end{bmatrix} = \begin{bmatrix} 1 & u_1 & u_1^2 \\ 1 & u_2 & u_2^2 \\ 1 & u_3 & u_3^2 \end{bmatrix} \begin{bmatrix} \mathbf{a}_1 \\ \mathbf{a}_2 \\ \mathbf{a}_3 \end{bmatrix} = \begin{bmatrix} 1 & 0 & 0 \\ 1 & 1/2 & 1/4 \\ 1 & 1 & 1 \end{bmatrix} \begin{bmatrix} \mathbf{a}_1 \\ \mathbf{a}_2 \\ \mathbf{a}_3 \end{bmatrix} \tag{5.31}$$

Simultaneous solution yields

$$\begin{aligned} \mathbf{a}_1 &= \mathbf{P}_1 \\ \mathbf{a}_2 &= -3\mathbf{P}_1 + 4\mathbf{P}_2 - \mathbf{P}_3 \\ \mathbf{a}_3 &= 2\mathbf{P}_1 - 4\mathbf{P}_2 + 2\mathbf{P}_3 \end{aligned} \tag{5.32}$$

which can be written out as $a_{1x} = x_1$, $a_{1y} = y_1$, . . . , $a_{3z} = 2z_1 - 4z_2 + 2z_3$. Variation of u in Eq. (5.30) over the range 0 to 1 with the $\mathbf{a}$'s determined from (5.32) describes a continuous curve that connects the three arbitrary points in space.

Although the parametric quadratic may find use in smoothing or blending data, other parametric functions have additional properties that make them much more versatile. Three particular forms—parametric cubic, Bezier, and B-spline—find wide use in engineering applications for generation of arbitrary curves and sculptured surfaces. Any of these forms can generate accurate approximations to conic sections.

EXAMPLE 5.4

Use the parametric quadratic curve that passes through the points (0, 0, 0), (1, 2, 1), and (3, 2, 2) to determine the coordinates of the point equally between (in the parametric sense) the first two points. Repeat the process with the first and second points interchanged.

Solution. It is convenient to select $u = 0, 1/2, 1$ for the three given points. From Eq. (5.32),

$$\mathbf{a}_1 = \mathbf{P}_1 = [0 \quad 0 \quad 0]$$

$$\mathbf{a}_2 = 4\mathbf{P}_2 - \mathbf{P}_3 = [1 \quad 6 \quad 2]$$

$$\mathbf{a}_3 = -4\mathbf{P}_2 + 2\mathbf{P}_3 = [2 \quad -4 \quad 0]$$

Use of Eq. (5.29) with $u = 0.25$ yields

$$\begin{aligned}[\mathbf{P}_4] &= [\mathbf{a}_1] + [\mathbf{a}_2]u + [\mathbf{a}_3]u^2 \\ &= [0 \quad 0 \quad 0] + [1 \quad 6 \quad 2](0.25) + [2 \quad -4 \quad 0](0.25)^2 \\ &= [0.375 \quad 1.250 \quad 0.500]\end{aligned}$$

When $\mathbf{P}_1$ and $\mathbf{P}_2$ are interchanged, $\mathbf{P}_2^* = \mathbf{P}_1$ and $\mathbf{P}_1^* = \mathbf{P}_2$. Thus,

$$\mathbf{a}_1^* = \mathbf{P}_1^* = [1 \quad 2 \quad 1]$$

$$\mathbf{a}_2^* = -3\mathbf{P}_1^* - \mathbf{P}_3 = [-6 \quad -8 \quad -5]$$

$$\mathbf{a}_3^* = 2\mathbf{P}_1^* + 2\mathbf{P}_3 = [8 \quad 8 \quad 6]$$

Subsequently, the interpolation yields

$$\begin{aligned}[\mathbf{P}_4^*] &= [\mathbf{a}_1^*] + [\mathbf{a}_2^*]u + [\mathbf{a}_3^*]u^2 \\ &= [1 \quad 2 \quad 1] + [-6 \quad -8 \quad -5](0.25) + [8 \quad 8 \quad 6](0.25)^2 \\ &= [0 \quad 0.5 \quad 0.125]\end{aligned}$$

which is greatly different from the first result. The two ways of ordering the points are plotted in the accompanying figure.

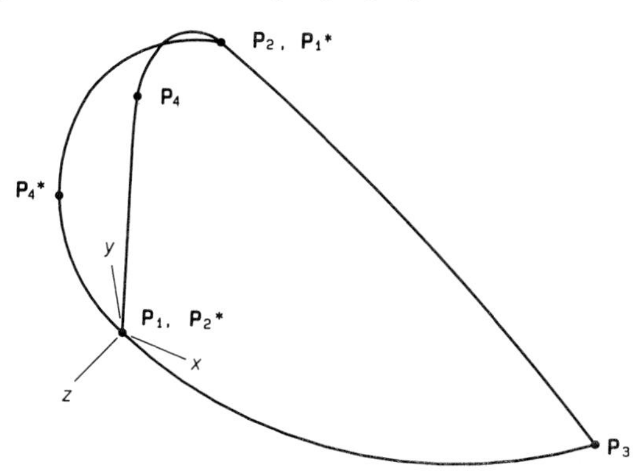

5.3 PARAMETRIC CUBIC CURVES

The parametric parabola in Eqs. (5.29) through (5.32) passes through three points in space but does not permit the specification of the position and slope of the two end points. For four conditions to be specified, four vector-valued constants are needed in the parametric polynomial, or

$$\mathbf{P}(u) = \mathbf{a}_1 + \mathbf{a}_2 u + \mathbf{a}_3 u^2 + \mathbf{a}_4 u^3$$

$$= [1 \quad u \quad u^2 \quad u^3]\begin{bmatrix}\mathbf{a}_1\\ \mathbf{a}_2\\ \mathbf{a}_3\\ \mathbf{a}_4\end{bmatrix} = [1 \quad u \quad u^2 \quad u^3]\begin{bmatrix}a_{1x} & a_{1y} & a_{1z}\\ a_{2x} & a_{2y} & a_{2z}\\ a_{3x} & a_{3y} & a_{3z}\\ a_{4x} & a_{4y} & a_{4z}\end{bmatrix}$$

$$= [1 \quad u \quad u^2 \quad u^3][\mathbf{A}] \tag{5.33}$$

where u is a dimensionless parameter $0 \le u \le 1$, $\mathbf{P}(u) = [x(u) \quad y(u) \quad z(u)]$, and $\mathbf{a}_i$ are the vector-valued algebraic coefficients. This equation is known as the *algebraic form* of the parametric cubic curve. Two more useful forms, the *point form* and the *geometric form,* which use a combination of the points and slopes along the line, are developed next.[36]

Point Form

A simple application of the parametric cubic curve, fitting a smooth curve through four points in space, leads to four simultaneous linear equations in four unknowns. Conveniently, the four points are parametrized at $u = 0$, 1/3, 2/3, and 1. As in Example 5.4, the result is affected by the order in which the points are used. The problem of determining the constants $\mathbf{a}_i$ is stated as

$$\begin{bmatrix}\mathbf{P}(0)\\ \mathbf{P}(1/3)\\ \mathbf{P}(2/3)\\ \mathbf{P}(1)\end{bmatrix} = \begin{bmatrix}1 & 0 & 0 & 0\\ 1 & 1/3 & 1/9 & 1/27\\ 1 & 2/3 & 4/9 & 8/27\\ 1 & 1 & 1 & 1\end{bmatrix}\begin{bmatrix}\mathbf{a}_1\\ \mathbf{a}_2\\ \mathbf{a}_3\\ \mathbf{a}_4\end{bmatrix} \tag{5.34}$$

Inversion of the square matrix yields

$$[\mathbf{A}] = \begin{bmatrix}\mathbf{a}_1\\ \mathbf{a}_2\\ \mathbf{a}_3\\ \mathbf{a}_4\end{bmatrix} = \begin{bmatrix}1 & 0 & 0 & 0\\ -11/2 & 9 & -9/2 & 1\\ 9 & -45/2 & 18 & -9/2\\ -9/2 & 27/2 & -27/2 & 9/2\end{bmatrix}\begin{bmatrix}\mathbf{P}(0)\\ \mathbf{P}(1/3)\\ \mathbf{P}(2/3)\\ \mathbf{P}(1)\end{bmatrix} \tag{5.35}$$

Viewing and modeling transformations such as rotation and translation are applied to the point form of the parametric cubic curve.

EXAMPLE 5.5

Fit the parametric cubic approximation for a semicircular arc through four coplanar points which are equally spaced along the circumference. Use (a) polar (ρ, θ) coordinates and (b) Cartesian (x, y) coordinates.

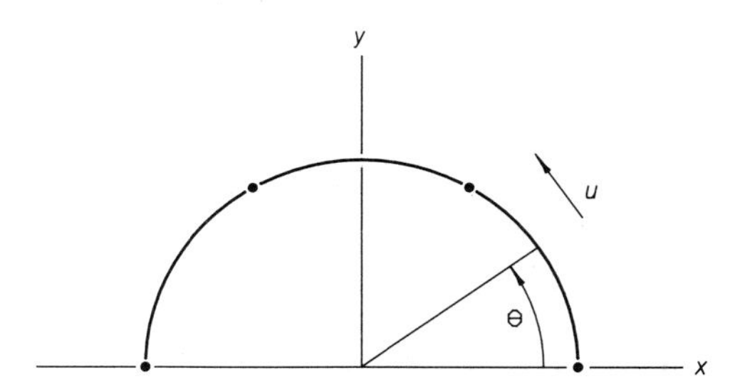

Solution. (**a**) Since r is a constant, it is only necessary to parametrize θ. The four equally spaced points around the arc are $(r, 0)$, $(r, \pi/3)$, $(r, 2\pi/3)$, and (r, π) in polar coordinates. The curve is evaluated at $u = 0$, $1/3$, $2/3$, and 1. By inspection,

$$\theta = \pi u, \qquad \rho = r = \text{const}$$

which produces a perfect circular arc. In polar coordinates, only a constant and a linear term in u are necessary to describe the circular arc. The elements of the matrix [**A**] of Eq. (5.33) are $a_{1_\rho} = r$, $a_{1\theta} = 0$, $a_{2_\rho} = 0$, $a_{2\theta} = \pi$, $a_{3_\rho} = a_{3\theta} = a_{4_\rho} = a_{4\theta} = 0$.

(**b**) In Cartesian coordinates, the four points equally spaced in x and the corresponding values of y which lie on the circular arc are

u	x	y
0	r	0
1/3	$r/2$	$r\sqrt{3}/2$
2/3	$-r/2$	$r\sqrt{3}/2$
1	$-r$	0

The equations in the form of Eq. (5.34) are

$$\begin{bmatrix} 1 & 0 & 0 & 0 \\ 1 & 1/3 & 1/9 & 1/27 \\ 1 & 2/3 & 4/9 & 8/27 \\ 1 & 1 & 1 & 1 \end{bmatrix} \begin{bmatrix} \mathbf{a}_1 \\ \mathbf{a}_2 \\ \mathbf{a}_3 \\ \mathbf{a}_4 \end{bmatrix} = r \begin{bmatrix} 1 & 0 \\ 1/2 & \sqrt{3}/2 \\ -1/2 & \sqrt{3}/2 \\ -1 & 0 \end{bmatrix}$$

The solution from Eq. (5.35) is

$$[\mathbf{A}] = r \begin{bmatrix} 1 & 0 \\ 0.25 & 3.897 \\ -6.75 & -3.897 \\ 4.5 & 0 \end{bmatrix}$$

which gives the cubic equations

$$x = (1 + 0.25u - 6.75u^2 + 4.5u^3)r$$

$$y = (3.897u - 3.897u^2)r$$

This particular parametric representation of a circular arc departs somewhat from the correct geometry. For example, at the point $x = 0$, $u = 0.5$ and

$$y = (3.897(0.5) - 3.897(0.5)^2)r = 0.9743r$$

which is 2.57% less than the correct value. While an error of this magnitude would be imperceptible in a plotted circle, a more accurate approximation is available by adaptation of the geometric form which is described below.

Geometric Form

The geometric form of the cubic curve defines a curve by specifying the two end points and the two tangent vectors at the ends of the curve. From the algebraic form, Eq. (5.33), the parametric slope at any point along the line is

$$\begin{aligned} \frac{d}{du}\mathbf{P}(u) &= \mathbf{P}'(u) = \mathbf{P}_u \\ &= [0 \quad 1 \quad 2u \quad 3u^2][\mathbf{A}] \\ &= \mathbf{a}_2 + 2\mathbf{a}_3 u + 3\mathbf{a}_4 u^2 \end{aligned} \tag{5.36}$$

where $0 \leq u \leq 1$. Note that the subscript u denotes the operation of differentiation, while subscripts x, y, and z denote Cartesian components.

The four conditions to define the curve are arbitrarily selected as the coordinates of the two end points and the parametric slopes at the two end points. With Eq. (5.33) defining the coordinates at $u = 0$ and $u = 1$ and Eq. (5.36) defining the parametric slopes at these same points,

$$\begin{bmatrix} \mathbf{P}(0) \\ \mathbf{P}(1) \\ \mathbf{P}'(0) \\ \mathbf{P}'(1) \end{bmatrix} = \begin{bmatrix} 1 & 0 & 0 & 0 \\ 1 & 1 & 1 & 1 \\ 0 & 1 & 0 & 0 \\ 0 & 1 & 2 & 3 \end{bmatrix} \begin{bmatrix} \mathbf{a}_1 \\ \mathbf{a}_2 \\ \mathbf{a}_3 \\ \mathbf{a}_4 \end{bmatrix} = \begin{bmatrix} \mathbf{b}_1 \\ \mathbf{b}_2 \\ \mathbf{b}_3 \\ \mathbf{b}_4 \end{bmatrix} = [\mathbf{B}] \tag{5.37}$$

where $[\mathbf{B}]$ is the matrix of geometric coefficients. Inversion yields

$$[\mathbf{A}] = \begin{bmatrix} \mathbf{a}_1 \\ \mathbf{a}_2 \\ \mathbf{a}_3 \\ \mathbf{a}_4 \end{bmatrix} = \begin{bmatrix} 1 & 0 & 0 & 0 \\ 0 & 0 & 1 & 0 \\ -3 & 3 & -2 & -1 \\ 2 & -2 & 1 & 1 \end{bmatrix} [\mathbf{B}] \tag{5.38}$$

The foregoing two equations are the conversions between the algebraic and geometric forms.

A useful extension of Eq. (5.38) which permits the evaluation of any point on the parametric cubic curve is obtained by combining Eqs. (5.33) and (5.38). Thus,

$$\mathbf{P}(u) = [1 \quad u \quad u^2 \quad u^3]\begin{bmatrix} 1 & 0 & 0 & 0 \\ 0 & 0 & 1 & 0 \\ -3 & 3 & -2 & -1 \\ 2 & -2 & 1 & 1 \end{bmatrix}[\mathbf{B}]$$

$$= [1 \quad u \quad u^2 \quad u^3]\begin{bmatrix} \mathbf{P}(0) \\ \mathbf{P}'(0) \\ -3\mathbf{P}(0) + 3\mathbf{P}(1) - 2\mathbf{P}'(0) - \mathbf{P}'(1) \\ 2\mathbf{P}(0) - 2\mathbf{P}(1) + \mathbf{P}'(0) + \mathbf{P}'(1) \end{bmatrix} \tag{5.39}$$

where $0 \leq u \leq 1$. This also can be written as

$$\mathbf{P}(u) = [\mathbf{F}][\mathbf{B}]$$
$$= [F_1(u) \quad F_2(u) \quad F_3(u) \quad F_4(u)][\mathbf{P}(0) \quad \mathbf{P}(1) \quad \mathbf{P}'(0) \quad \mathbf{P}'(1)]^T \tag{5.40}$$

where
$$F_1(u) = 1 - 3u^2 + 2u^3$$
$$F_2(u) = 3u^2 - 2u^3$$
$$F_3(u) = u - 2u^2 + u^3$$
$$F_4(u) = -u^2 + u^3$$

with $0 \leq u \leq 1$.

Conversion from geometric form to point form is needed prior to performing viewing and modeling transformations. Substitution of $u = 0, \frac{1}{3}, \frac{2}{3}$, and 1 into Eq. (5.40) yields

$$\begin{bmatrix} \mathbf{P}(0) \\ \mathbf{P}(1/3) \\ \mathbf{P}(2/3) \\ \mathbf{P}(0) \end{bmatrix} = \begin{bmatrix} 1 & 0 & 0 & 0 \\ 20/27 & 7/27 & 4/27 & -2/27 \\ 7/27 & 20/27 & 2/27 & -4/27 \\ 0 & 1 & 0 & 0 \end{bmatrix}[\mathbf{B}] \tag{5.41}$$

The inverse process, conversion from point form to geometric form, is given by

$$[\mathbf{B}] = \begin{bmatrix} \mathbf{P}(0) \\ \mathbf{P}(1) \\ \mathbf{P}'(0) \\ \mathbf{P}'(1) \end{bmatrix} = \begin{bmatrix} 1 & 0 & 0 & 0 \\ 0 & 0 & 0 & 1 \\ -11/2 & 9 & -9/2 & 1 \\ -1 & 9/2 & -9 & 11/2 \end{bmatrix}\begin{bmatrix} \mathbf{P}(0) \\ \mathbf{P}(1/3) \\ \mathbf{P}(2/3) \\ \mathbf{P}(1) \end{bmatrix} \tag{5.42}$$

The application of the geometric form is actually more intuitive than the algebraic form. The first two rows, $\mathbf{P}(0)$, and $\mathbf{P}(1)$, of the geometric coefficient matrix define the two end points of the line. The other two rows, $\mathbf{P}'(0)$ and $\mathbf{P}'(1)$

define the *parametric slopes* at the end points. From calculus, the geometric and parametric slopes are related as

$$\frac{dy}{dx} = \frac{dy/du}{dx/du} = \frac{P'_y}{P'_x} \tag{5.43}$$

Note that here the *ratio* of P'_y to P'_x affects the geometric slope. If the magnitudes of P'_x and P'_y are changed but the ratio of the two is unchanged, the geometric slopes of the end points are unchanged. However, the shape of the curve is affected. As shown in the example of Figure 5.5, the geometric form of the parametric cubic curve can be modeled as a stiff steel wire for the line with guides at the end points to fix the geometric slopes. As more wire is fed through the guides, the curve bulges up and up. Eventually, a condition is reached where the wire doubles back on itself.

The parametric slopes for a straight line in geometric form are written as

$$\mathbf{P}'(0) = \mathbf{P}'(1) = \mathbf{P}(1) - \mathbf{P}(0) \tag{5.44}$$

This relationship yields even parameterization, which means that there is a linear relationship between u and the geometric distance between the two end points of the line. Two forms of the geometric coefficient matrix for a straight line, where one is not evenly parametrized, are shown in Figure 5.5.

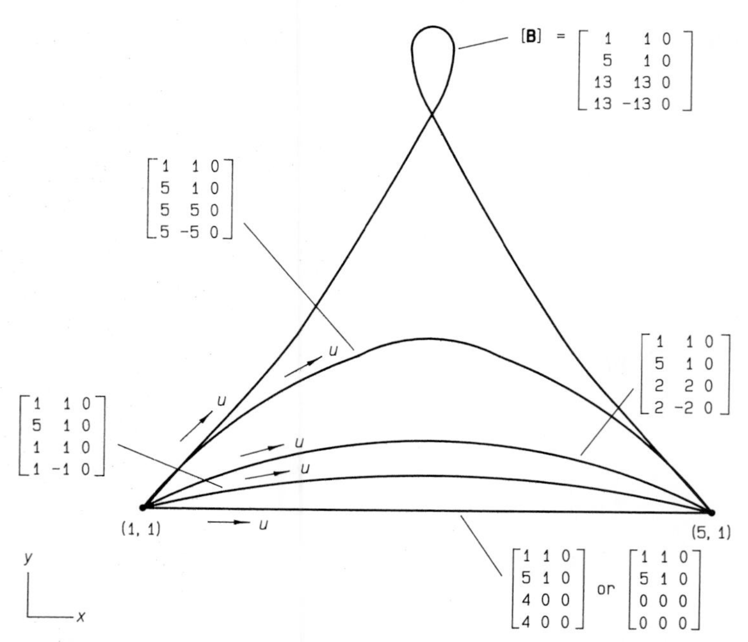

Figure 5.5 Effect of parametric slope terms on curves with the same end points and the same geometric slopes at the end points. The parametric slopes are given in the geometric coefficient matrices.

Approximation of Conic Sections

The geometric form of the parametric cubic curve leads to very good approximations[36] for construction of conic sections and is easier to use than the methods of Section 5.2. In the conic section curve shown in Figure 5.6, construction lines define the tangents to the end points of the curve and the midpoint of the line connecting the two end points. Because a conic section is a planar curve, the tangents intersect as shown at a point in the same plane designated as $\mathbf{P}_E$. The ratio of the distances DC to EC—denoted by ρ—determines the type of conic curve produced. A shallow curve, that is, one for which $0 \leq \rho < 0.5$, is an ellipse or possibly a circle. A parabola is defined by $\rho = 0.5$. For $0.5 < \rho \leq 1$, a hyperbola results.

As has been shown in the example of Figure 5.5, the magnitude of the parametric slopes of the end points affects the shape of the curve produced. The directions of the end point slopes, of course, are the lines AE and BE, which are described with the given points as $\mathbf{P}_E - \mathbf{P}_A$ and $\mathbf{P}_B - \mathbf{P}_E$, where $\mathbf{P}_A = \mathbf{P}(0)$ and $\mathbf{P}_B = \mathbf{P}(1)$. In conjunction with the parameter ρ, the geometric coefficients for a conic section can be shown[36] to be

$$[\mathbf{B}] = [\mathbf{P}_A \quad \mathbf{P}_B \quad 4\rho(\mathbf{P}_E - \mathbf{P}_A) \quad 4\rho(\mathbf{P}_B - \mathbf{P}_E)]^T \tag{5.45}$$

where $0 \leq \rho < 0.5$ defines a family of ellipses, $\rho = 0.5$ defines a parabola, and $0.5 < \rho \leq 1$ defines a family of hyperbolas.

The approximation of a circular arc is important in engineering applications. The foregoing can be extended to a circular arc which subtends less than 180° with the aid of the geometrical constructions shown in Figure 5.7. The parameter ρ, which is the ratio CD/DE, is given by

$$\rho = \frac{R(1 - \cos\phi)}{(R\tan\phi)\sin\phi} = \frac{\cos\phi}{1 + \cos\phi} \tag{5.46}$$

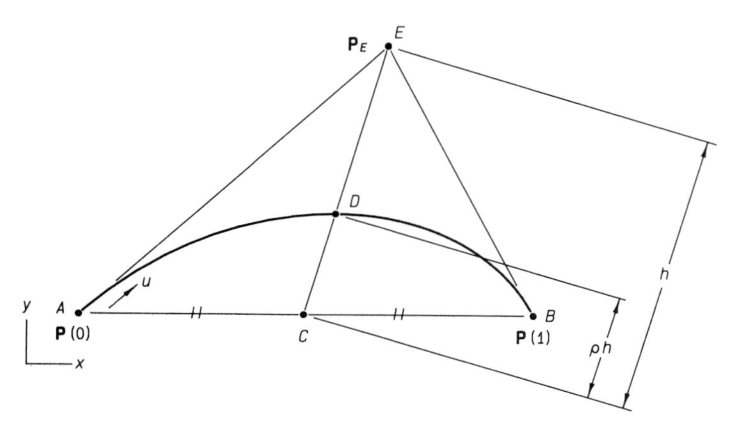

Figure 5.6 Definition of a conic section curve for approximation by a parametric cubic curve. Lines AE and BE are tangent to the ends of the curve. Point C bisects the line AB which connects the end points of the given curve.

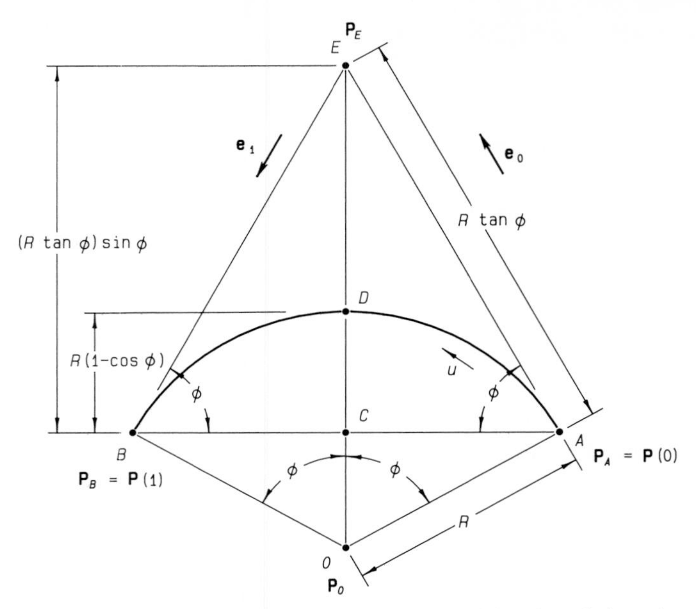

Figure 5.7 Approximation of circular arc with subtended angle 2ϕ by means of Eq. (5.45). The direction of parametrization has been reversed from that in Fig. 5.6 for counterclockwise rotation to be positive.

where ϕ is *half* the angle subtended by the circular arc and $0 < \phi < 90°$. Combination of Eqs. (5.40), (5.45), and (5.46) yields

$$\mathbf{P}(u) = [\mathbf{F}] \begin{bmatrix} \mathbf{P}_A \\ \mathbf{P}_B \\ \dfrac{4 \cos \phi}{1 + \cos \phi} (\mathbf{P}_E - \mathbf{P}_A) \\ \dfrac{4 \cos \phi}{1 + \cos \phi} (\mathbf{P}_B - \mathbf{P}_E) \end{bmatrix} \tag{5.47a}$$

Unit vectors $\mathbf{e}_0$ and $\mathbf{e}_1$ in the directions of $(\mathbf{P}_E - \mathbf{P}_A)$ and $(\mathbf{P}_B - \mathbf{P}_E)$ are shown in Figure 5.7. Alternatively, Eq. (5.47a) can be written in terms of the unit vectors as

$$\mathbf{P}(u) = [\mathbf{F}] \begin{bmatrix} \mathbf{P}_A \\ \mathbf{P}_B \\ \dfrac{4\, R \sin \phi}{1 + \cos \phi} \mathbf{e}_0 \\ \dfrac{4\, R \sin \phi}{1 + \cos \phi} \mathbf{e}_1 \end{bmatrix} \tag{5.47b}$$

Because it is more convenient to specify a circular arc in terms of its end points and the center of the circle, the unit vectors in Eq. (5.47b) are replaced as follows. The unit vector normal to the plane defined by $\mathbf{P}_A$, $\mathbf{P}_B$, and $\mathbf{P}_O$ is

$$\mathbf{n} = \frac{(\mathbf{P}_A - \mathbf{P}_O) \times (\mathbf{P}_B - \mathbf{P}_O)}{R^2 \sin 2\phi}$$

Unit vectors for use in Eq. (5.47b) are thus

$$\mathbf{e}_0 = \frac{\mathbf{n} \times (\mathbf{P}_A - \mathbf{P}_O)}{R} \quad \text{and} \quad \mathbf{e}_1 = \frac{\mathbf{n} \times (\mathbf{P}_B - \mathbf{P}_O)}{R} \tag{5.48}$$

If the circle is in the x-y plane, $\mathbf{n}$ is already known to be $\mathbf{k}$. In this case, Eq. (5.47b) may be specialized to

$$\mathbf{P}(u) = [x(u) \quad y(u)]$$

$$= [\mathbf{F}] \begin{bmatrix} x_A & y_A \\ x_B & y_B \\ \dfrac{-4 \sin \phi}{1 + \cos \phi}(y_A - y_O) & \dfrac{4 \sin \phi}{1 + \cos \phi}(x_A - x_O) \\ \dfrac{-4 \sin \phi}{1 + \cos \phi}(y_B - y_O) & \dfrac{4 \sin \phi}{1 + \cos \phi}(x_B - x_O) \end{bmatrix} \tag{5.49}$$

where (x_A, y_A), (x_B, y_B), and (x_O, y_O) are the components of $\mathbf{P}_A$, $\mathbf{P}_B$, and $\mathbf{P}_O$, respectively. As before, $0 < \phi < 90°$.

The accuracy of the circles approximated by Eqs. (5.47) or (5.49) depends on the angle subtended. For a subtended angle 2ϕ which approaches 180°, the maximum deviation between the approximation for the radius and the true radius is about 1.8%. For a subtended angle $2\phi = 90°$, the maximum deviation is about 0.03%, while for $2\phi = 30°$, the maximum deviation is less than 0.00004%. The circular arc fits exactly at $u = 0$, 1/2, and 1, and the deviation, always toward a larger circle, occurs between these points. A slight adjustment[36] can be made to these points so that the circle is slightly smaller with the maximum deviation reduced to about half the given values. In either case, construction of a parametric approximation to a whole circle with four arcs should be accurate enough for most engineering purposes.

EXAMPLE 5.6

A circular arc in the x-y plane subtends 120° counterclockwise. The center is at the point (2, 3) and the arc starts at the point (4, 3). Use Eq. (5.49) to form a parametric cubic approximation. Determine the percentage of deviation from a true circle at the parametric quarter point.

Solution. For the given circular arc, $R = 2$. From trigonometry, the coordinates of $\mathbf{P}_B$ are $[x_B \quad y_B] = [1 \quad 3 + \sqrt{3}]$. Use of Eqs. (5.49) and (5.40) yields

$$\mathbf{P}(u) = [\mathbf{F}][\mathbf{B}]$$

$$= [(1 - 3u^2 + 2u^3) \quad (3u^2 - 2u^3) \quad (u - 2u^2 + u^3) \quad (-u^2 + u^3)]$$

$$\times \begin{bmatrix} x_A & y_A \\ x_B & y_B \\ -c(y_A - y_O) & c(x_A - x_O) \\ -c(y_B - y_O) & c(x_B - x_O) \end{bmatrix}$$

where $c = 4 \sin \phi/(1 + \cos \phi)$.

Evaluation of $[\mathbf{B}]$ yields

$$[\mathbf{B}] = \begin{bmatrix} 4 & 3 \\ 1 & 3 + \sqrt{3} \\ 0 & 8\sqrt{3}/3 \\ -4 & -4\sqrt{3}/3 \end{bmatrix}$$

At the parametric quarter point, $u = 1/4$,

$$\mathbf{P}(1/4) = \begin{bmatrix} \frac{27}{32} & \frac{5}{32} & \frac{9}{64} & \frac{-3}{64} \end{bmatrix} [\mathbf{B}]$$

$$= [3.718\,750\,000 \qquad 4.028\,405\,168]$$

The radius at this point is 2.002 928, which is 0.146% higher than the true radius of 2. This is much better accuracy than that of the circular arc generated from four points in Example 5.5.

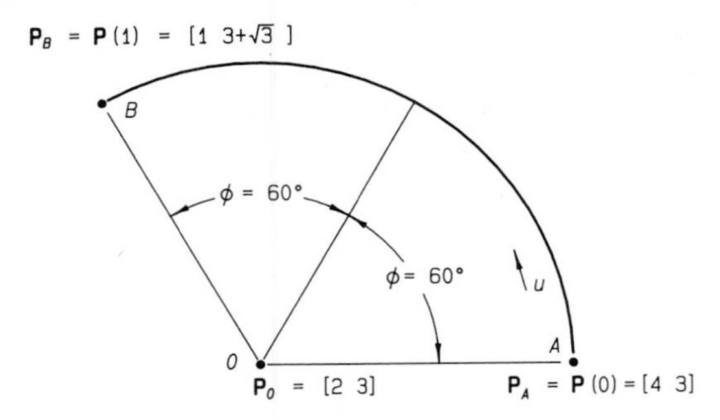

Cubic Spline

Since slopes can be specified at the end points with the geometric form of the parametric cubic curve, it is possible to join two or more curves at *knots* (connecting points) to form a continuous curve. The *cubic spline* is a piecewise set of cubic curves passing through specified points—the knots—with slope continuity everywhere including the knots.

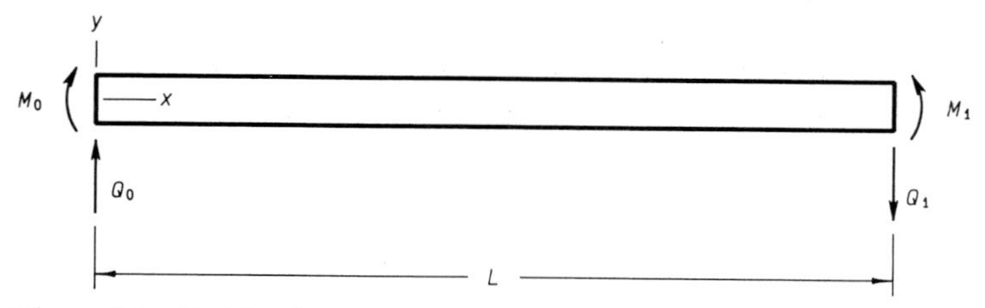

Figure 5.8 End loading on an elastic beam.

The elastic curve from mechanics of materials beam theory is familiar to engineers. The differential equation of the elastic curve of a beam of Figure 5.8 with prescribed moment M_0 and prescribed force Q_0 on the end where $x = 0$ is

$$\frac{d^2y}{dx^2} = \frac{Q_0x + M_0}{EI} \tag{5.50}$$

where EI is the stiffness. Double integration of Eq. (5.50) produces

$$y = \frac{Q_0x^3}{6EI} + \frac{M_0x^2}{2EI} + C_1x + C_2 \tag{5.51}$$

where C_1 and C_2 are constants to be determined from boundary conditions. Clearly, the equation of the elastic curve is the same as a nonparametric cubic polynomial, Eq. (5.2).

The model presented by the elastic beam is very useful in understanding the continuity in piecewise cubic splines.[16] A beam with multiple supports is shown in Figure 5.9, where, at the supports or "knots," there is continuity in displacement, slope, and moment. The knots, then, are where beams described by individual equations of the form of Eq. (5.51) are connected. The same continuity is achieved in the mathematical curve known as the cubic spline.

The foregoing concept can be extended to construction of a parametric cubic spline from piecewise segments of parametric cubic curves. From Eq. (5.39) the equation of the ith segment of a parametric cubic curve is

$$\begin{aligned}\mathbf{P}(u) &= \mathbf{P}_i + \mathbf{P}'_i u + (3(\mathbf{P}_{i+1} - \mathbf{P}_i) - 2\mathbf{P}'_i - \mathbf{P}'_{i+1})u^2 \\ &\quad + (2(\mathbf{P}_i - \mathbf{P}_{i+1}) + \mathbf{P}'_i + \mathbf{P}'_{i+1})u^3\end{aligned} \tag{5.52}$$

where $u = 0$ at point i and $u = 1$ at point $i + 1$. The analogy with the elastic beam shown in Figure 5.9 shows that displacement, slope, and moment continuity are needed at the knots. Continuity in the parametric second derivatives is tantamount to continuity in the geometric second derivative; see Eq. (5.50). When the second derivative of Eq. (5.52) (the subscripted terms are end conditions and are treated as constants) is evaluated with $u = 1$ and the same second derivative with

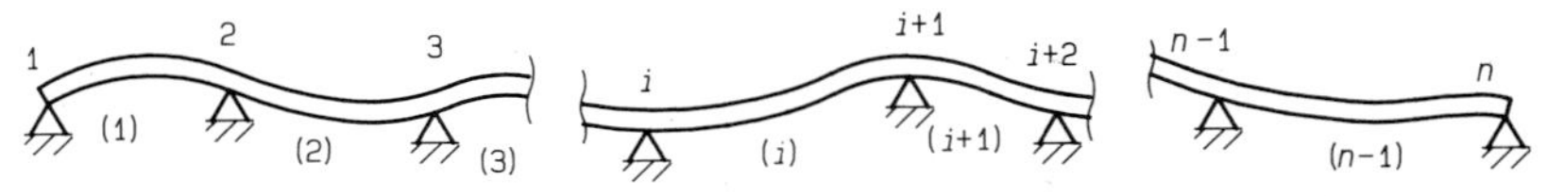

Figure 5.9 Continuous beam model with n supports and $n - 1$ spans. This suggests a cubic spline with n knots and $n - 1$ segments.

all i's replaced by $i + 1$ is evaluated with $u = 0$, equating the two expresses the required second derivative continuity at the knot $i + 1$,

$$2(-3\mathbf{P}_i + 3\mathbf{P}_{i+1} - 2\mathbf{P}'_i - \mathbf{P}'_{i+1}) + 6(2\mathbf{P}_i - 2\mathbf{P}_{i+1} + \mathbf{P}'_i + \mathbf{P}'_{i+1}) = 2(-3\mathbf{P}_{i+1} + 3\mathbf{P}_{i+2} - 2\mathbf{P}'_{i+1} - \mathbf{P}'_{i+2}) \tag{5.53}$$

This simplifies to

$$\mathbf{P}'_i + 4\mathbf{P}'_{i+1} + \mathbf{P}'_{i+2} = 3\mathbf{P}_{i+2} - 3\mathbf{P}_i \tag{5.54}$$

For the $n - 2$ interior knots, Eq. (5.54) indicates

$$\begin{bmatrix} 4 & 1 & 0 & 0 & \cdots & 0 & 0 \\ 1 & 4 & 1 & 0 & \cdots & 0 & 0 \\ 0 & 1 & 4 & 1 & \cdots & 0 & 0 \\ \cdots & \cdots & \cdots & \cdots & \cdots & \cdots & \cdots \\ 0 & 0 & \cdots & 0 & 1 & 4 & 1 \\ 0 & 0 & \cdots & 0 & 0 & 1 & 4 \end{bmatrix} \begin{bmatrix} \mathbf{P}'_2 \\ \mathbf{P}'_3 \\ \mathbf{P}'_4 \\ \vdots \\ \mathbf{P}'_{n-2} \\ \mathbf{P}'_{n-1} \end{bmatrix} = \begin{bmatrix} 3\mathbf{P}_3 - 3\mathbf{P}_1 - \mathbf{P}'_1 \\ 3\mathbf{P}_4 - 3\mathbf{P}_2 \\ 3\mathbf{P}_5 - 3\mathbf{P}_3 \\ \vdots \\ 3\mathbf{P}_{n-1} - 3\mathbf{P}_{n-3} \\ 3\mathbf{P}_n - 3\mathbf{P}_{n-2} - \mathbf{P}'_n \end{bmatrix} \tag{5.55}$$

which may be solved for the slopes at the interior points $\mathbf{P}'_2$ through $\mathbf{P}'_{n-1}$. This completes the construction of a piecewise continuous set of cubic splines through n knots with specified end slopes.

Other conditions besides slope may be specified on the free ends of the spline. The most common is the moment-free or relaxed end where

$$\mathbf{P}''_1 = 0 \;\; \text{(left end)} \quad \text{and} \quad \mathbf{P}''_n = 0 \;\; \text{(right end)} \tag{5.56}$$

The second derivative of Eq. (5.33) set equal to zero yields for the left end

$$2\mathbf{a}_3 = 0 \;\; \text{(left end)} \tag{5.57}$$

In view of Eq. (5.38),

$$\mathbf{a}_3 = -3\mathbf{P}_1 + 3\mathbf{P}_2 - 2\mathbf{P}'_1 - \mathbf{P}'_2 = 0 \;\; \text{(left end)} \tag{5.58}$$

Since now $\mathbf{P}_1$ and $\mathbf{P}'_n$ are unknown rather than being specified, the relationship given in Eq. (5.56) must hold, or

$$\begin{aligned} 2\mathbf{P}'_1 + \mathbf{P}'_2 &= 3\mathbf{P}_2 - 3\mathbf{P}_1 \;\; \text{(left end)} \\ 2\mathbf{P}'_n + \mathbf{P}'_{n-1} &= 3\mathbf{P}_n - 3\mathbf{P}_{n-1} \;\; \text{(right end)} \end{aligned} \tag{5.59}$$

Next, Eqs. (5.59) must be incorporated in Eq. (5.55) to provide for the simultaneous solution of the two additional unknowns. This yields

$$\begin{bmatrix} 2 & 1 & 0 & 0 & 0 & \cdots & & & 0 & 0 \\ 1 & 4 & 1 & 0 & 0 & \cdots & & & 0 & 0 \\ 0 & 1 & 4 & 1 & 0 & \cdots & & & 0 & 0 \\ 0 & 0 & 1 & 4 & 1 & \cdots & & & 0 & 0 \\ \cdots & & & & & & & & & \\ 0 & 0 & \cdots & & 0 & 1 & 4 & 1 & & 0 \\ 0 & 0 & \cdots & & & 0 & 1 & 4 & & 1 \\ 0 & 0 & \cdots & & & & 0 & 1 & & 2 \end{bmatrix} \begin{bmatrix} \mathbf{P}'_1 \\ \mathbf{P}'_2 \\ \mathbf{P}'_3 \\ \vdots \\ \\ \mathbf{P}'_{n-2} \\ \mathbf{P}'_{n-1} \\ \mathbf{P}'_n \end{bmatrix} = \begin{bmatrix} 3\mathbf{P}_2 - 3\mathbf{P}_1 \\ 3\mathbf{P}_3 - 3\mathbf{P}_1 \\ 3\mathbf{P}_4 - 3\mathbf{P}_2 \\ \vdots \\ \\ 3\mathbf{P}_{n-1} - 3\mathbf{P}_{n-3} \\ 3\mathbf{P}_n - 3\mathbf{P}_{n-2} \\ 3P_n - 3\mathbf{P}_{n-1} \end{bmatrix} \tag{5.60}$$

Using $\mathbf{P}'_i$ determined by Eq. (5.60), Eq. (5.52) is applied to generate an interpolating curve passing through n arbitrary points.

The cubic spline is used in some CAD systems to fit a smooth curve through a selected set of points. In presentation graphics, a 2-D version of the cubic spline is useful for smoothing x-y plots and contour plots constructed from a small number of data. The cubic spline presented here is a *parametric* cubic spline, which is more versatile than the more common natural (nonparametric) cubic spline presented in Section 8.4. Among the disadvantages of the natural cubic spline are the inability to handle infinite slopes and the fact that it is only a 2-D curve.

EXAMPLE 5.7

Fit a parametric cubic spline through the five data

$$[\mathbf{P}] = \begin{bmatrix} 0 & 0 & 0 \\ 1 & 0 & 0 \\ 2 & 1 & 1 \\ 3 & 2 & 2 \\ 4 & 3 & 3 \end{bmatrix}$$

Solution. Relaxed ends are assumed. Application of Eq. (5.60) with $n = 5$ produces the system

$$\begin{bmatrix} 2 & 1 & 0 & 0 & 0 \\ 1 & 4 & 1 & 0 & 0 \\ 0 & 1 & 4 & 1 & 0 \\ 0 & 0 & 1 & 4 & 1 \\ 0 & 0 & 0 & 1 & 2 \end{bmatrix} \begin{bmatrix} \mathbf{P}'_1 \\ \mathbf{P}'_2 \\ \mathbf{P}'_3 \\ \mathbf{P}'_4 \\ \mathbf{P}'_5 \end{bmatrix} = \begin{bmatrix} 3\mathbf{P}_2 - 3\mathbf{P}_1 \\ 3\mathbf{P}_3 - 3\mathbf{P}_1 \\ 3\mathbf{P}_4 - 3\mathbf{P}_2 \\ 3\mathbf{P}_5 - 3\mathbf{P}_3 \\ 3\mathbf{P}_5 - 3\mathbf{P}_4 \end{bmatrix} = \begin{bmatrix} 3 & 0 & 0 \\ 6 & 3 & 3 \\ 6 & 6 & 6 \\ 6 & 6 & 6 \\ 3 & 3 & 3 \end{bmatrix}$$

The solutions for the parametric slopes are

$$\begin{bmatrix} \mathbf{P}'_1 \\ \mathbf{P}'_2 \\ \mathbf{P}'_3 \\ \mathbf{P}'_4 \\ \mathbf{P}'_5 \end{bmatrix} = \begin{bmatrix} 1 & -0.268 & -0.268 \\ 1 & 0.536 & 0.536 \\ 1 & 1.125 & 1.125 \\ 1 & 0.964 & 0.964 \\ 1 & 1.018 & 1.018 \end{bmatrix}$$

Equation (5.52) is then used to generate the spline, where i takes the range 1 to 4. The resulting curve (below) has first and second derivative continuity at the knots. The oscillation between knots 2 and 4 is characteristic of the parametric cubic curve when a sudden chain in direction is present. A simple method of straightening such curves is given in Section 8.4.

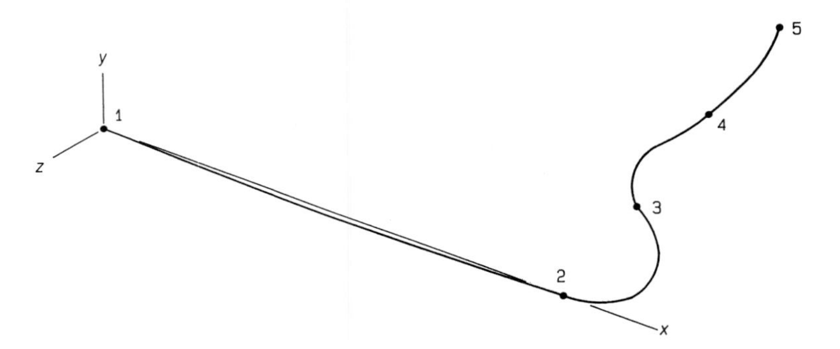

5.4 BEZIER CURVES

The parametric cubic curves of the prvious section are classified as *interpolating* polynomials because they fit exactly through the given control points. Alternatively, curves classified as *approximating* polynomials only pass near the given control points.

In the early 1960s P. Bezier[22] developed a versatile system of curves that combines features of both interpolating and approximating polynomials. Bezier's curves are part of the computer-aided design system of the French automobile manufacturer Renault and subsequently have been incorporated in many other systems. The features of Bezier curves include

The first and last control point define the exact end points of the curve—like interpolating polynomials.

Intermediate control points influence the path of the curve without introducing unwanted oscillations—like approximating polynomials.

The first two and last two control points define lines which are tangent to the beginning and the end of the curve. Thus, with the ability to specify the direction at the beginning and the end of the curve, multiple curves can be blended together.

Bezier curves are strong in intuitive appeal. In the original work of Bezier, curves of any arbitrary degree are described. However, cubic or quartic Bezier curves represent a balance between versatility and tractability. A useful format for a cubic Bezier curve is

$$\mathbf{P}(u) = (1 - u)^3\mathbf{P}_1 + 3u(1 - u)^2\mathbf{P}_2 + 3u^2(1 - u)\mathbf{P}_3 + u^3\mathbf{P}_4$$

or

$$\mathbf{P}(u) = [(1 - u)^3 \quad 3u(1 - u)^2 \quad 3u^2(1 - u) \quad u^3]\,[\mathbf{P}_1 \quad \mathbf{P}_2 \quad \mathbf{P}_3 \quad \mathbf{P}_4]^T \tag{5.61}$$

where the parameter $0 \leq u \leq 1$, and the control points $\mathbf{P}_1$, $\mathbf{P}_2$, $\mathbf{P}_3$, and $\mathbf{P}_4$ define a Bezier polygon. Note that this does not necessarily have to be a planar figure. The series of Bezier curves and the corresponding Bezier polygons in Figure 5.10 show how repositioning the two intermediate control points changes the shape of the curve.

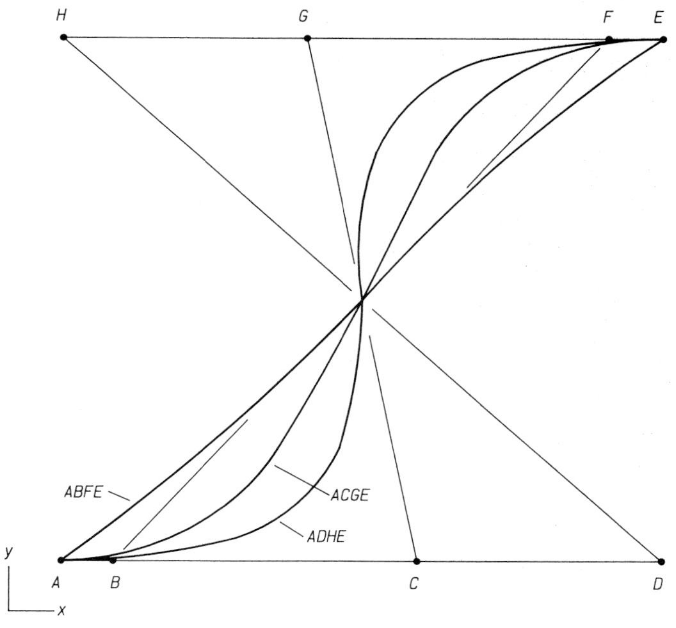

Figure 5.10 Effect of control point placement on shape of Bezier curves. The intermediate control points are moved so that the initial and final tangents of the curves remain the same. Note how the control points "pull" the curve while maintaining smoothness.

The quartic Bezier curve has five control points instead of the four in the cubic Bezier curve. Here,

$$\mathbf{P}(u) = [(1-u)^4 \quad 4u(1-u)^3 \quad 6u^2(1-u)^2 \quad 4u^3(1-u) \quad u^4] \times [\mathbf{P}_1 \quad \mathbf{P}_2 \quad \mathbf{P}_3 \quad \mathbf{P}_4 \quad \mathbf{P}_5]^T \tag{5.62}$$

Although quartic Bezier curves give a little more control in shape, they are not as easy to use.

Examination of Eqs. (5.61) and (5.62) shows that two (or three) of the intermediate control points may be coincident, which will produce a sharper bend in the curve. If the first and last control points are coincident, a closed curve results.

The process of blending Bezier curves is intuitive. Slope continuity is maintained by having three collinear control points, the middle one being common to two curves, as shown in Figure 5.11. As higher degree of continuity is accomplished with quartic Bezier curves by aligning five control points.

It has been seen with parametric cubic curves that the magnitude of the parametric derivatives at the end points affects the shape of the curve, while the ratios of the parametric derivatives affect the geometric slopes at the end points. The same control exists with the spacing of the two end control points in Bezier curves. The parametric derivative of a cubic Bezier curve from Eq. (5.61) is

$$\mathbf{P}'(u) = [-3(u-1)^2 \quad 3(3u^2-4u+1) \quad -3u(3u-2) \quad 3u^2] \times [\mathbf{P}_1 \quad \mathbf{P}_2 \quad \mathbf{P}_3 \quad \mathbf{P}_4]^T \tag{5.63}$$

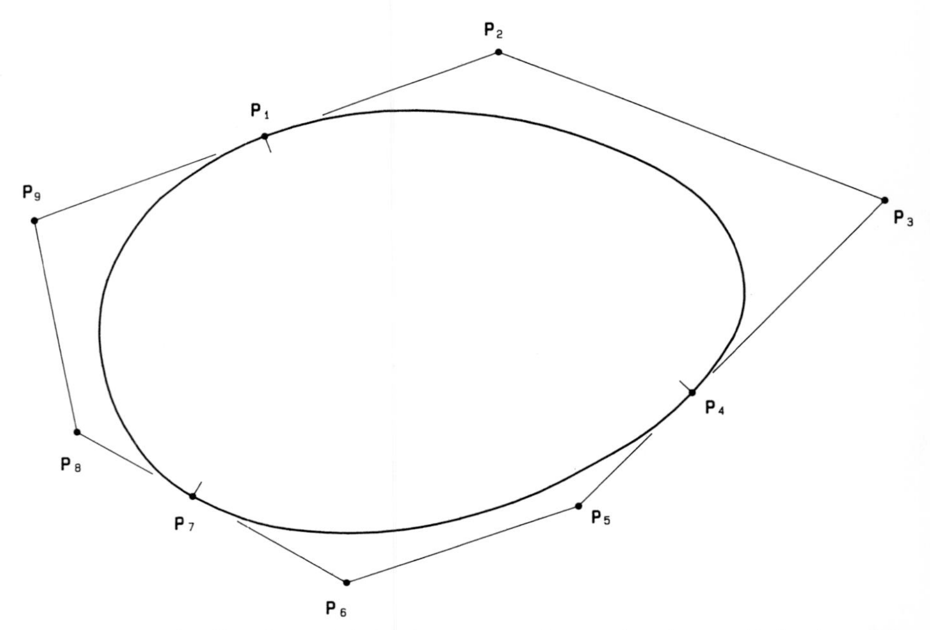

Figure 5.11 Blending three cubic Bezier curves to form a closed curve. End points of the three separate curves are $\mathbf{P}_1$, $\mathbf{P}_4$, and $\mathbf{P}_7$, where each end point is common to two curves.

which, evaluated at $u = 0$ and $u = 1$, produces

$$\mathbf{P}'(0) = [-3 \quad 3 \quad 0 \quad 0][\mathbf{P}_1 \quad \mathbf{P}_2 \quad \mathbf{P}_3 \quad \mathbf{P}_4]^T$$
$$\mathbf{P}'(1) = [0 \quad 0 \quad -3 \quad 3][\mathbf{P}_1 \quad \mathbf{P}_2 \quad \mathbf{P}_3 \quad \mathbf{P}_4]^T \tag{5.64}$$

This indicates that the magnitude and direction of the parametric derivatives at the two end points are related to the associated control points by

$$\mathbf{P}'(0) = 3(\mathbf{P}_2 - \mathbf{P}_1) \quad \text{and} \quad \mathbf{P}'(1) = 3(\mathbf{P}_4 - \mathbf{P}_3) \tag{5.65}$$

With this equation, the cubic Bezier curve can be converted to a parametric cubic curve in geometric form.

Bezier curves are very well suited for use in interactive design applications. Input is from points only, with intuitive control of initial and final slope. The curves may be blended, open, or closed, without wild oscillations. It happens that the Bezier curves are special cases of B-spline curves, treated in the next section.

EXAMPLE 5.8

A circular arc in the x-y plane subtends an angle of 60° counterclockwise. Its center is at the point (2, 3, 0) and the arc starts at the point (4, 3, 0). Represent this arc with a cubic Bezier curve.

Solution. The arc is first written in terms of the parameter u, which has the range 0 to 1. Thus, the x and y coordinates of a circular arc in the x-y plane with radius r and subtended angle ϕ (as measured from the x axis) are described by

$$x = x_0 + r \cos \phi u \quad \text{and} \quad y = y_0 + r \sin \phi u$$

where x_0 and y_0 are the center of the arc. For the data given, $x_0 = 2$, $y_0 = 3$, $r = 2$, and $\phi = \pi/3$ rad. Thus,

$$x = 2 + 2 \cos\left(\frac{\pi}{3} u\right) \quad \text{and} \quad y = 3 + 2 \sin\left(\frac{\pi}{3} u\right)$$

The end point is therefore

$$\mathbf{P}(1) = \mathbf{P}_4 = [3 \quad 4.732 \quad 0]$$

The slopes are, at $u = 0$,

$$\frac{dx}{du} = 0 \quad \text{and} \quad \frac{dy}{du} = 2\frac{\pi}{3} = 2.094$$

and, at $u = 1$,

$$\frac{dx}{du} = \frac{-\pi\sqrt{3}}{3} = -1.814, \qquad \frac{dy}{du} = \frac{\pi}{3} = 1.047$$

This specification of slopes permits finding the intermediate control points by means of Eq. (5.65). Thus,

$$\mathbf{P}'(0) = 3(\mathbf{P}_2 - \mathbf{P}_1)$$

$$[0 \quad 2.094 \quad 0] = 3[x_2 - 4 \quad y_2 - 3 \quad 0]$$

and

$$\mathbf{P}'(1) = 3(\mathbf{P}_4 - \mathbf{P}_3)$$

$$[-1.814 \quad 1.047 \quad 0] = 3[3 - x_3 \quad 4.732 - y_3 \quad 0]$$

Solving these equations gives

$$\mathbf{P}_2 = [4.000 \quad 3.698 \quad 0] \quad \text{and} \quad \mathbf{P}_3 = [3.605 \quad 4.383 \quad 0]$$

which results in the Bezier polygon and curve shown.

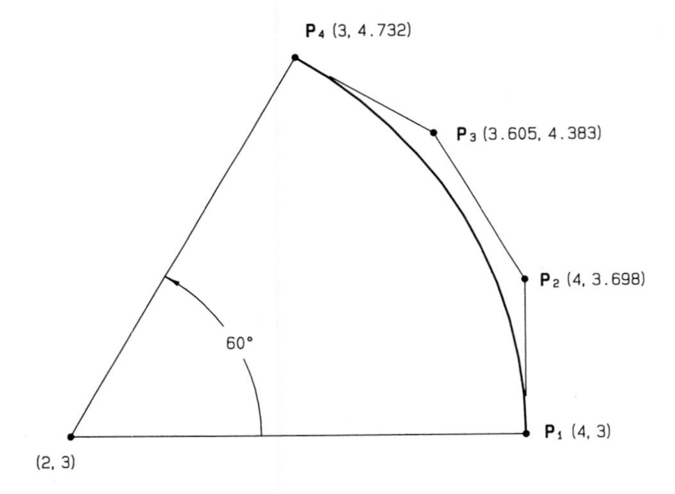

5.5 B-SPLINE CURVES

As seen in the previous section, cubic Bezier curves, each defined by four control points, can be blended to produce a smooth curve through more than four control points. However, the additional control points must be supplied in groups of three, with the knot between individual Bezier curves sharing a control point. One advantage of blended Bezier curves is that of local control—moving a control point affects the curve only in the vicinity of the changed point.

The advantages of an approximating function with local control and the versatility of using an arbitrary number of control points are available with the family of curves known as B-splines.[16,36,50] These curves are generated by multiplying an approximating function, which is in terms of the parameter u ($0 \leq u \leq 1$), times a matrix which contains a subset of the control points in the vicinity of the B-spline curve. Considered here are *quadratic* and *cubic* B-spline curves, so classified according to the order of the approximating function. B-splines are also classified as *nonperiodic* and *periodic* according to the way in which control points at the beginning and end are utilized.

Nonperiodic Form

As applied to B-spline curves, the term *nonperiodic* means that special forms of the approximating function are used to begin and terminate the curves. The approximating function for the central part of the curve is reused. For *quadratic* B-spline curves with n control points, there are $n - 2$ curve segments joined by $n - 3$ knots.

A central segment of a quadratic B-spline curve is given by

$$\mathbf{P}_i(u) = \begin{bmatrix} \dfrac{(1-u)^2}{2} & \dfrac{-2u^2 + 2u + 1}{2} & \dfrac{u^2}{2} \end{bmatrix} \begin{bmatrix} \mathbf{P}_i \\ \mathbf{P}_{i+1} \\ \mathbf{P}_{i+2} \end{bmatrix} \tag{5.66}$$

where the notation $\mathbf{P}_i(u)$ denotes the ith segment; $\mathbf{P}_i$, $\mathbf{P}_{i+1}$, and so on are control points; $0 \leq u \leq 1$ and $2 \leq i \leq n - 3$ (for the nonperiodic form). Substitution of $u = 0$ shows that the ith segment of the curve starts at the knot $\mathbf{P}_i(0) = (\mathbf{P}_i + \mathbf{P}_{i+1})/2$ and substitution of $u = 1$ shows that the segment ends at the knot $\mathbf{P}_i(1) = (\mathbf{P}_{i+1} + \mathbf{P}_{i+2})/2$. Segments from the central part of the B-spline join as shown in Figure 5.12. Special segments are needed to make the B-spline begin on the first control point and end on the last control point.

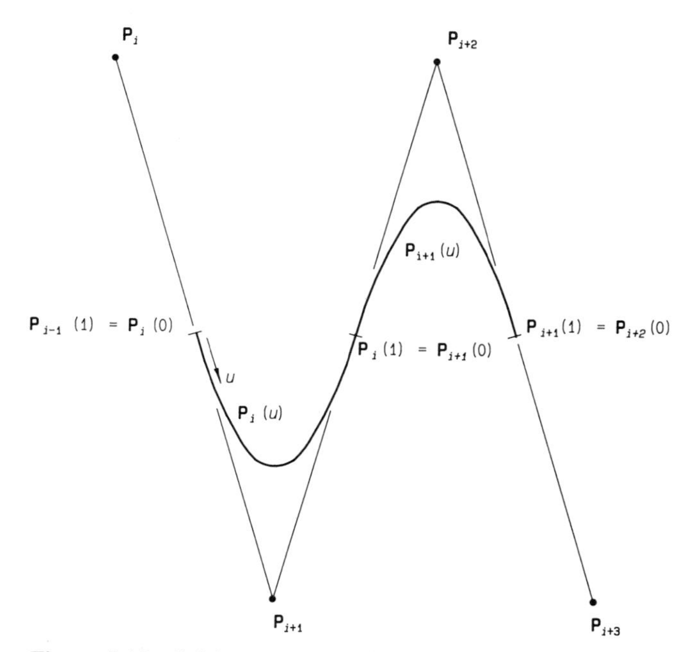

Figure 5.12 Joining two central segments of a quadratic B-spline curve. Knots for central segments are half-way between the control points.

The first segment in the nonperiodic curve is

$$\mathbf{P}_1(u) = \begin{bmatrix} (1-u)^2 & \dfrac{-3u^2+4u}{2} & \dfrac{u^2}{2} \end{bmatrix} \begin{bmatrix} \mathbf{P}_1 \\ \mathbf{P}_2 \\ \mathbf{P}_3 \end{bmatrix} \tag{5.67}$$

where $0 \le u \le 1$. It can be verified that the first segment starts at the knot $\mathbf{P}_1(0) = \mathbf{P}_1$ and goes to the knot $\mathbf{P}_1(1) = (\mathbf{P}_2 + \mathbf{P}_3)/2$, as shown in Figure 5.13(a). The last segment of the curve has the same form as Eq. (5.67), where u is replaced by $(1 - u)$ and the order of the terms in the row matrix is reversed, giving

$$\mathbf{P}_{n-2}(u) = \begin{bmatrix} \dfrac{(1-u)^2}{2} & \dfrac{-3u^2+2u+1}{2} & u^2 \end{bmatrix} \begin{bmatrix} \mathbf{P}_{n-2} \\ \mathbf{P}_{n-1} \\ \mathbf{P}_n \end{bmatrix} \tag{5.68}$$

This segment starts at the knot $\mathbf{P}_{n-2}(0) = (\mathbf{P}_{n-2} + \mathbf{P}_{n-1})/2$ and goes to the last control point, $\mathbf{P}_{n-2}(1) = \mathbf{P}_n$ as shown in Figure 5.13(b).

The foregoing B-spline segments connect with first derivative (slope) continuity, as demonstrated by differentiating Eq. (5.66),

$$\mathbf{P}_i'(u) = [u-1 \quad -2u+1 \quad u] \begin{bmatrix} \mathbf{P}_i \\ \mathbf{P}_{i+1} \\ \mathbf{P}_{i+2} \end{bmatrix} \tag{5.69}$$

where $0 \le u \le 1$. Here, for the ith segment, the slope at $u = 1$ is $\mathbf{P}_i'(1) = -\mathbf{P}_{i+1} + \mathbf{P}_{i+2}$. For the $i + 1$st segment, the slope at $u = 0$ is the same. It may be

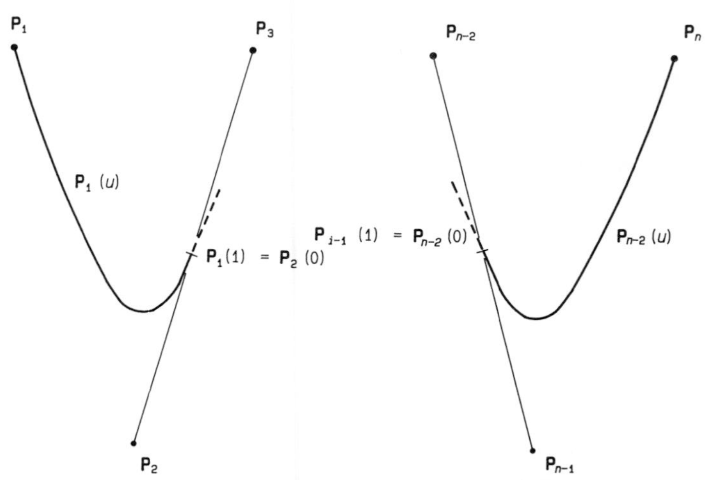

Figure 5.13 Special segments of nonperiodic quadratic B-splines. (a) Beginning segment; (b) last segment.

verified that the slope at $u = 1$ derived from Eq. (5.67) matches Eq. (5.69) with $u = 0$ and $i = 2$. Similarly, the slope at $u = 0$ from Eq. (5.68) matches Eq. (5.69) with $u = 1$ and $i = n - 3$.

A *cubic* B-spline curve uses four control points to define each segment, where an additional order of continuity contributes to making smoother-appearing curves than the quadratic B-spline. In the cubic B-spline, n control points produce a spline of $n - 3$ segments joined by $n - 4$ knots. However, a nonperiodic cubic B-spline requires two special segments to start and two special segments to end. The central segments are

$$\mathbf{P}_i(u) = \begin{bmatrix} \dfrac{(1-u)^3}{6} & \dfrac{3u^3 - 6u^2 + 4}{6} & \dfrac{-3u^3 + 3u^2 + 3u + 1}{6} & \dfrac{u^3}{6} \end{bmatrix} \begin{bmatrix} \mathbf{P}_i \\ \mathbf{P}_{i+1} \\ \mathbf{P}_{i+2} \\ \mathbf{P}_{i+3} \end{bmatrix} \tag{5.70}$$

where $0 \le u \le 1$ and $3 \le i \le n - 5$ (for the nonperiodic form). The ith segment connects the knot $\mathbf{P}_i(0) = \mathbf{P}_i/6 + 2\mathbf{P}_{i+1}/3 + \mathbf{P}_{i+2}/6$ to the knot $\mathbf{P}_i(1) = \mathbf{P}_{i+1}/6 + 2\mathbf{P}_{i+2}/3 + \mathbf{P}_{i+3}/6$.

The first of the two special segments needed to start a nonperiodic cubic B-spline curve is

$$\mathbf{P}_1(u) = \begin{bmatrix} (1-u)^3 & \dfrac{21u^3 - 54u^2 + 36u}{12} & \dfrac{-11u^3 + 18u^2}{12} & \dfrac{u^3}{6} \end{bmatrix} \begin{bmatrix} \mathbf{P}_1 \\ \mathbf{P}_2 \\ \mathbf{P}_3 \\ \mathbf{P}_4 \end{bmatrix} \tag{5.71}$$

where $0 \le u \le 1$. This connects the knot $\mathbf{P}_1(0) = \mathbf{P}_1$ to the knot $\mathbf{P}_1(1) = \mathbf{P}_2/4 + 7\mathbf{P}_3/12 + \mathbf{P}_4/6$. The second segment is

$$\mathbf{P}_2(u) = \begin{bmatrix} \dfrac{(1-u)^3}{4} & \dfrac{7u^3 - 15u^2 + 3u + 7}{12} & \dfrac{-3u^3 + 3u^2 + 3u + 1}{6} & \dfrac{u^3}{6} \end{bmatrix} \begin{bmatrix} \mathbf{P}_2 \\ \mathbf{P}_3 \\ \mathbf{P}_4 \\ \mathbf{P}_5 \end{bmatrix} \tag{5.72}$$

where $0 \le u \le 1$. This generates a curve from the knot $\mathbf{P}_2(0) = \mathbf{P}_2/4 + 7\mathbf{P}_3/12 + \mathbf{P}_4/6$ to the knot $\mathbf{P}_2(1) = \mathbf{P}_3/6 + 2\mathbf{P}_4/3 + \mathbf{P}_5/6$. Thus, this segment makes the transition between the first segment, Eq. (5.71), and the first central segment, Eq. (5.70).

In a similar fashion, two special segments are needed to make the transition from the last central segment to the last control point, $\mathbf{P}_n$. For the next-to-last segment,

$$\mathbf{P}_{n-4}(u) = \begin{bmatrix} \dfrac{(1-u)^3}{6} & \dfrac{3u^3 - 6u^2 + 4}{6} & \dfrac{-7u^3 + 6u^2 + 6u + 2}{12} & \dfrac{u^3}{4} \end{bmatrix} \begin{bmatrix} \mathbf{P}_{n-4} \\ \mathbf{P}_{n-3} \\ \mathbf{P}_{n-2} \\ \mathbf{P}_{n-1} \end{bmatrix} \tag{5.73}$$

where $0 \le u \le 1$. For the last segment,

$$\mathbf{P}_{n-3}(u) = \begin{bmatrix} \dfrac{(1-u)^3}{6} & \dfrac{11u^3 - 15u^2 - 3u + 7}{12} & \dfrac{-7u^3 + 3u^2 + 3u + 1}{4} & u^3 \end{bmatrix} \begin{bmatrix} \mathbf{P}_{n-3} \\ \mathbf{P}_{n-2} \\ \mathbf{P}_{n-1} \\ \mathbf{P}_{n} \end{bmatrix} \tag{5.74}$$

where $0 \le u \le 1$.

From the foregoing two equations, it can be ascertained that the last three knots are located at the points $(\mathbf{P}_{n-4}/6 + 2\mathbf{P}_{n-3}/3 + \mathbf{P}_{n-2}/6)$, $(\mathbf{P}_{n-3}/6 + 7\mathbf{P}_{n-2}/12 + \mathbf{P}_{n-1}/4)$, and $\mathbf{P}_n$.

A quadratic B-spline passes a little closer to the control points than does the cubic B-spline, as demonstrated in Figure 5.14. In applications where the smoothness is critical, the cubic B-spline is preferred because of the continuity in the second derivatives at the knots.

In B-splines, two or more control points may be coincident. This "pulls" the curve closer to the control points (see Figure 5.15). Local control is illustrated in Figure 5.16, where one control point is moved to three different locations.

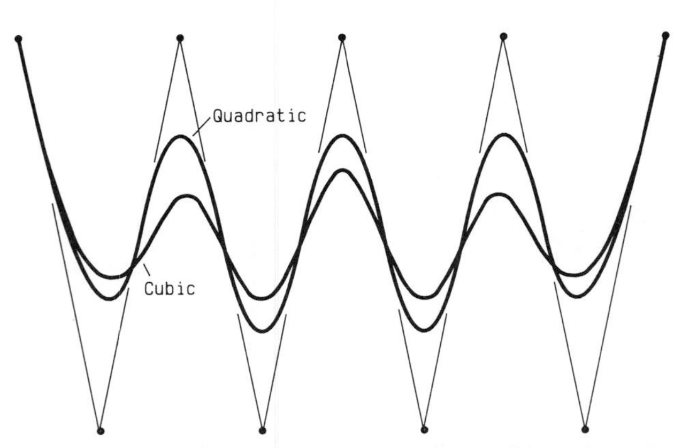

Figure 5.14 Comparison of cubic and quadratic nonperiodic B-spline curves.

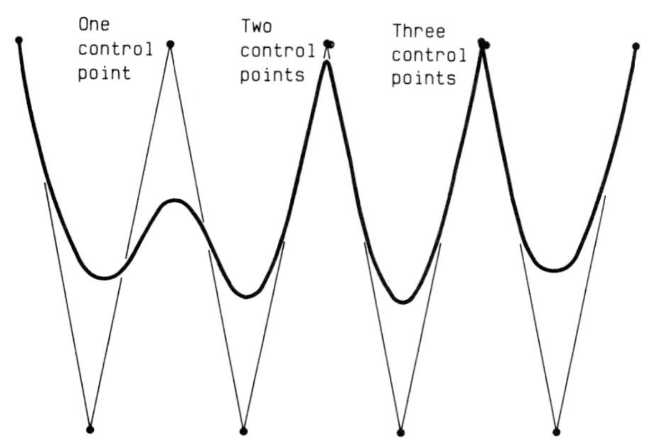

Figure 5.15 Nonperiodic cubic B-splines resulting from one, two, or three control points occupying the same location.

Periodic Form

The foregoing family of B-spline curves is called *nonperiodic* because of the use of special segments at the ends. The special segments, of course, cause the resulting curve to coincide with the first and the last control point. A *periodic* B-spline curve is much simpler to construct because only the function for central segments is needed. A periodic quadratic B-spline curve is generated from Eq. (5.66) with $1 \leq i \leq n - 2$ and a periodic cubic B-spline curve is generated from Eq. (5.70) with $1 \leq i \leq n - 3$.

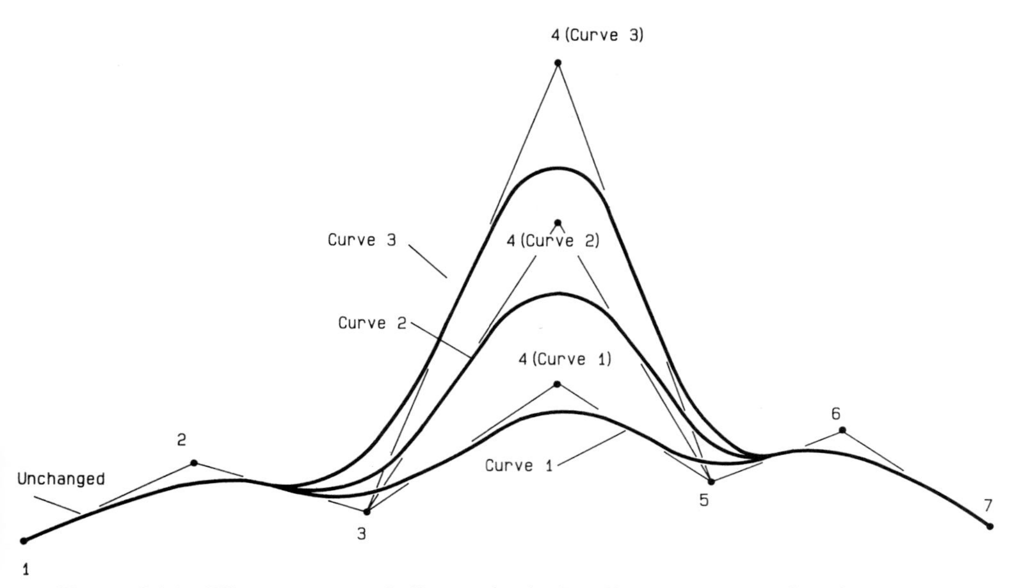

Figure 5.16 Effect on nonperiodic quadratic B-spline curves resulting from three different placements of control point $\mathbf{P}_4$.

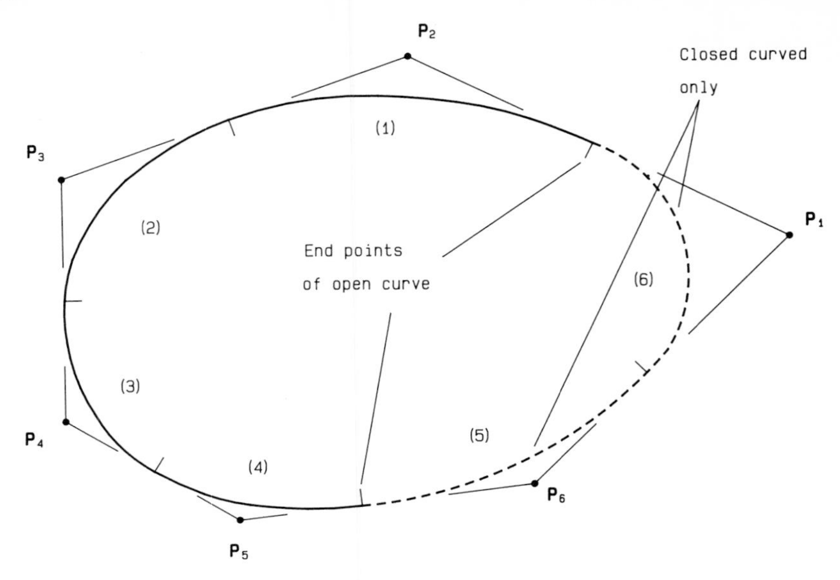

Open B-spline

Segment	Control points		
(1)	1	2	3
(2)	2	3	4
(3)	3	4	5
(4)	4	5	6

Closed B-spline

Segment	Control points		
(1)	1	2	3
(2)	2	3	4
(3)	3	4	5
(4)	4	5	6
(5)	5	6	1
(6)	6	1	2

Figure 5.17 Open and closed quadratic periodic B-spline curves produced from the same six control points, using Eq. (5.66). Segment numbers are in parentheses. The open curve has four segments and does not start or terminate at the end control points, $\mathbf{P}_1$ and $\mathbf{P}_6$. The closed curve has six segments, joined with continuous parametric slopes.

As seen in Figure 5.17, n control points produce $n - 2$ segments for the open curve. The resulting curve does not meet the first and last control points, a characteristic of open periodic B-spline curves.

For a closed curve, the control points are reused as shown in Figure 5.17. Here it is seen that the number of segments now equals the number of control points. The resulting closed curve is continuous to the degree provided by the order (quadratic, cubic, etc.) of the B-spline selected.

EXAMPLE 5.9

Fit a quadratic nonperiodic B-spline curve to the same circular arc as in Example 5.8. Use four control points, and check the accuracy of the radius given by the B-spline approximation where $u = 0.5$ on the first segment.

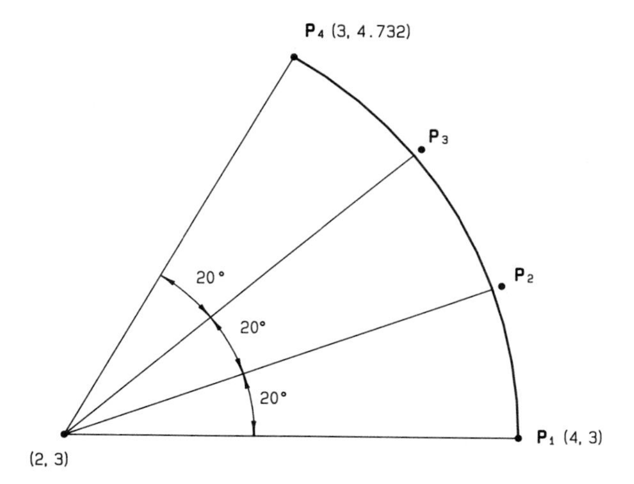

Solution. Symmetry suggests that the $n = 4$ control points be spaced in 20° increments as shown above. Furthermore, four control points allow for only two segments, both special segments for beginning and ending the B-spline. Because the two segments join at the point $(\mathbf{P}_2 + \mathbf{P}_3)/2$, these two control points should be located such that a connecting line is tangent to the given circular arc. This can be accomplished with the construction shown below.

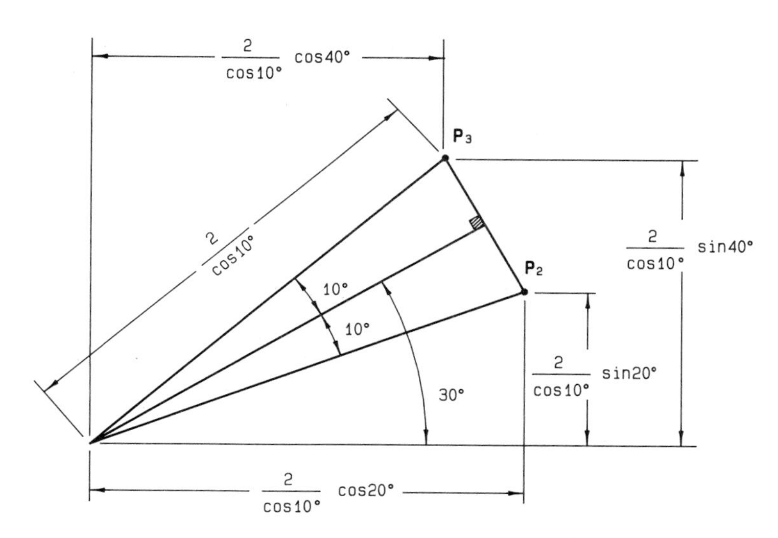

The computation for the two intermediate control points yields

$$\mathbf{P}_2 = (2 + 2\cos 20°/\cos 10°, \quad 3 + 2\sin 20°/\cos 10°)$$

$$= (3.908378, \quad 3.694593)$$

$$\mathbf{P}_3 = (2 + 2\cos 40°/\cos 10°, \quad 3 + 2\sin 40°/\cos 10°)$$

$$= (3.555724, \quad 4.305407)$$

The B-spline curve is generated by Eqs. (5.67) and (5.68),

$$\mathbf{P}_1(u) = \begin{bmatrix} (1-u)^2 & \dfrac{-3u^2+4u}{2} & \dfrac{u^2}{2} \end{bmatrix} \begin{bmatrix} 4 & 3 \\ 3.908378 & 3.694593 \\ 3.555724 & 4.305407 \end{bmatrix}$$

and

$$\mathbf{P}_{n-2}(u) = \begin{bmatrix} \dfrac{(1-u)^2}{2} & \dfrac{-3u^2+2u+1}{2} & u^2 \end{bmatrix} \begin{bmatrix} 3.908378 & 3.694593 \\ 3.555724 & 4.305407 \\ 3 & 4.732051 \end{bmatrix}$$

When the equation for segment 1 is evaluated at $u = 0.5$, it is seen that $\mathbf{P}(0.5) = [3.887202 \quad 3.597297]$. The radius at this point is

$$\sqrt{(3.887202 - 2)^2 + (3.597297 - 3)^2} = 1.979468$$

which is 1.03% smaller than the actual radius of the arc.

5.6 CLOSURE

Conic sections, parametric cubic curves and splines, Bezier curves, and B-splines find widespread use in geometric modeling applications. Advantages in accuracy and compactness make any of these preferable to a list of straight line elements for representation of curves. Viewing and modeling transformations such as rotation and translation can be applied to the control points which define the Bezier and B-spline curves, as well as to the point form of the parametric cubic. Surfaces and solids, an important part of the geometric modeling found in computer-aided drafting and solid modeling systems, are described in the next chapter.

Many other interpolating and approximating methods[16,59] are also suited for geometric modeling. Another application of curvilinear representations is in smoothing data plots, as described in Sections 8.4 and 8.5.

PROBLEMS

5.1. Determine the coefficients c_i in the equation

$$v = c_1 + c_2 t^{0.5} + c_3 t$$

in order that the (t, v) data pairs pass through the points (0, 0), (1, 22.3), and (4, 9.6).

5.2. Read the values of y for $x = 2.0$, 6.0, and 10.0 meters from the graph of Figure 5.1. Use Eq. (5.7) to determine the constants c_1, c_2, and c_3 which fit a parabola to the data. Compare the values of these coefficients with those values computed directly from Eq. (5.4).

5.3. Determine the equation of a circle in the x-y plane which passes through the three points $(2, 3)$, $(-1, 4)$, and $(3, -1)$.

5.4. It is specified that the straight line segments (chords) approximating a plotted circle by the method of Section 5.2 should not have a length greater than some quantity c. Derive an expression which relates n, the number of segments, to c and r, the radius of the circle. Note that n must be an integer value.

5.5. Modify Eqs. (5.25) through (5.28) to generate parabolas which are symmetrical to a line parallel to the y axis.

5.6. The end points of a parabola that is symmetrical to the line $y = -1$ are $(2.5, 0)$ and $(4, -3)$. Determine the equation of the parabola in the form of Eq. (5.25) and the range of the parameter u.

5.7. What is the minimum number of points needed to completely specify an ellipse? In general, an ellipse need not be centered at the origin or have its semimajor and semiminor axes parallel to the coodinate axes. Substantiate your answer.

5.8. An ellipse with its semimajor and semiminor axes coincident with the x and y coordinate axes passes through the points $(3.670, 0.886)$ and $(1.701, 1.561)$. Determine the lengths of the semimajor and semiminor axes.

5.9. Determine the algebraic coefficients of a parametric quadratic curve which connects the points $(0, 1, 0)$, $(1, 2, 4)$, and $(2, 4, 5)$. Assume the three given points are parametrized at $u = 0, 1/2$, and 1, respectively.

5.10. Repeat Problem 5.9, but assume that the three points are parametrized at $u = 0$, $1/3$, and 1.

5.11. A parametric quadratic curve passes through the points $(0, 0)$, $(4, 3)$, and $(4, -3)$, which are parametrized at $u = 0, 1/2$, and 1, respectively. Determine the geometric slope dy/dx at each end.

5.12. A straight line connects the point $(1, 0, 2)$ to the point $(3, -1, 5)$. Write the equation for the parametric cubic representation of this line in (a) algebraic form (*Hint:* $\mathbf{a}_3 = \mathbf{a}_4 = 0$), and (b) geometric form. (c) What happens to the representation of the line if $\mathbf{a}_3 \neq \mathbf{a}_4 \neq 0$?

5.13. The points $(2, 3)$ and $(5, 0)$ are to be connected by a parametric cubic representation of a straight line. Demonstrate that the use of $\mathbf{P}'(0) = \mathbf{P}'(1) = (0, 0)$ produces nonuniform parametrization.

5.14. A parametric cubic curve connects the points $(0, 0)$, $(1, 3)$, $(2, 2)$, and $(3, -3)$. What conic section does this curve approximate?

5.15. Four equally spaced data $y = f(x)$ as tabulated are to be connected with a parametric cubic curve. Determine the geometric coefficient matrix, and determine the values of x and y predicted at $u = 0.5$.

x	y
0	4.0
1	2.0
2	1.0
3	0.0

5.16. A parametric cubic curve connects the points $(0, 1)$ and $(4, 1)$. The parametric slopes $\mathbf{P}'(0) = (t, 2)$ and $\mathbf{P}'(1) = (t, -2)$, where t is dx/du at the two end points. Determine the minimum value of t that will cause the curve to double back on itself.

Hint: By symmetry, the geometric slope becomes infinite at $u = 0.5$ at the critical value of t.

5.17. Determine the geometric coefficient matrix for a parametric cubic approximation of a parabola which has its end points at $(-2, 0)$ and $(5, 0)$ and geometric slopes at the end points which make angles of 60° and 120°, respectively, with the x axis.

5.18. Verify the derivation of Eq. (5.49) from Eq. (5.45).

5.19. The simple way to parametrize a circular arc shown in Example 5.5(a) may be generalized as

$$x = r\cos\phi u + x_0$$

$$y = r\sin\phi u + y_0$$

where $0 \le u \le 1$, r is the radius, and the arc begins at $x = r + x_0$, $y = y_0$ and is swept out counterclockwise through a total angle of ϕ. Form the geometric coefficient matrix by setting $\mathbf{P}'(0) = [x_u(0) \quad y_u(0)]$ and $\mathbf{P}'(1) = [x_u(1) \quad y_u(1)]$ where $x_u \equiv dx/du$, etc. Compare the resulting parametric cubic representation of a circular arc with the same data as used in Example 5.6.

5.20. A parametric cubic spine in the x-y plane is to connect the points

$$[\mathbf{P}] = \begin{bmatrix} 0 & 0 \\ 3 & -2 \\ 7 & -1 \\ 10 & 0 \end{bmatrix}$$

where the points are numbered 1, 2, 3, and 4 in order. The end fixity of the spline is specified by $\mathbf{P}'_1 = \mathbf{P}'_4 = [1 \quad 0]$. Determine the unknown parametric slopes $\mathbf{P}'_2$ and $\mathbf{P}'_3$. Is the spline used here the same as a parametric cubic curve which is defined by four points? Why or why not?

5.21. A plane cubic Bezier curve is described by the four control points

$$[\mathbf{P}] = \begin{bmatrix} 0 & 0 \\ 6 & 8 \\ 12 & 8 \\ 18 & 0 \end{bmatrix}$$

where the first and last points are the end points of the curve. Determine whether these points generate a parabola.

5.22 The geometric coefficient matrix for a parametric cubic curve is

$$\begin{bmatrix} 2 & 3 \\ 4 & 0 \\ 3 & 2 \\ 3 & -4 \end{bmatrix}$$

Determine the control points for a cubic Bezier curve which will most closely match the given curve.

5.23. Smooth the points given in Problem 5.20 with a nonperiodic quadratic B-spline curve.

5.24. The control points for a nonperiodic quadratic B-spline curve are (0, 0), (1, 2), (4, 2), (5, 3). Determine the value of y when (a) $x = 1$ and (b) $x = 3$.

5.25 Use the control points in Problem 5.24 to construct a periodic quadratic B-spline. For this curve, determine (a) the value of y when $x = 1$ and (b) the maximum value of y.

5.26. How would the straight line connecting the points (0, 0, 0) to (5, 2, 3) best be described in the form of a quadratic B-spline?

5.27. Show how a circle can be approximated by a closed, periodic quadratic B-spline with four control points. Check the accuracy of the approximation halfway between the knots. Repeat for six control points.

5.28. Demonstrate that the cubic B-spline, Eq. (5.70), has continuous first and second derivatives but discontinuous third derivatives at the knots.

PROJECTS

NOTE: A graphics tools library is recommended to keep the amount of programming down. In some cases, the projects can serve as a basis for user familiarization with existing graphics systems.

5.1. Develop a program which traces out the trajectory of a projectile in real time. Allow the user to input the initial velocity v_0 and initial angle θ_0, initial position (x_0, y_0), and maximum time t_{max}. Your program should determine the necessary window size so that the trajectory fills the screen. Generate the trajectory with Eqs. (5.5). The use of even increments of time makes a more realistic simulation.

5.2. Test the nonparametric explicit form of a circle

$$y = \sqrt{r^2 - x^2}$$

to generate circles centered at the origin of coordinates of varying degrees of accuracy. Use constant steps of x. Plot the circles large enough that the straight line segments are easily visible.

5.3. Develop an interactive program that fits a circle through any three points input by a user. It is preferable to use the locator for input. The amount of work may be reduced by using a library routine such as CIRCLE in Appendix B.

5.4. Modify the subroutine in Example 5.3 to draw circular arcs. The inputs to the subroutine should match those for the ARC routine in Appendix B. Write a main program which can supply data to test your subroutine. User input can be either by locator or by string.

5.5. Develop an interactive program that constructs fillet radii between two intersecting line segments. The first user input should be the fillet radius (string). The next input should be the three points defining the two lines (locator), where the intermediate point is the intersection between the two lines. The line segments should be trimmed (relimited) to blend with the fillet. What happens when the three input points lie on the same line?

5.6. Develop and test a subroutine that uses recursion to draw ellipses. A main program should handle input to include the coordinates of the center, the angle ϕ between the x axis and the semimajor axis, and the lengths a and b.

5.7. Develop and test a program which accepts the x-y-z coordinates of three points as user input and draws a smooth parametric quadratic curve between them. The program should accept subsequent user input of a viewpoint followed by redisplay of the curve.

5.8. Repeat Project 5.7, except that the program defines a parametric cubic curve defined by four points.

5.9. Develop a program that fits a 2-D parametric cubic spline through an arbitrary number of data. Use the relaxed end condition. Test your program on a data set as assigned.

5.10. Repeat Project 5.9 using a nonperiodic quadratic B-spline.

5.11. Write an interactive program that fits a cubic Bezier curve through any four points input by locator. Demonstrate blended Bezier curves with your program.

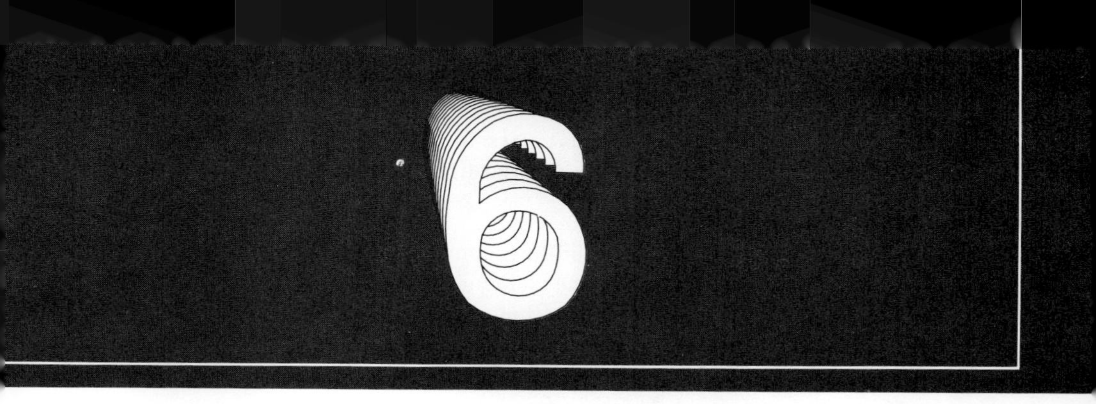

Surfaces and Solids

A curve can be generated by sweeping a point through space, a surface by sweeping a line through space, and a solid by sweeping a surface through space. Thus, the hierarchy exists, for example, that a cube can be generated by translating a point to form a straight line, translating the line to form a square, and translating the square to form the cube. Furthermore, depending on the trajectory, the lines, surfaces, and solids may have any desired curvature. The two trajectories of most practical use describe translation and rotation.

For spatial representation of surfaces and solids, it is convenient to use vector-valued parametric functions, just as has been done for space curves. A curve—for example, Eq. (5.33)—is given by the vector-valued parametric function

$$\mathbf{P}(u) = [x(u) \quad y(u) \quad z(u)] \tag{6.1}$$

where $\mathbf{P}$ is a vector to a point in x-y-z space and u is a parameter. A surface is described by a bivariate (two-parameter) vector-valued function,

$$\mathbf{P}(u, v) = [x(u, v) \quad y(u, v) \quad z(u, v)] \tag{6.2}$$

where u, v are the two parameters which sweep out the surface. It follows that a solid can be described with a trivariate vector-valued function,

$$\mathbf{P}(u, v, w) = [x(u, v, w) \quad y(u, v, w) \quad z(u, v, w)] \tag{6.3}$$

where u, v, w are the three parameters which sweep out the solid.

Among the advantages of using vector-valued parametric functions are:

Representations are independent of the coordinate system, which avoids such problems as infinite slopes.

Multivalued surfaces can be unambiguously represented.

The resulting structures are compatible with the three-dimensional transformations (including perspective) developed in Chapter 4.

Real-world surfaces and solids can be formed by joining together any number of functions with appropriate matching of coordinates and slopes at the boundaries.

The general approach of building complex solids by the assemblage of simpler elements is the basis of finite element analysis. Here, assemblies of plane elements such as triangles are used to model a plane region or a faceted shell, assemblies of bricks or wedges are used to model a solid, and so forth. The combination of simple surface or solid elements possesses the versatility required for objects of any degree of complexity. Intuitively, increasing fineness by using more but smaller elements improves the accuracy of the approximation.

In the developments in the sections which follow, the *synthesis* or construction of surfaces and solids is emphasized. Synthesis in this context is the process of determining a surface which fits certain specified points and meets other criteria such as having specified slopes at certain points.

6.1 BILINEAR SURFACES

Three points in space define a triangle, which of course is a planar figure. Linear interpolation between the three points produces points only within the triangle. Similarly, four points which lie in the same plane define a planar quadrilateral. Here, linear interpolation between the four points yields points within the plane region bounded by the quadrilateral.

Linear interpolation between four points which do not lie in the same plane describes a curved surface in space, a *bilinear surface* or a *bilinear patch*. The four points appear as vertices or corner points. Shown in Figure 6.1 is a bilinear surface which has ruling lines at quarter intervals to facilitate visibility. The four corners are given as the points matrix

$$[\mathbf{P}] = \begin{bmatrix} \mathbf{P}(0,0) \\ \mathbf{P}(1,0) \\ \mathbf{P}(0,1) \\ \mathbf{P}(1,1) \end{bmatrix} \tag{6.4}$$

where the parameters u, v of Eq. (6.2) have the range 0 to 1 for convenience in manipulation.

Linear interpolation between the four corner points gives

$$\begin{aligned} \mathbf{P}(u,v) = {} & (1-u)(1-v)\mathbf{P}(0,0) + u(1-v)\mathbf{P}(1,0) \\ & + (1-u)v\mathbf{P}(0,1) + uv\mathbf{P}(1,1) \end{aligned} \tag{6.5}$$

which may be written in matrix notation as

$$\mathbf{P}(u,v) = [(1-u)(1-v) \quad u(1-v) \quad (1-\mathrm{u})v \quad uv] \begin{bmatrix} \mathbf{P}(0,0) \\ \mathbf{P}(1,0) \\ \mathbf{P}(0,1) \\ \mathbf{P}(1,1) \end{bmatrix} \tag{6.6}$$

where $0 \le u \le 1$ and $0 \le v \le 1$.

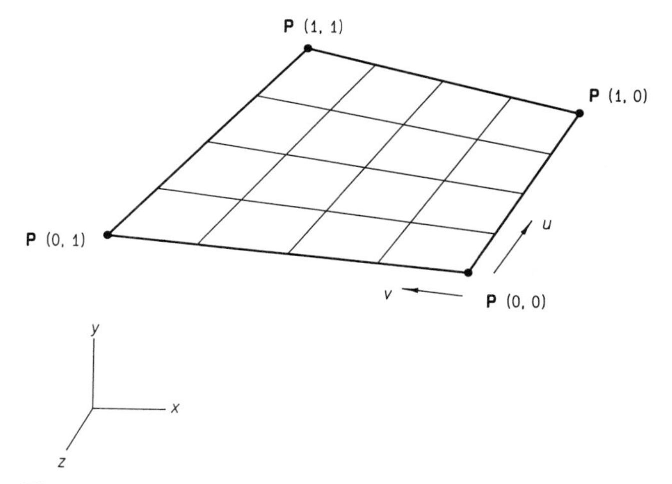

Figure 6.1 Bilinear patch.

The very simple bilinear surface does not provide the necessary flexibility to control slopes or to fit any more than the four corner points. This disadvantage can be partially overcome by the use of a large number of bilinear patches to describe a curved surface, as in the following example.

EXAMPLE 6.1

Demonstrate the use of four bilinear surface patches to approximate the surface of the conical frustum shown below. Discuss the fit at the midpoint of a patch. Repeat the process with an eight-patch approximation.

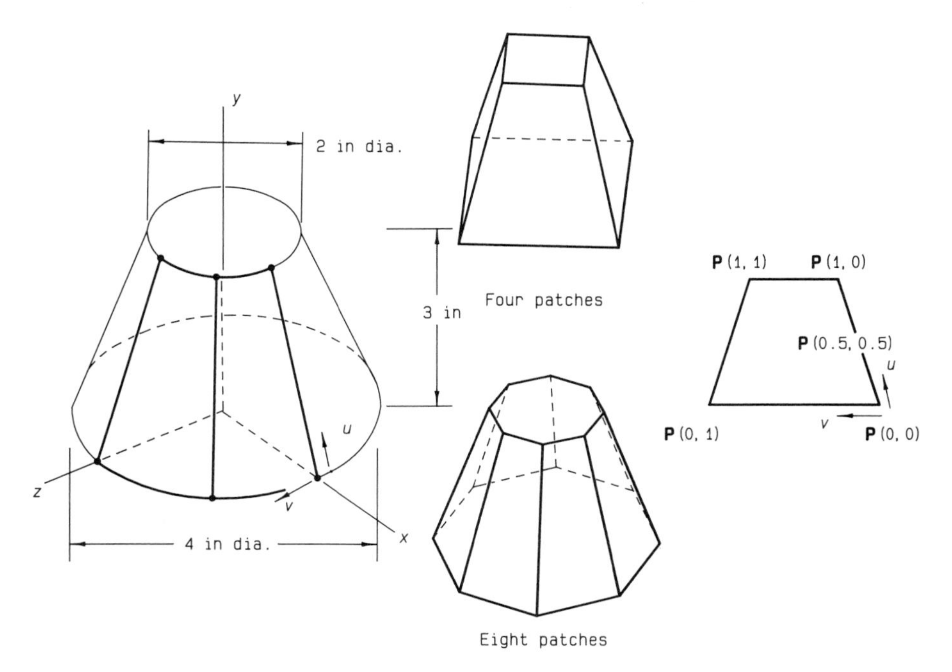

Solution. One typical patch of a four-patch approximation has corners

$$\mathbf{P}(1, 1) = (0, 3, 1) \qquad \mathbf{P}(1, 0) = (1, 3, 0)$$

$$\mathbf{P}(0, 1) = (0, 0, 2) \qquad \mathbf{P}(0, 0) = (2, 0, 0)$$

Hence by Eq. (6.6),

$$\mathbf{P}(u, v) = [(1 - u)(1 - v) \quad u(1 - v) \quad (1 - u)v \quad uv]\begin{bmatrix} 2 & 0 & 0 \\ 1 & 3 & 0 \\ 0 & 0 & 2 \\ 0 & 3 & 1 \end{bmatrix}$$

For $u = v = 0.5$

$$\mathbf{P}(0.5, 0.5) = [0.75 \quad 1.5 \quad 0.75]$$

which does not compare well with the actual coordinates (1.0607, 1.5, 1.0607) at the midpoint of the part of the conical surface represented by this patch. This problem is essentially due to the flatness of the bilinear patch.

When the patch is made half as large, the point just considered will be predicted exactly if it is on the edge of the new patch. The corner points of a typical bilinear patch for the eight-patch model are

$$\mathbf{P}(1, 1) = (0.707, 3, 0.707) \qquad \mathbf{P}(1, 0) = (1, 3, 0)$$

$$\mathbf{P}(0, 1) = (1.414, 0, 1.414) \qquad \mathbf{P}(0, 0) = (2, 0, 0)$$

which when substituted into Eq. (6.6) and evaluated at $u = v = 0.5$ gives

$$[\mathbf{P}(0.5 \quad 0.5)] = [1.280 \quad 1.500 \quad 0.530]$$

This compares more favorably than the four-patch approximation with the actual coordinates of (1.3858, 1.5, 0.5740) at the midpoint of the region represented by this patch.

It is clear that subdivision of the surface into even more bilinear surface patches will produce an even better approximation.

6.2 COONS PATCHES

It has been seen that the bilinear surface patch is formed by linear interpolation between four points in space. A logical extension, credited to S. Coons,[50] is the generation of a surface by linear interpolation between four arbitrary curves which bound the patch, Figure 6.2. Linear interpolation between the boundary curves $\mathbf{P}(0, v)$, $\mathbf{P}(u, 0)$, $\mathbf{P}(1, v)$, and $\mathbf{P}(u, 1)$ produces

$$\mathbf{f}(u, v) = (1 - v)\mathbf{P}(u, 0) + u\mathbf{P}(1, v) + v\mathbf{P}(u, 1) + (1 - u)\mathbf{P}(0, v) \qquad (6.7)$$

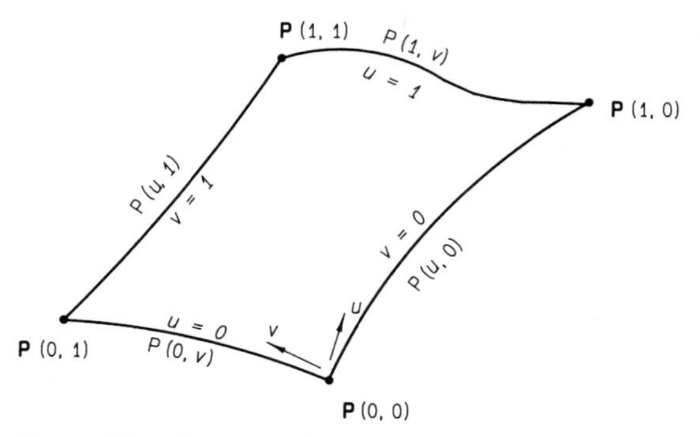

Figure 6.2 Coons patch.

which is *not* the desired result because this equation predicts the wrong values at the corners ($u, v = 0, 1$),

$$\mathbf{f}(0, 0) = \mathbf{P}(0, 0) + \mathbf{P}(0, 0) = 2\mathbf{P}(0, 0)$$

$$\mathbf{f}(1, 0) = 2\mathbf{P}(1, 0), \quad \text{etc.}$$

The Coons patch is formed by a modification of Eq. (6.7) where the corners are subtracted, or

$$\begin{aligned}\mathbf{P}(u, v) &= (1 - v)\mathbf{P}(u, 0) + u\mathbf{P}(1, v) + v\mathbf{P}(u, 1) \\ &\quad + (1 - u)\mathbf{P}(0, v) - (1 - u)(1 - v)\mathbf{P}(0, 0) \\ &\quad - u(1 - v)\mathbf{P}(1, 0) - (1 - u)v\mathbf{P}(0, 1) - uv\mathbf{P}(1, 1) \qquad (6.8)\end{aligned}$$

where $0 \le u \le 1$ and $0 \le v \le 1$. This gives the correct results on the boundaries as well as at the corners. For computational purposes, it is more convenient to write this equation as

$$\begin{aligned}\mathbf{P}(u, v) &= [(1 - v) \quad u \quad v \quad (1 - u)] \begin{bmatrix} \mathbf{P}(u, 0) \\ \mathbf{P}(1, v) \\ \mathbf{P}(u, 1) \\ \mathbf{P}(0, v) \end{bmatrix} \\ &\quad - [(1 - u)(1 - v) \quad u(1 - v) \quad (1 - u)v \quad uv] \begin{bmatrix} \mathbf{P}(0, 0) \\ \mathbf{P}(1, 0) \\ \mathbf{P}(0, 1) \\ \mathbf{P}(1, 1) \end{bmatrix} \\ &= [(1 - u) \quad u \quad 1] \begin{bmatrix} -\mathbf{P}(0, 0) & -\mathbf{P}(0, 1) & \mathbf{P}(0, v) \\ -\mathbf{P}(1, 0) & -\mathbf{P}(1, 1) & \mathbf{P}(1, v) \\ \mathbf{P}(u, 0) & \mathbf{P}(u, 1) & 0 \end{bmatrix} \begin{bmatrix} (1 - v) \\ v \\ 1 \end{bmatrix} \\ &= [\mathbf{f}(u)][\mathbf{B}_c(u, v)][\mathbf{f}(v)] \qquad (6.9)\end{aligned}$$

where $0 \leq u \leq 1$ and $0 \leq v \leq 1$. This general form of surface representation, which has a square vector-valued matrix that is pre- and postmultiplied by blending matrices, finds application for the curved surface patches in the sections that follow.

6.3 PARAMETRIC CUBIC PATCHES

A useful and versatile curved surface patch[36,42,50] results from having parametric cubic (PC) polynomials as the four boundary curves. It is recalled that the geometric form of the PC curve, as used in the cubic spline, permits specification of the location and the slope at the two end points of the line. Hence, with the control of slopes at the edges, PC patches have the possibility of being smoothly blended. To deal with the partial derivatives with respect to the parametric variables, the notation is introduced where

$$\frac{\partial \mathbf{P}(0, v)}{\partial u} = \mathbf{P}_u(0, v), \qquad \frac{\partial \mathbf{P}(u, 1)}{\partial v} = \mathbf{P}_v(u, 1), \quad \text{etc.} \tag{6.10}$$

A patch with four edges described by PC curves is shown in Figure 6.3. At the four corners, the partial derivatives at the end points of the lines correspond to the parametric slopes $\mathbf{P}'(0)$ and $\mathbf{P}'(1)$ of the geometric form of the PC curve, Eq. (5.40). Thus, 12 vector-valued terms—four end points and eight parametric slopes—describe the four curves that bound the patch. It turns out that a total of 16 vector-valued terms is required to completely describe the PC patch. This array, the geometric coefficient matrix, can be written as

$$[\mathbf{B}_p] = \begin{bmatrix} \mathbf{P}(0,0) & \mathbf{P}(0,1) & \mathbf{P}_v(0,0) & \mathbf{P}_v(0,1) \\ \mathbf{P}(1,0) & \mathbf{P}(1,1) & \mathbf{P}_v(1,0) & \mathbf{P}_v(1,1) \\ \mathbf{P}_u(0,0) & \mathbf{P}_u(0,1) & \mathbf{P}_{uv}(0,0) & \mathbf{P}_{uv}(0,1) \\ \mathbf{P}_u(1,0) & \mathbf{P}_u(1,1) & \mathbf{P}_{uv}(1,0) & \mathbf{P}_{uv}(1,1) \end{bmatrix} \tag{6.11}$$

where the elements containing second partial derivatives are the *twist vectors* at the four corners. Twist vectors control the internal shape of the surface, Figure 6.3, by affecting one of the two parametric slopes along the edges of the patch.

To illustrate how twist vectors work, consider the edge of the patch which is defined by curve 1 of Figure 6.3, where v = constant. One of the parametric slopes, $\mathbf{P}_u(u, 0)$, is readily obtained from the derivative of Eq. (5.40) with the geometric coefficients taken from the first column of Eq. (6.11),

$$\mathbf{P}_u(u,0) = [(-6u + 6u^2) \quad (6u - 6u^2) \quad (1 - 4u + 3u^2)(-2u + 3u^2)] \begin{bmatrix} \mathbf{P}(0,0) \\ \mathbf{P}(1,0) \\ \mathbf{P}_u(0,0) \\ \mathbf{P}_u(1,0) \end{bmatrix} \tag{6.12}$$

This partial derivative, illustrated in Figure 6.4 for an arbitrary point on curve 1, controls the shape of the edge itself. The other partial derivative of curve 1,

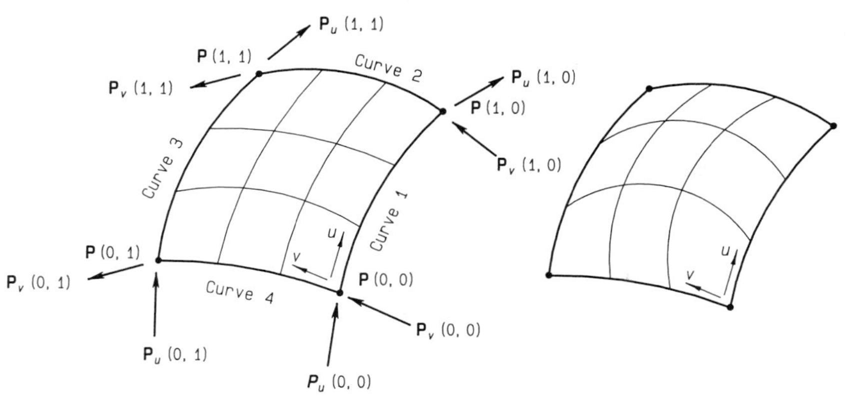

Figure 6.3 Parametric cubic patches with same edges but different twist vectors.

$\mathbf{P}_v(u, 0)$, is the one that controls the internal shape of the patch. To define the function $\mathbf{P}_v(u, 0)$, a new PC curve, the "auxiliary," is used. The end points of this are

$$\begin{aligned} \mathbf{P}_v(0, 0)_{\text{curve 1}} &= \mathbf{P}_v(0, 0)_{\text{curve 4}} \\ \mathbf{P}_v(1, 0)_{\text{curve 1}} &= \mathbf{P}_v(1, 0)_{\text{curve 2}} \end{aligned} \tag{6.13}$$

The partial derivatives at the end points of the auxiliary, which dictate the other two conditions needed to describe a PC curve, are

$$\begin{aligned} \frac{\partial}{\partial u} \mathbf{P}_v(0, 0) &= \mathbf{P}_{uv}(0, 0) \\ \frac{\partial}{\partial u} \mathbf{P}_v(1, 0) &= \mathbf{P}_{uv}(1, 0) \end{aligned} \tag{6.14}$$

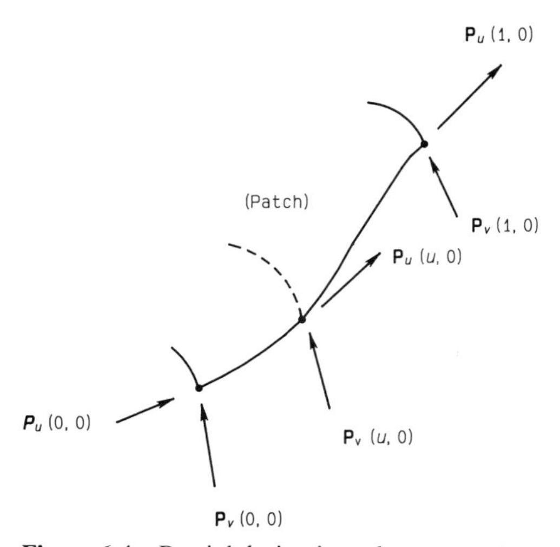

Figure 6.4 Partial derivatives along curve 1.

These two second partial derivatives are the twist vectors at the end points of curve 1. The four vector-valued functions, which represent the third column of Eq. (6.11), produce the PC curve in Figure 6.5.

The twist vectors are found in similar fashion for the other three edges. At the corners, twist vectors are shared. This means, for example, that at $\mathbf{P}(0, 0)$, the derivative of the curve 1 auxiliary must match that of the curve 4 auxiliary, or

$$\frac{\partial \mathbf{P}_u(0, 0)}{\partial v} = \frac{\partial \mathbf{P}_v(0, 0)}{\partial u}$$

Thus, the elements of the matrix of geometric coefficients $[\mathbf{B}_p]$ for a PC patch in Eq. (6.11) may be interpreted as

$$[\mathbf{B}_p] = \begin{array}{l} \\ \text{Curve 4} \\ \text{Curve 2} \\ \text{Curve 4 aux} \\ \text{Curve 2 aux} \end{array} \begin{array}{c} \begin{array}{cccc} \text{Curve 1} & \text{Curve 3} & \text{Curve 1 aux} & \text{Curve 3 aux} \end{array} \\ \begin{bmatrix} \mathbf{P}(0, 0) & \mathbf{P}(0, 1) & \mathbf{P}_v(0, 0) & \mathbf{P}_v(0, 1) \\ \mathbf{P}(1, 0) & \mathbf{P}(1, 1) & \mathbf{P}_v(1, 0) & \mathbf{P}_v(1, 1) \\ \mathbf{P}_u(0, 0) & \mathbf{P}_u(0, 1) & \mathbf{P}_{uv}(0, 0) & \mathbf{P}_{uv}(0, 1) \\ \mathbf{P}_u(1, 0) & \mathbf{P}_u(1, 1) & \mathbf{P}_{uv}(1, 0) & \mathbf{P}_{uv}(1, 1) \end{bmatrix} \end{array} \tag{6.15}$$

This matrix can be partitioned for visualization of the meaning of the 16 coefficients as

4 corner points	Tangent vectors with respect to v
Tangent vectors with respect to u	Twist vectors at 4 corners

An aid in the determination of twist vectors is the plot of the auxiliary, Figure 6.5. In general, as the magnitude of the twist vectors increases, the patch bulges out.

The simplified *F-patch*[22,50] has the four twist vectors set equal to zero. With the minor drawback of possible nonuniform parametrization, F-patches can

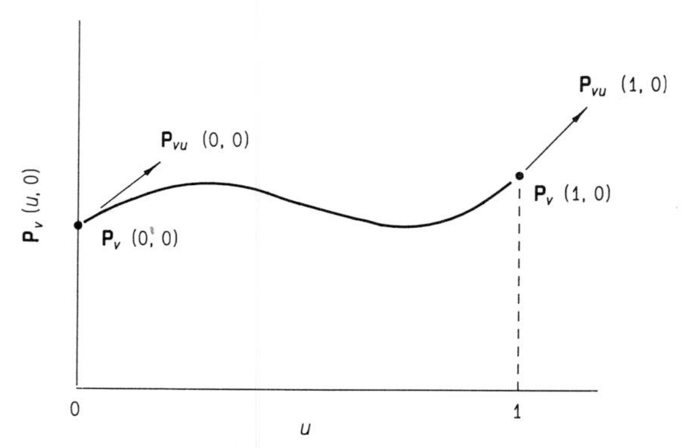

Figure 6.5 Plot of curve 1 auxiliary.

represent flat surfaces and curved surfaces swept out by straight line generators. However, for generally curved patches, nonzero twist vectors are required.

Everything is now in place to define $\mathbf{P}(u, v)$ for any u and v within the region defined for the patch, $0 \le u \le 1$ and $0 \le v \le 1$, as shown in Figure 6.6. The PC curve for u = constant which is manifolded (i.e., curved to fit on the surface) to the patch can be constructed with

$$\begin{bmatrix} \mathbf{P}(0, v) \\ \mathbf{P}(1, v) \\ \mathbf{P}_u(0, v) \\ \mathbf{P}_u(1, v) \end{bmatrix} = [\mathbf{B}_p] \begin{bmatrix} F_1(v) \\ F_2(v) \\ F_3(v) \\ F_4(v) \end{bmatrix} \tag{6.16}$$

where $[\mathbf{B}_p]$ is the square matrix in Eq. (6.15), and the elements of $[\mathbf{F}(v)]$ are, from Eq. (5.40),

$$\begin{aligned} F_1(v) &= 1 - 3v^2 + 2v^3 \\ F_2(v) &= 3v^2 - 2v^3 \\ F_3(v) &= v - 2v^2 + v^3 \\ F_4(v) &= -v^2 + v^3 \qquad (0 \le v \le 1) \end{aligned}$$

Next, combination of Eqs. (6.16) and (5.40) yields

$$\mathbf{P}(u, v) = [F_1(u) \quad F_2(u) \quad F_3(u) \quad F_4(u)] \quad [\mathbf{B}_p] \begin{bmatrix} F_1(v) \\ F_2(v) \\ F_3(v) \\ F_4(v) \end{bmatrix} \tag{6.17}$$

which may be shortened to

$$\mathbf{P}(u, v) = [\mathbf{F}(u)][\mathbf{B}_p][\mathbf{F}(v)]^T \tag{6.18}$$

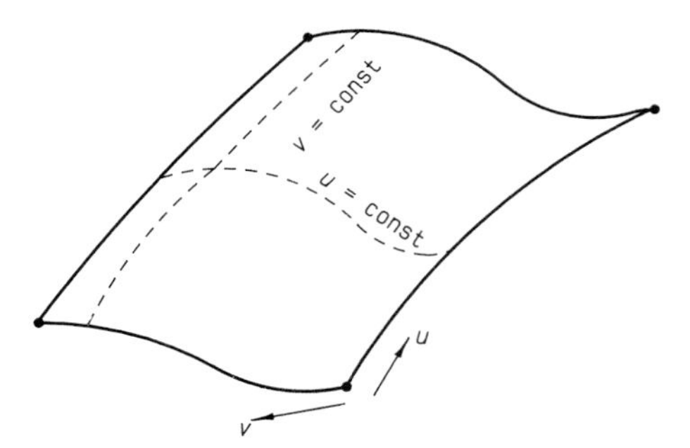

Figure 6.6 Definition of the point (u, v) in the PC patch. The dashed isoparametric curves (u = constant and v = constant) aid in visualization.

The partial derivatives are simply

$$\mathbf{P}_u(u, v) = [\mathbf{F}'(u)][\mathbf{B}_p][\mathbf{F}(v)]^T \tag{6.19a}$$

$$\mathbf{P}_v(u, v) = [\mathbf{F}(u)][\mathbf{B}_p][\mathbf{F}'(v)]^T \tag{6.19b}$$

where

$$[\mathbf{F}(u)] = [(1 - 3u^2 + 2u^3) \quad (3u^2 - 2u^3) \quad (u - 2u^2 + u^3) \quad (-u^2 + u^3)]$$

$$[\mathbf{F}'(u)] = [(-6u + 6u^2) \quad (6u - 6u^2) \quad (1 - 4u + 3u^2) \quad (-2u + 3u^2)] \qquad 0 \le u \le 1$$

and

$$[\mathbf{F}(v)]^T = [(1 - 3v^2 + 2v^3) \quad (3v^2 - 2v^3) \quad (v - 2v^2 + v^3) \quad (-v^2 + v^3)]^T$$

$$[\mathbf{F}'(v)]^T = [(-6v + 6v^2) \quad (6v - 6v^2) \quad (1 - 4v + 3v^2) \quad (-2v + 3v^2)]^T \qquad 0 \le v \le 1$$

Just as for PC curves, there are point and algebraic forms for the PC patch. For the point form, the parameters u, v take the arbitrary values 0, $\frac{1}{3}$, $\frac{2}{3}$, 1, so that the patch is defined by the 16 points

$$\begin{matrix} \mathbf{P}(0, 0) & \mathbf{P}(0, 1/3) & \mathbf{P}(0, 2/3) & \mathbf{P}(0, 1) \\ \mathbf{P}(1/3, 0) & \mathbf{P}(1/3, 1/3) & \mathbf{P}(1/3, 2/3) & \mathbf{P}(1/3, 1) \\ \mathbf{P}(2/3, 0) & \mathbf{P}(2/3, 1/3) & \mathbf{P}(2/3, 2/3) & \mathbf{P}(2/3, 1) \\ \mathbf{P}(1, 0) & \mathbf{P}(1, 1/3) & \mathbf{P}(1, 2/3) & \mathbf{P}(1, 1) \end{matrix} \tag{6.20}$$

The 16 points define four curves, in the same way that Eqs. (5.33) and (5.35) define one curve with four points. The modeling and viewing transformations of Chapter 4 can be applied to the point form to rotate or translate the patch.

The algebraic format is

$$\mathbf{P}(u, v) = \sum_{i=1}^{4} \sum_{j=1}^{4} \mathbf{a}_{ij} u^{i-1} v^{j-1}$$

$$= [1 \quad u \quad u^2 \quad u^3] \begin{bmatrix} \mathbf{a}_{11} & \mathbf{a}_{12} & \mathbf{a}_{13} & \mathbf{a}_{14} \\ \mathbf{a}_{21} & \mathbf{a}_{22} & \mathbf{a}_{23} & \mathbf{a}_{24} \\ \mathbf{a}_{31} & \mathbf{a}_{32} & \mathbf{a}_{33} & \mathbf{a}_{34} \\ \mathbf{a}_{41} & \mathbf{a}_{42} & \mathbf{a}_{43} & \mathbf{a}_{44} \end{bmatrix} \begin{bmatrix} 1 \\ v \\ v^2 \\ v^3 \end{bmatrix} \tag{6.21}$$

where $0 \le u \le 1$, $0 \le v \le 1$, and the 16 elements of $[\mathbf{A}]$ define the patch. One way to obtain these elements is by combination of Eqs. (5.35) and (6.20). Furthermore, a conversion between geometric form and algebraic form is possible by setting Eq. (6.18) equal to Eq. (6.21). Since it usually is the objective to blend the PC patches by matching slopes at their edges, the geometric form, Eq. (6.18), is the one most used.

EXAMPLE 6.2

A flat rectangular patch lies in the x-y plane with vertices located as shown. Determine the $[\mathbf{B}_p]$ matrix, and determine the world coordinates of a point at the parametric location $u = \frac{1}{3}$, $v = \frac{1}{3}$. Repeat, using an F-patch.

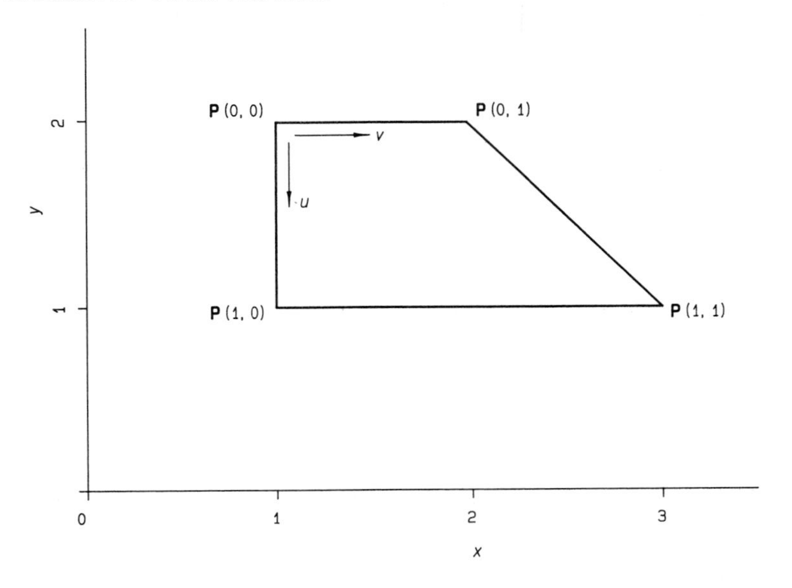

Solution. Because the patch is flat and in the x-y plane, all z coordinates are identically zero. The x-y coordinates of the corners as indicated in the figure are first entered into the proper locations of Eq. (6.15). Since the edges are straight, Eq. (5.44) is convenient for determining the slope vectors at the corners. Because the auxiliaries for the four edges are straight lines, each twist vector is simply the difference between the slope vectors at the corresponding corners. The resulting expression is

$$[\mathbf{B}_p] = \begin{bmatrix} (1, 2) & (2, 2) & (1, 0) & (1, 0) \\ (1, 1) & (3, 1) & (2, 0) & (2, 0) \\ (0, -1) & (1, -1) & (1, 0) & (1, 0) \\ (0, -1) & (1, -1) & (1, 0) & (1, 0) \end{bmatrix}$$

Use of Eq. (6.17) yields

$$x\left(\frac{1}{3}, \frac{1}{3}\right) = \frac{1}{27^2}\,[20 \quad 7 \quad 4 \quad -2] \begin{bmatrix} 1 & 2 & 1 & 1 \\ 1 & 3 & 2 & 2 \\ 0 & 1 & 1 & 1 \\ 0 & 1 & 1 & 1 \end{bmatrix} \begin{bmatrix} 20 \\ 7 \\ 4 \\ -2 \end{bmatrix}$$

$$= \frac{13}{9} = 1.444$$

and

$$y\left(\frac{1}{3}, \frac{1}{3}\right) = \frac{1}{27^2}[20 \quad 7 \quad 4 \quad -2]\begin{bmatrix} 2 & 2 & 0 & 0 \\ 1 & 1 & 0 & 0 \\ -1 & -1 & 0 & 0 \\ -1 & -1 & 0 & 0 \end{bmatrix}\begin{bmatrix} 20 \\ 7 \\ 4 \\ -2 \end{bmatrix}$$

$$= \frac{5}{3} = 1.667$$

These values agree exactly with that predicted by linear interpolation, Eq. (6.6).

The F-patch assumption changes only the value of x. The new value of x is computed from

$$x\left(\frac{1}{3}, \frac{1}{3}\right) = \frac{1}{27^2}[20 \quad 7 \quad 4 \quad -2]\begin{bmatrix} 1 & 2 & 1 & 1 \\ 1 & 3 & 2 & 2 \\ 0 & 1 & 0 & 0 \\ 0 & 1 & 0 & 0 \end{bmatrix}\begin{bmatrix} 20 \\ 7 \\ 4 \\ -2 \end{bmatrix}$$

$$= \frac{1049}{27^2} = 1.439$$

Thus, for the F-patch, it is shown that the parametrization is not linear. Ordinarily, this is not a problem.

EXAMPLE 6.3

A parametric cubic patch is to be manifolded onto the surface of a sphere with a radius of 10 m. The edges of the patch are to be on the equator, a line of 0° longitude, a line of 45° north lattitude, and a line of 30° longitude.

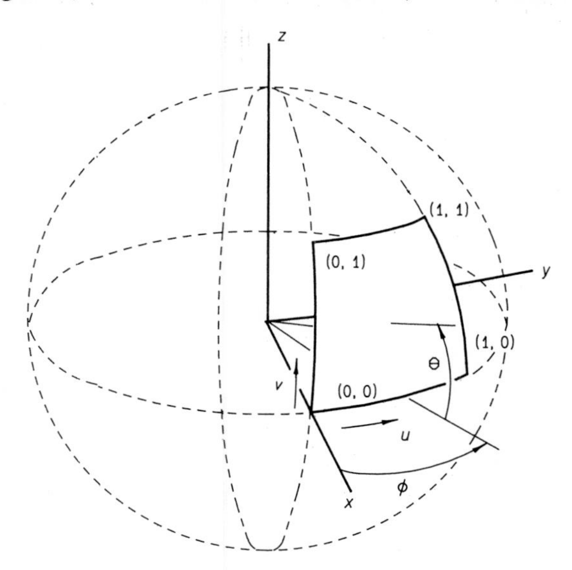

Determine the geometric coefficients. Ascertain the accuracy of the approximation at the parametric midpoints.

Solution. The origin of the coordinates is taken at the center of the sphere and the patch as shown. The Cartesian, spherical, and local (u, v) coordinates are related by

$$x = r \cos \phi \cos \theta = r \cos\left(\frac{\pi}{6} u\right) \cos\left(\frac{\pi}{4} v\right)$$

$$y = r \sin \phi \cos \theta = r \sin\left(\frac{\pi}{6} u\right) \cos\left(\frac{\pi}{4} v\right)$$

$$z = r \sin \theta = r \sin\left(\frac{\pi}{4} v\right)$$

where $r = 10$ m. Note the ranges $0 \le u \le 1$ and $0 \le v \le 1$ sweep out the desired region. The corners of the patch (Cartesian coordinates) are located at

$$\mathbf{P}(0,1) = [7.071 \quad 0 \quad 7.071] \qquad \mathbf{P}(1,1) = [6.124 \quad 3.536 \quad 7.071]$$

$$\mathbf{P}(0,0) = [10 \quad 0 \quad 0] \qquad \mathbf{P}(1,0) = [8.66 \quad 5 \quad 0]$$

The Cartesian components of the parametric slopes are found by differentiation,

$$x_u = -\frac{\pi}{6} r \sin\left(\frac{\pi}{6} u\right) \cos\left(\frac{\pi}{4} v\right)$$

$$x_v = -\frac{\pi}{4} r \cos\left(\frac{\pi}{6} u\right) \sin\left(\frac{\pi}{4} v\right) \qquad x_{uv} = \frac{\pi^2}{24} r \sin\left(\frac{\pi}{6} u\right) \sin\left(\frac{\pi}{4} v\right)$$

$$y_u = \frac{\pi}{6} r \cos\left(\frac{\pi}{6} u\right) \cos\left(\frac{\pi}{4} v\right)$$

$$y_v = -\frac{\pi}{4} r \sin\left(\frac{\pi}{6} u\right) \sin\left(\frac{\pi}{4} v\right) \qquad y_{uv} = -\frac{\pi^2}{24} r \cos\left(\frac{\pi}{6} u\right) \sin\left(\frac{\pi}{4} v\right)$$

$$z_u = 0$$

$$z_v = \frac{\pi}{4} r \cos\left(\frac{\pi}{4} v\right) \qquad z_{uv} = 0$$

Although x_u, x_v, y_u, and so forth could be determined more accurately by differenting elements of Eq. (5.47b), the method above simplifies determination of the twist vectors. Substitution of u, v at 0, 1 yields numerical values for the slope and twist vectors. Substitution into Eq. (6.15) gives

$$[\mathbf{B}_x] = \begin{bmatrix} 10 & 7.071 & 0 & -5.554 \\ 8.660 & 6.124 & 0 & -4.810 \\ 0 & 0 & 0 & 0 \\ -2.618 & -1.851 & 0 & 1.454 \end{bmatrix}$$

$$[\mathbf{B}_y] = \begin{bmatrix} 0 & 0 & 0 & 0 \\ 5 & 3.536 & 0 & -2.777 \\ 5.236 & 3.702 & 0 & -2.908 \\ 4.534 & 3.206 & 0 & -2.518 \end{bmatrix}$$

$$[\mathbf{B}_z] = \begin{bmatrix} 0 & 7.071 & 7.854 & 5.554 \\ 0 & 7.071 & 7.854 & 5.554 \\ 0 & 0 & 0 & 0 \\ 0 & 0 & 0 & 0 \end{bmatrix}$$

where $[\mathbf{B}_x]$, $[\mathbf{B}_y]$, and $[\mathbf{B}_z]$ are components of the vector-valued matrix of geometric coefficients $[\mathbf{B}_p]$.

At the parametric midpoints $u = 0.5$, $v = 0.5$, the point on the PC patch is given by Eq. (6.18) as

$$\begin{aligned} \mathbf{P}(1/2, 1/2) &= [\mathbf{F}(1/2)][\mathbf{B}_p][\mathbf{F}(1/2)] \\ &= [1/2 \quad 1/2 \quad 1/8 \quad -1/8][\mathbf{B}_p][1/2 \quad 1/2 \quad 1/8 \quad -1/8]^T \\ &= [8.914 \quad 2.389 \quad 3.823] \end{aligned}$$

The radius at this point is $\sqrt{8.914^2 + 2.389^2 + 3.823^2} = 9.989$ m, which is 1.1% less than the actual value of 10 m.

6.4 BEZIER SURFACES

Just as parametric cubic curves are extended to parametric cubic patches, Bezier curves may be extended to Bezier surface[36,39,50] patches. While both kinds of patches pass through the four corner points, all the other vector-valued parameters are control points with the Bezier patch. Using the placement of these points to specify edge slopes is more intuitive than determinating the parametric slopes and twist vectors for the PC patch. As a result, the Bezier patch is easier to use, because the control points themselves approximate the location of the desired surface.

Starting with the cubic Bezier curve, Eq. (5.61), an expansion similar to Eq. (6.18) yields

$$\mathbf{P}(u, v) = [(1-u)^3 \quad 3u(1-u)^2 \quad 3u^2(1-u) \quad u^3]\,[\mathbf{P}_B]\begin{bmatrix}(1-v)^3 \\ 3v(1-v)^2 \\ 3v^2(1-v) \\ v^3\end{bmatrix} \tag{6.22}$$

where $0 \leq u \leq 1$, $0 \leq v \leq 1$, and the 16 Bezier control points are given by the following:

$$[\mathbf{P}_B] = \begin{bmatrix}\mathbf{P}_{11} & \mathbf{P}_{12} & \mathbf{P}_{13} & \mathbf{P}_{14} \\ \mathbf{P}_{21} & \mathbf{P}_{22} & \mathbf{P}_{23} & \mathbf{P}_{24} \\ \mathbf{P}_{31} & \mathbf{P}_{32} & \mathbf{P}_{33} & \mathbf{P}_{34} \\ \mathbf{P}_{41} & \mathbf{P}_{42} & \mathbf{P}_{43} & \mathbf{P}_{44}\end{bmatrix}$$

A Bezier surface patch is shown in Figure 6.7. Note that only the four corner points—$\mathbf{P}_{11}$, $\mathbf{P}_{14}$, $\mathbf{P}_{41}$, and $\mathbf{P}_{44}$—are on the surface itself. The other points of the matrix of control points, or "net," serve the same purposes as the remaining terms of the PC patch. For example, points $\mathbf{P}_{21}$ and $\mathbf{P}_{11}$ define the initial slope of the curved edged $v = 0$. Thus, the eight control points on the edges (but not on the corners) serve the same purpose as the parametric slope vectors of the PC patch. The four interior control points—$\mathbf{P}_{22}$, $\mathbf{P}_{32}$, $\mathbf{P}_{23}$, and $\mathbf{P}_{33}$—function like the twist vectors in the PC patch.

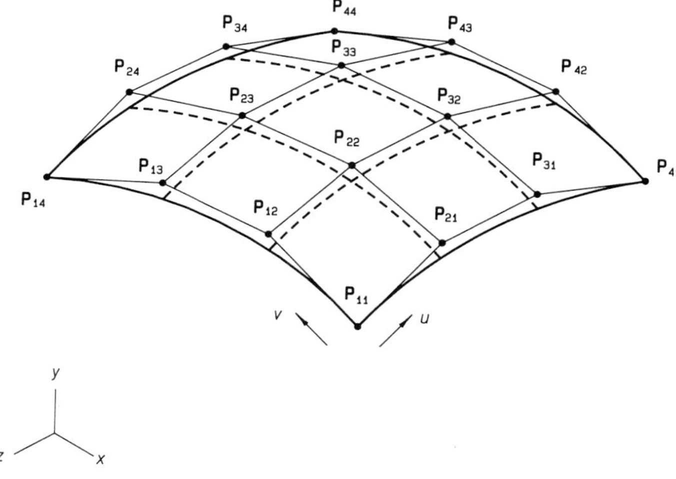

Figure 6.7 Net of control points for a cubic Bezier patch.

Bezier surfaces can also be generated with higher-order functions. Extension of the quartic Bezier curve of Eq. (5.62) leads to a Bezier patch with 25 control points,

$$\mathbf{P}(u, v) = [(1-u)^4 \quad 4u(1-u)^3 \quad 6u^2(1-u)^2 \quad 4u^3(1-u) \quad u^4]\,[\mathbf{P}_B]\begin{bmatrix}(1-v)^4\\ 4v(1-v)^3\\ 6v^2(1-v)^2\\ 4v^3(1-v)\\ v^4\end{bmatrix} \tag{6.23}$$

where $0 \le u \le 1$, $0 \le v \le 1$, and the 25 Bezier control points are given by

$$[\mathbf{P}_B] = \begin{bmatrix}\mathbf{P}_{11} & \mathbf{P}_{12} & \mathbf{P}_{13} & \mathbf{P}_{14} & \mathbf{P}_{15}\\ \mathbf{P}_{21} & \mathbf{P}_{22} & \mathbf{P}_{23} & \mathbf{P}_{24} & \mathbf{P}_{25}\\ \mathbf{P}_{31} & \mathbf{P}_{32} & \mathbf{P}_{33} & \mathbf{P}_{34} & \mathbf{P}_{35}\\ \mathbf{P}_{41} & \mathbf{P}_{42} & \mathbf{P}_{43} & \mathbf{P}_{44} & \mathbf{P}_{45}\\ \mathbf{P}_{51} & \mathbf{P}_{52} & \mathbf{P}_{53} & \mathbf{P}_{54} & \mathbf{P}_{55}\end{bmatrix}$$

In the 5×5 array of Bezier control points, the middle control point $\mathbf{P}_{33}$ can be adjusted without affecting the slopes around the edge of the patch. This may be an advantage when two or more patches share edges.

Bezier surfaces are not limited to square arrays of control points. For example, a surface patch which is cubic in one direction and quartic in the other is

$$\mathbf{P}(u, v) = [(1-u)^3 \quad 3u(1-u)^2 \quad 3u^2(1-u) \quad u^3]\,[\mathbf{P}_B]\begin{bmatrix}(1-v)^4\\ 4v(1-v)^3\\ 6v^2(1-v)^2\\ 4v^3(1-v)\\ v^4\end{bmatrix} \tag{6.24}$$

where $0 \le u \le 1$, $0 \le v \le 1$, and the 20 Bezier control points are given by

$$[\mathbf{P}_B] = \begin{bmatrix}\mathbf{P}_{11} & \mathbf{P}_{12} & \mathbf{P}_{13} & \mathbf{P}_{14} & \mathbf{P}_{15}\\ \mathbf{P}_{21} & \mathbf{P}_{22} & \mathbf{P}_{23} & \mathbf{P}_{24} & \mathbf{P}_{25}\\ \mathbf{P}_{31} & \mathbf{P}_{32} & \mathbf{P}_{33} & \mathbf{P}_{34} & \mathbf{P}_{35}\\ \mathbf{P}_{41} & \mathbf{P}_{42} & \mathbf{P}_{43} & \mathbf{P}_{44} & \mathbf{P}_{45}\end{bmatrix}$$

Blending Bezier patches with slope continuity requires that (1) control points on the common edges be shared and (2) three control points—one on the

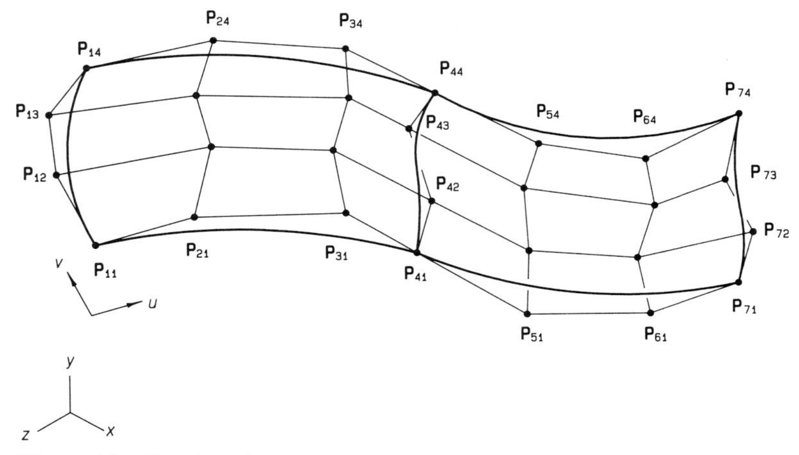

Figure 6.8 Two blended Bezier patches. Control points $\mathbf{P}_{41}$, $\mathbf{P}_{42}$, $\mathbf{P}_{43}$, and $\mathbf{P}_{44}$ are shared by both patches. Slope continuity between the two patches is maintained by having each group of three control points which cross the shared edge ($\mathbf{P}_{31}$, $\mathbf{P}_{41}$, $\mathbf{P}_{51}$ etc.) lie on straight lines.

edge and ones on either side of the edge—form a straight line. Two blended cubic Bezier patches are shown in Figure 6.8. The next order of continuity requires that five control points be aligned, which is best managed with higher-order Bezier surfaces.

For design work, the Bezier surface patch is easy to use. In blending Bezier patches, the following should be observed:

Control points must be used in the correct multiples. For example, an $n \times m$ array of cubic Bezier patches requires a $(4 + 3(n - 1))$ by $(4 + 3(m - 1))$ array of control points.

Proper placement of control points is necessary to maintain slope continuity across patch boundaries.

The effect of moving a single control point depends on what kind of control point it is.

The advantages of the intuitive use of Bezier surfaces without these limitations are available in B-spline surfaces, described in the next section.

EXAMPLE 6.4

A vaulted roof structure with four edges formed by circular arcs has the dimensions shown. The longitudinal line along the top of the structure is straight. Describe a smooth surface with two blended cubic Bezier patches.

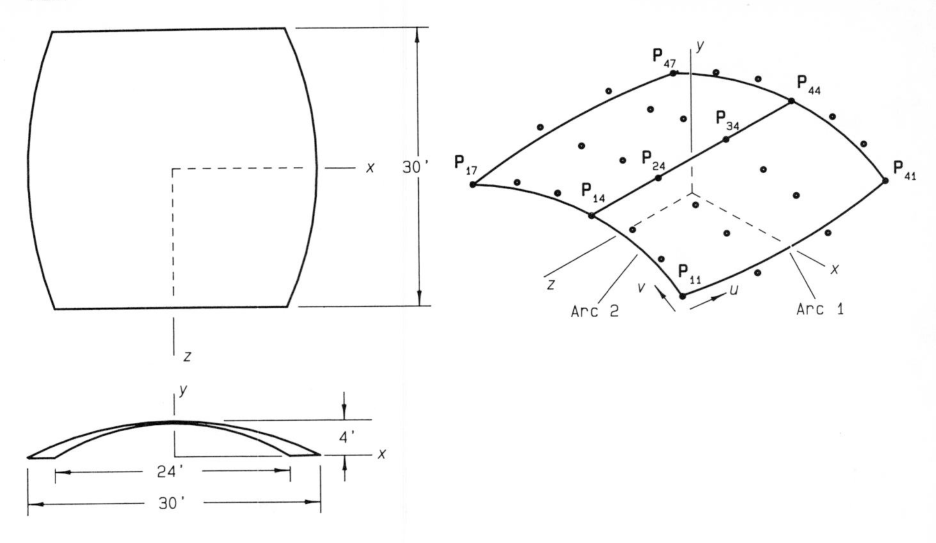

Solution. Shown above on the right is an isometric projection of the structure with the 28 Bezier control points. The axes have been placed so that the eight control points in the first octant can be used to generate the rest by mirror operations. The two corner control points of known location in this octant are $\mathbf{P}_{11} = [12 \quad 0 \quad 15]$ and $\mathbf{P}_{14} = [0 \quad 4 \quad 15]$. The top edge, where the two patches are blended, has four equally spaced control points, making $\mathbf{P}_{24} = [0 \quad 4 \quad 5]$.

The control points on the patch edges will be determined to approximate the circular arcs. The bottom edge, which is in the x-z plane, is labeled arc 1; the front edge, which is parallel to the x-y plane, is labeled arc 2. The radius r_1 is given by the right triangle relationship,

$$r_1^2 = (r_1 - 3)^2 + 15^2 \qquad r_1 = 39 \text{ ft}$$

and the angle

$$\phi_1 = \sin^{-1}(15/59) = 22.620° = 0.39479 \text{ rad}$$

The radius r_2 is

$$r_2^2 = (r_2 - 4)^2 + 12^2 \qquad r_2 = 20 \text{ ft}$$

and the angle

$$\phi_2 = \sin^{-1}(12/20) = 36.870° = 0.64350 \text{ rad}$$

The location of the control points on the two circular edges is aided by parametrizing the arcs in u or v, where $0 \le u \le 1$ and $0 \le v \le 1$. In the first octant, there is only half a patch, so the range is modified to $0 \le u \le \frac{1}{2}$.

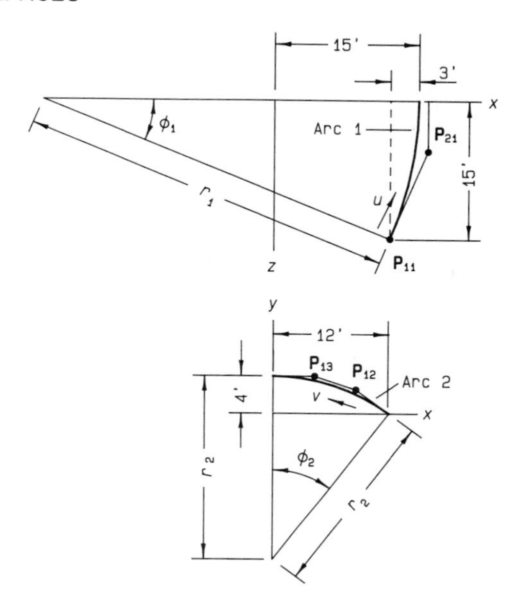

With the aid of the figure above, the parametric representations and corresponding parametric derivatives are

Arc 1:

$$x = 39 \cos (\phi_1(2u - 1)) - 24 \qquad x_u = -78\phi_1 \sin (\phi_1(2u - 1))$$

$$y = 0 \qquad y_u = 0$$

$$z = 39 \sin (\phi_1(2u - 1)) \qquad z_u = 78\phi_1 \cos (\phi_1(2u - 1))$$

Arc 2:

$$x = 20 \sin (\phi_2(1 - v)) \qquad x_v = -20\phi_2 \cos (\phi_2(1 - v))$$

$$y = 20 \cos (\phi_2(1 - v)) - 16 \qquad y_v = 20\phi_2 \sin (\phi_2(1 - v))$$

$$z = 0 \qquad z_v = 0$$

From Eq. (5.65) the control points to produce these circular arcs are

$$\mathbf{P}_u(0, 0) = 3(\mathbf{P}_{21} - \mathbf{P}_{11})$$

or

$$\mathbf{P}_{21} = [12 \quad 0 \quad 15] + \frac{[x_u(0) \quad y_u(0) \quad z_u(0)]}{3}$$

$$= [12 \quad 0 \quad 15] + \frac{[11.844 \quad 0 \quad -28.425]}{3} = [15.948 \quad 0 \quad 5.525]$$

$$\mathbf{P}_v(0, 0) = 3(\mathbf{P}_{12} - \mathbf{P}_{11})$$

or

$$\mathbf{P}_{12} = [12 \quad 0 \quad 15] + \frac{[x_v(0) \quad y_v(0) \quad z_v(0)]}{3}$$

$$= [12 \quad 0 \quad 15] + \frac{[-10.296 \quad 7.722 \quad 0]}{3} = [8.568 \quad 2.574 \quad 15]$$

$$\mathbf{P}_v(0, 1) = 3(\mathbf{P}_{14} - \mathbf{P}_{13})$$

or

$$\mathbf{P}_{13} = [0 \quad 4 \quad 15] - \frac{[x_v(1) \quad y_v(1) \quad z_v(1)]}{3}$$

$$= [0 \quad 4 \quad 15] - \frac{[-12.870 \quad 0 \quad 0]}{3} = [4.290 \quad 4 \quad 15]$$

The control point $\mathbf{P}_{23}$ is selected to line up with control point $\mathbf{P}_{24}$ in a way that ensures slope continuity at the top and with control point $\mathbf{P}_{13}$ to have uniform curvature along the top. This is done by giving $\mathbf{P}_{13}$ a translation of -10 ft in the z direction, or

$$\mathbf{P}_{23} = [4.290 \quad 4 \quad 5]$$

The remaining control point in the first octant can be placed somewhat arbitrarily to give a good shape. The curvature along the bottom edge will be approximately the same if control point $\mathbf{P}_{22}$ is the same x and y distances from $\mathbf{P}_{21}$ as $\mathbf{P}_{12}$ is from $\mathbf{P}_{11}$. The z for $\mathbf{P}_{22}$ (z_{22}) is arbitrarily selected as halfway between z_{21} and z_{23}. Thus,

$$\mathbf{P}_{22} = [12.516 \quad 2.574 \quad 5.263]$$

Now the matrices of control points can be completed with the use of mirror transformations. For patch 1, the eight points found in the first octant are mirrored across the x-y plane to complete the matrix,

$$[\mathbf{P}_{B1}] = \begin{bmatrix} \mathbf{P}_{11} & \mathbf{P}_{12} & \mathbf{P}_{13} & \mathbf{P}_{14} \\ \mathbf{P}_{21} & \mathbf{P}_{22} & \mathbf{P}_{23} & \mathbf{P}_{24} \\ \mathbf{P}_{31} & \mathbf{P}_{32} & \mathbf{P}_{33} & \mathbf{P}_{34} \\ \mathbf{P}_{41} & \mathbf{P}_{42} & \mathbf{P}_{43} & \mathbf{P}_{44} \end{bmatrix}$$

$$[\mathbf{P}_{B1}] = \begin{bmatrix} (12, 0, 15) & (8.568, 2.574, 15) & (4.290, 4, 15) & (0, 4, 15) \\ (15.948, 0, 5.525) & (12.516, 2.574, 5.263) & (4.290, 4, 5) & (0, 4, 5) \\ (15.948, 0, -5.525) & (12.516, 2.574, -5.263) & (4.290, 4, -5) & (0, 4, -5) \\ (12, 0, -15) & (8.568, 2.574, -15) & (4.290, 4, -15) & (0, 4, -15) \end{bmatrix}$$

Patch 2 is a mirror of patch 1, where the control points are renumbered as shown, and the x's are all negative,

$$[\mathbf{P}_{B2}] = \begin{bmatrix} \mathbf{P}_{14} & \mathbf{P}_{15} & \mathbf{P}_{16} & \mathbf{P}_{17} \\ \mathbf{P}_{24} & \mathbf{P}_{25} & \mathbf{P}_{26} & \mathbf{P}_{27} \\ \mathbf{P}_{34} & \mathbf{P}_{35} & \mathbf{P}_{36} & \mathbf{P}_{37} \\ \mathbf{P}_{44} & \mathbf{P}_{45} & \mathbf{P}_{46} & \mathbf{P}_{47} \end{bmatrix}$$

$$[\mathbf{P}_{B2}] = \begin{bmatrix} (0, 4, 15) & (-4.290, 4, 15) & (-8.568, 2.574, 15) & (-12, 0, 15) \\ (0, 4, 5) & (-4.290, 4, 5) & (-12.516, 2.574, 5.263) & (-15.948, 0, 5.525) \\ (0, 4, -5) & (-4.290, 4, -5) & (-12.516, 2.574, -5.263) & (-15.948, 0, -5.525) \\ (0, 4, -15) & (-4.290, 4, -15) & (-8.568, 2.574, -15) & (-12, 0, -15) \end{bmatrix}$$

6.5 B-SPLINE SURFACES

Surfaces can be generated from B-spline blending functions in a manner analogous to that of generating surfaces from functions associated with parametric cubic and Bezier curves.[39] B-spline surfaces are defined by an array of control points which define patches parametrized in u and v, where $0 \le u \le 1$ and $0 \le v \le 1$ for each individual patch. B-spline surfaces are classified as *approximating* functions, since the control points do not necessarily lie on the surface.

Quadratic, cubic, or higher-order blending functions can be used for B-spline surfaces; the blending function associated with u can be of different order from that associated with v. Attention here will be directed to the simple, but useful, quadratic B-spline surfaces which have only first-derivative continuity at the patch boundaries. Cubic B-spline surfaces, which also have second-derivative continuity at the patch boundaries, may be the choice for applications requiring greater apparent smoothness.

In the same manner as B-spline curves, B-spline surfaces are classified as *periodic* or *nonperiodic*. Periodic blending functions constitute the central regions of both classifications and are the basis for closed surfaces. Nonperiodic blending functions are chosen when the edges of a surface are to be defined exactly by control points. A tubular surface—that is, a surface closed in one direction and open in the other—can be developed with periodic blending in one parametric direction and nonperiodic blending in the other.

The periodic blending functions for quadratic B-spline surfaces have the same form as in Eq. (5.66). Extension to surfaces, analogous to the techniques of Eq. (6.17) or (6.22), is in terms of the u and v parametric directions. For central patches,

$$\mathbf{P}_{ij}(u, v) = [\mathbf{N}(u)][\mathbf{P}_2][\mathbf{N}(v)]^T \tag{6.25}$$

where

$$[\mathbf{N}(u)] = \left[\frac{(1-u)^2}{2} \quad \frac{-2u^2+2u+1}{2} \quad \frac{u^2}{2}\right] \qquad 0 \le u \le 1$$

$[\mathbf{N}(v)]$ is similarly defined with $0 \le v \le 1$, and

$$[\mathbf{P}_2] = \begin{bmatrix} \mathbf{P}_{i,j} & \mathbf{P}_{i,j+1} & \mathbf{P}_{i,j+2} \\ \mathbf{P}_{i+1,j} & \mathbf{P}_{i+1,j+1} & \mathbf{P}_{i+1,j+2} \\ \mathbf{P}_{i+2,j} & \mathbf{P}_{i+2,j+1} & \mathbf{P}_{i+2,j+2} \end{bmatrix}$$

the set of nine control points for the ijth B-spline patch, shown in Figure 6.9.

In the case of nonperiodic B-spline surfaces, special blending functions are used for edge patches. Quadratic surfaces require a total of eight specialized forms of edge patches—four corners and four sides. The layout and notation for the various special patches are shown in Figure 6.10.

The specialized quadratic blending functions for edge and corner patches are the same as those found in Eqs. (5.67) and (5.68). For leading edges, that is, edges of the surface where $u = 0$ or $v = 0$, the form is

$$[\mathbf{N}_1(u)] = \left[(1-u)^2 \quad \frac{-3u^2+4u}{2} \quad \frac{u^2}{2}\right] \tag{6.26}$$

where $0 \le u \le 1$ (or $0 \le v \le 1$).

For edges of the surface where $u = 1$ or $v = 1$, the notation n denotes the trailing edge in the u parametric direction, and m denotes the trailing edge in the v parametric direction. The blending functions in both directions have the form

$$[\mathbf{N}_n(u)] = \left[\frac{(1-u)^2}{2} \quad \frac{-3u^2+2u+1}{2} \quad u^2\right] \tag{6.27}$$

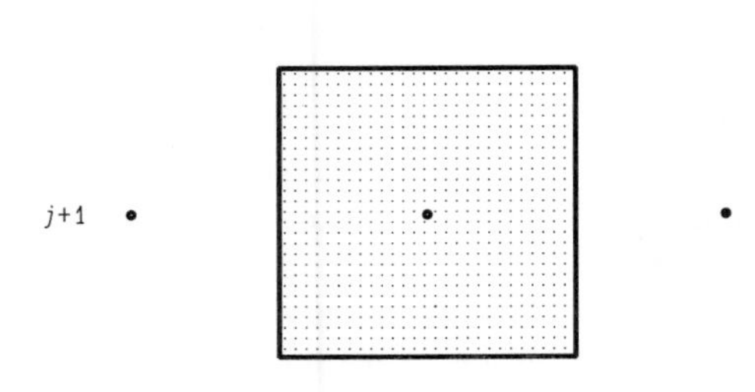

Figure 6.9 Central patch described by nine control points for quadratic B-spline surface.

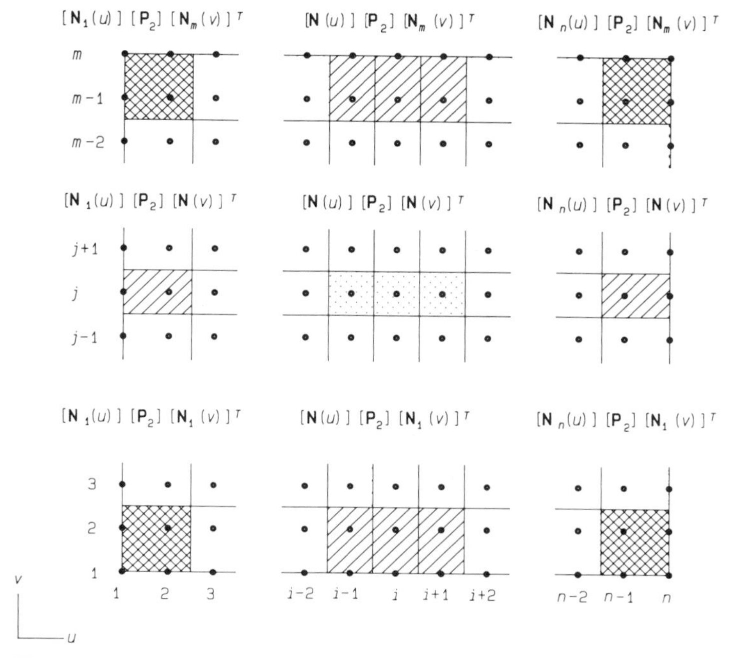

Figure 6.10 Arrangement of quadratic B-spline patches to form a nonperiodic B-spline surface.

where $0 \leq u \leq 1$ (or $0 \leq v \leq 1$, with the blending function designated as $[\mathbf{N}_m(v)]$).

The foregoing scheme for combination of patches to form continuous surfaces can be extended to the use of cubic B-spline functions. Here, a total of 24 forms of special patches are needed in addition to central patches. From Eq. (5.70), the central patch is

$$\mathbf{P}_{ij}(u, v) = [\mathbf{N}(u)][\mathbf{P}_3][\mathbf{N}(v)] \tag{6.28}$$

where

$$[\mathbf{N}(u)] = \left[\frac{(1-u)^3}{6} \quad \frac{3u^3 - 6u^2 + 4}{6} \quad \frac{-3u^3 + 3u^2 + 3u + 1}{6} \quad \frac{u^3}{6}\right] \qquad 0 \leq u \leq 1$$

$[\mathbf{N}(v)]$ is similarly defined with $0 \leq v \leq 1$, and

$$[\mathbf{P}_3] = \begin{bmatrix} \mathbf{P}_{i-1,j-1} & \mathbf{P}_{i-1,j} & \mathbf{P}_{i-1,j+1} & \mathbf{P}_{i-1,j+2} \\ \mathbf{P}_{i,j-1} & \mathbf{P}_{i,j} & \mathbf{P}_{i,j+1} & \mathbf{P}_{i,j+2} \\ \mathbf{P}_{i+1,j-1} & \mathbf{P}_{i+1,j} & \mathbf{P}_{i+1,j+1} & \mathbf{P}_{i+1,j+2} \\ \mathbf{P}_{i+2,j-1} & \mathbf{P}_{i+2,j} & \mathbf{P}_{i+2,j+1} & \mathbf{P}_{i+2,j+2} \end{bmatrix}$$

The relationship between the central patch and its control points is shown in Figure 6.11.

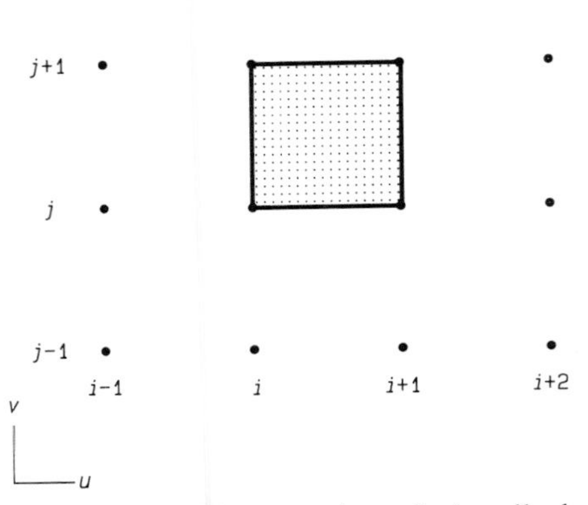

Figure 6.11 The central patch described by 16 control points for a cubic B-spline surface.

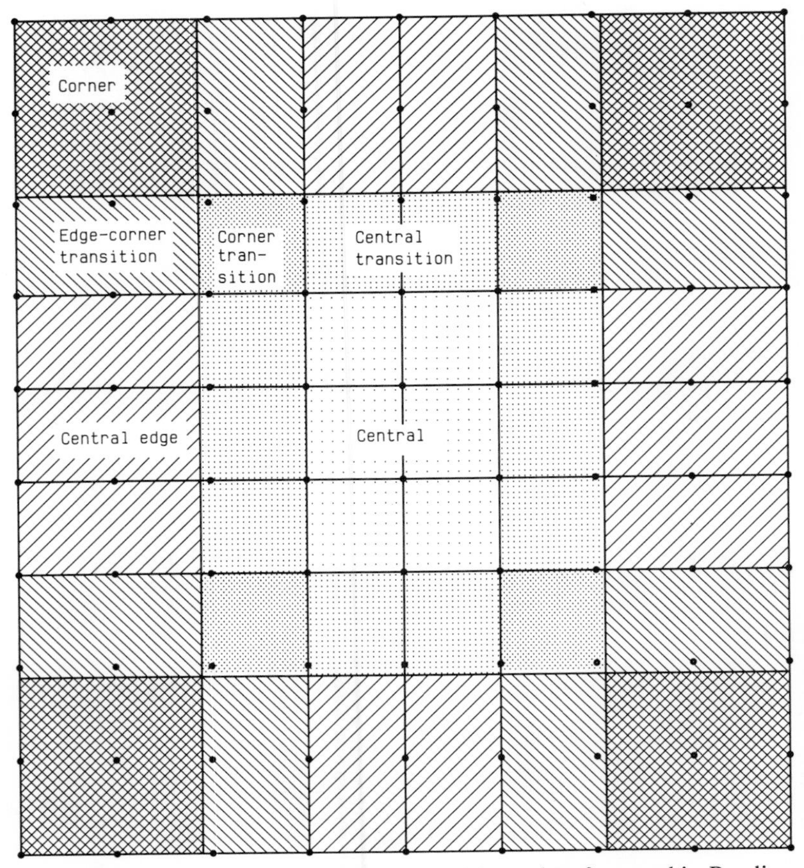

Figure 6.12 Arrangement of various special patches for a cubic B-spline surface.

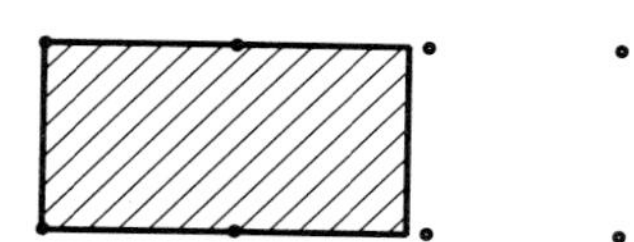

Figure 6.13 Control points for the typical edge patch used in construction of a cubic B-spline surface.

The 24 specialized forms of patches for a cubic B-splne surface follow the form suggested in Figure 6.12. The 24 different forms allow for special patches as follows: 4 corners, 4 central edges (top, bottom, left, right), 8 edge-corner transitions, 4 corner transitions, and 4 central transitions. With the exception of patches along the edges, the transition patches use the array of control points given by Eq. (6.28). For corners, the control points are the first four rows and columns including the edges. The other patches on the edges follow the pattern shown in Figure 6.13.

The cubic B-spline blending functions for each type of patch described in Figure 6.12 are related to Eqs. (5.71) through (5.74) in the same way that Eq. (6.28) is related to Eq. (5.70). Details are left as an exercise.

The simplest B-spline surfaces are periodic, where only one blending function is needed. Periodic B-spline surfaces can be closed by reusing control points in a manner analogous to that shown in Figure 5.17 for a B-spline curve. The polygons defined by the control points serve as an outer boundary to the resulting closed B-spline surface, Figure 6.14. The disadvantage of edges not being explicitly defined for periodic B-spline surfaces is balanced by simpler formulation.

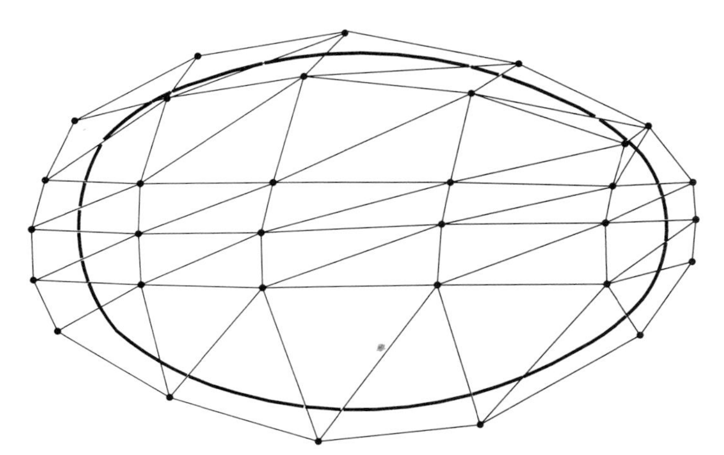

Figure 6.14 Closed convex B-spline surface. Triangles formed from control points provide an outer boundary for the surface.

Development of B-spline surfaces may be generalized[39] to permit expansion to quartic or higher-order surfaces. The increased degree of continuity between patches with increased order seldom justifies the additional complexity. In some cases, it may be advantageous to have a higher-order surface in one parametric direction than in the other. Furthermore, a surface may be periodic in one direction and nonperiodic in the other, as shown in the following example.

EXAMPLE 6.5

Approximate the thin-walled pipe elbow shown below with a quadratic B-spline surface. The surface should be described by four control points in each parametric direction. Check the accuracy of the approximation for the point on the actual surface $x = y = 300 \cos 45°$, $z = 0$.

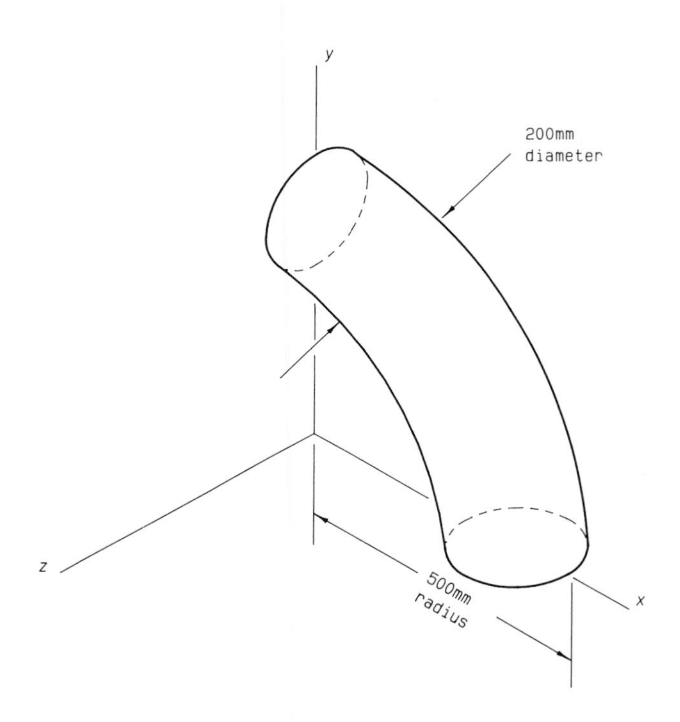

Solution. As seen in Figure 5.17, lines connecting control points in a constant parametric direction are tangent to the resulting quadratic B-spline curve. The end view in the x-z plane indicates the typical method used to place four control points on sections at 0°, 30°, 60°, and 90°. Control points use nonperiodic blending in the v parametric direction, which sweeps out the 90° bend of the elbow. In the u parametric direction, periodic blending is used in the circumferential direction with the control points placed to keep the cross section as near to 200 mm in diameter as possible. The locations of control points can be developed with the aid of the figure below.

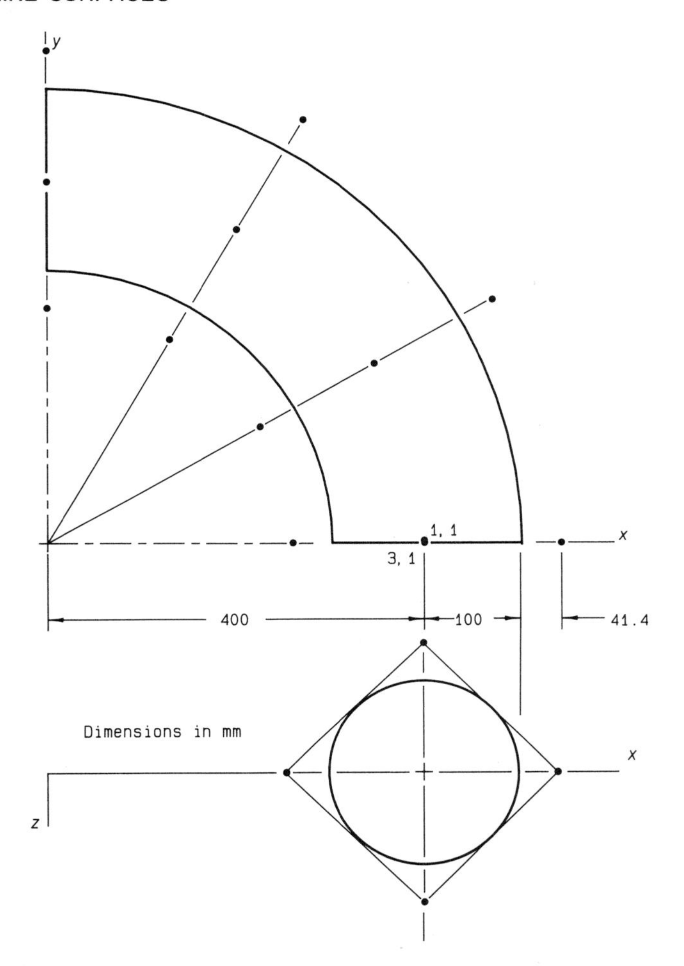

The Cartesian coordinates (mm) of the 16 control points are

$$[\mathbf{P}_{ij}] = \begin{bmatrix} (400, 0, 141.4) & (346, 250, 141.4) & (250, 346, 141.4) & (0, 400, 141.4) \\ (541, 0, 0) & (469, 271, 0) & (271, 469, 0) & (0, 541, 0) \\ (400, 0, -141.4) & (346, 250, -141.4) & (250, 346, -141.4) & (0, 400, -141.4) \\ (259, 0, 0) & (129.3, 224, 0) & (224, 129.3, 0) & (0, 259, 0) \end{bmatrix}$$

All eight curved patches on the B-spline surface use periodic blending for $[\mathbf{N}(u)]$ in Eq. 6.25. The four patches which start at the x-z plane use the nonperiodic blending for $[\mathbf{N}_1(v)]$ in Eq. (6.26). The four patches which end at the y-z plane use the blending for $[\mathbf{N}_1(v)]$ in Eq. (6.27). The array of patches, with hidden surfaces removed and with numbers for the patches in parentheses, is shown below. Since the equations for individual patches are essentially similar, only the one which contains the point to be checked will be expanded.

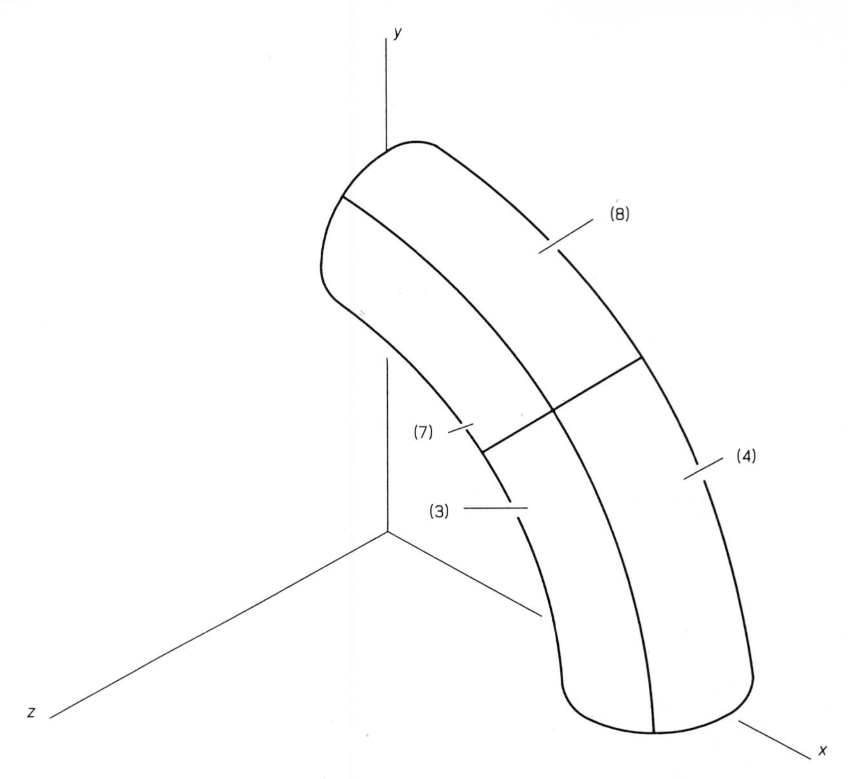

The point to be checked is on the boundary between patches (3) and (7), so either one may be used. The equation for patch (3) is

$$\mathbf{P}_{(3)}(u, v) = \begin{bmatrix} \dfrac{(1-u)^2}{2} & \dfrac{-2u^2+2u+1}{2} & \dfrac{u^2}{2} \end{bmatrix} \begin{bmatrix} \mathbf{P}_{3,1} & \mathbf{P}_{3,2} & \mathbf{P}_{3,3} \\ \mathbf{P}_{4,1} & \mathbf{P}_{4,2} & \mathbf{P}_{4,3} \\ \mathbf{P}_{1,1} & \mathbf{P}_{1,2} & \mathbf{P}_{1,3} \end{bmatrix} \begin{bmatrix} (1-v)^2 \\ (-3v^2+4v)/2 \\ v^2/2 \end{bmatrix}$$

At the point of interest on patch (3), $u = 0.5$ by symmetry and $v = 1.0$. Substitution of numerical values into the equation for patch (3) yields $\mathbf{P}_{(3)}(0.5, 1) = [207 \quad 207 \quad 0]$ mm to three places of accuracy. This is 2.4% less than the exact location $[300 \cos 45° \quad 300 \cos 45° \quad 0] = [212.13 \quad 212.13 \quad 0]$. One way accuracy could be improved is by using more control points and more patches. Alternatively, using cubic B-spline patches would also improve accuracy, but a minimum array of 45 control points would be needed—5 columns in the u direction by 9 rows in the v direction.

6.6 PARAMETRIC SOLIDS

Vector-valued parametric curves are described in terms of one parameter; vector-valued parametric surfaces (patches) are described in terms of two parameters. It follows that a *parametric solid,* or *hyperpatch,* can be described in terms of three parameters, usually $0 \le u \le 1$, $0 \le v \le 1$, and $0 \le w \le 1$. Holding any one of the three parameters constant describes a so-called isoparametric surface; holding any two of the three parameters constant describes an isoparametric curve. Should any one of the isoparametric values of u, v, or w be 0 or 1, the surface (or curve) will be on the outside surface of the hyperpatch.

An arbitrary point within a parametric solid can be given by the trivariate vector-valued function

$$\mathbf{P}(u, v, w) = [x(u, v, w) \quad y(u, v, w) \quad z(u, v, w)] \tag{6.29}$$

Representation of parametric cubic, Bezier, or B-spline curves is done with a list, or singly dimensioned array, of control points. Representation of surfaces is done with a rectangular or doubly dimensioned array of control points. It follows that a solid can be represented with a 3-D, or triply dimensioned, array of control points. The well-developed[42] parametric cubic hyperpatch has the three forms—algebraic, point, and geometric—all of which are defined by $4^3 = 64$ vector-valued functions, or $64 \times 3 = 192$ real numbers.

The algebraic form can be expressed compactly by the formula

$$\mathbf{P}(u, v, w) = \sum_{i=1}^{4} \sum_{j=1}^{4} \sum_{k=1}^{4} \mathbf{a}_{ijk} u^{i-1} v^{j-1} w^{k-1} \tag{6.30}$$

where the $\mathbf{a}_{ijk}$ are the algebraic coefficients and $0 \le u \le 1, 0 \le v \le 1, 0 \le w \le 1$.

Representation of the hyperpatch, Figure 6.15, in point form is useful for purposes of viewing and modeling transformations. The 64 points may be written in the array

$$\begin{matrix}
\mathbf{P}(0, 0, 0) & \mathbf{P}(0, 1/3, 0) & \mathbf{P}(0, 2/3, 0) & \mathbf{P}(0, 1, 0) \\
\mathbf{P}(1/3, 0, 0) & \mathbf{P}(1/3, 1/3, 0) & \mathbf{P}(1/3, 2/3, 0) & \mathbf{P}(1/3, 1, 0) \\
\mathbf{P}(2/3, 0, 0) & \mathbf{P}(2/3, 1/3, 0) & \mathbf{P}(2/3, 2/3, 0) & \mathbf{P}(2/3, 1, 0) \\
\mathbf{P}(1, 0, 0) & \mathbf{P}(1, 1/3, 0) & \mathbf{P}(1, 2/3, 0) & \mathbf{P}(1, 1, 0) \\
\\
\mathbf{P}(0, 0, 1/3) & \mathbf{P}(0, 1/3, 1/3) & \mathbf{P}(0, 2/3, 1/3) & \mathbf{P}(0, 1, 1/3) \\
\mathbf{P}(1/3, 0, 1/3) & \mathbf{P}(1/3, 1/3, 1/3) & \mathbf{P}(1/3, 2/3, 1/3) & \mathbf{P}(1/3, 1, 1/3) \\
\mathbf{P}(2/3, 0, 1/3) & \mathbf{P}(2/3, 1/3, 1/3) & \mathbf{P}(2/3, 2/3, 1/3) & \mathbf{P}(2/3, 1, 1/3) \\
\mathbf{P}(1, 0, 1/3) & \mathbf{P}(1, 1/3, 1/3) & \mathbf{P}(1, 2/3, 1/3) & \mathbf{P}(1, 1, 1/3) \\
\\
\mathbf{P}(0, 0, 2/3) & \mathbf{P}(0, 1/3, 2/3) & \mathbf{P}(0, 2/3, 2/3) & \mathbf{P}(0, 1, 2/3) \\
\mathbf{P}(1/3, 0, 2/3) & \mathbf{P}(1/3, 1/3, 2/3) & \mathbf{P}(1/3, 2/3, 2/3) & \mathbf{P}(1/3, 1, 2/3) \\
\mathbf{P}(2/3, 0, 2/3) & \mathbf{P}(2/3, 1/3, 2/3) & \mathbf{P}(2/3, 2/3, 2/3) & \mathbf{P}(2/3, 1, 2/3) \\
\mathbf{P}(1, 0, 2/3) & \mathbf{P}(1, 1/3, 2/3) & \mathbf{P}(1, 2/3, 2/3) & \mathbf{P}(1, 1, 2/3) \\
\\
\mathbf{P}(0, 0, 1) & \mathbf{P}(0, 1/3, 1) & \mathbf{P}(0, 2/3, 1) & \mathbf{P}(0, 1, 1) \\
\mathbf{P}(1/3, 0, 1) & \mathbf{P}(1/3, 1/3, 1) & \mathbf{P}(1/3, 2/3, 1) & \mathbf{P}(1/3, 1, 1) \\
\mathbf{P}(2/3, 0, 1) & \mathbf{P}(2/3, 1/3, 1) & \mathbf{P}(2/3, 2/3, 1) & \mathbf{P}(2/3, 1, 1) \\
\mathbf{P}(1, 0, 1) & \mathbf{P}(1, 1/3, 1) & \mathbf{P}(1, 2/3, 1) & \mathbf{P}(1, 1, 1)
\end{matrix} \tag{6.31}$$

where the rows (or the columns within each grouping) describe 16 curves (each described by four points) in point form, and the groups describe four patches ($w = 0, 1/3, 2/3, 1$) in point form. The points may be grouped in other ways, for example, by constant u or v instead of constant w. All the foregoing are examples of *parametrically parallel* curves and surfaces.

The third form is geometric. In a manner analogous to parametric cubic lines and surfaces, the parametric cubic solid is described by points and parametric derivatives on the surface. With reference to Figure 6.15, one possible way to write the geometric coefficient matrix is with the $4 \times 4 \times 4$ array $[\mathbf{B}_h]$, which is interpreted as

$$
\begin{array}{ccccc}
\begin{array}{c}\text{Bottom}\\ \text{surface}\\ v=0\end{array} & \begin{array}{c}\text{Top}\\ \text{surface}\\ v=1\end{array} & \begin{array}{c}\text{Rate of}\\ \text{change}\\ \text{at } v=0\end{array} & \begin{array}{c}\text{Rate of}\\ \text{change}\\ \text{at } v=1\end{array} & \\
\mathbf{P}(0,0,0) & \mathbf{P}(0,1,0) & \mathbf{P}_v(0,0,0) & \mathbf{P}_v(0,1,0) & \\
\mathbf{P}(1,0,0) & \mathbf{P}(1,1,0) & \mathbf{P}_v(1,0,0) & \mathbf{P}_v(1,1,0) & \text{Front surface} \\
\mathbf{P}_u(0,0,0) & \mathbf{P}_u(0,1,0) & \mathbf{P}_{uv}(0,0,0) & \mathbf{P}_{uv}(0,1,0) & w=0 \\
\mathbf{P}_u(1,0,0) & \mathbf{P}_u(1,1,0) & \mathbf{P}_{uv}(1,0,0) & \mathbf{P}_{uv}(1,1,0) & \\
& & & & \\
\mathbf{P}(0,0,1) & \mathbf{P}(0,1,1) & \mathbf{P}_v(0,0,1) & \mathbf{P}_v(0,1,1) & \\
\mathbf{P}(1,0,1) & \mathbf{P}(1,1,1) & \mathbf{P}_v(1,0,1) & \mathbf{P}_v(1,1,1) & \text{Rear surface} \\
\mathbf{P}_u(0,0,1) & \mathbf{P}_u(0,1,1) & \mathbf{P}_{uv}(0,0,1) & \mathbf{P}_{uv}(0,1,1) & w=1 \\
\mathbf{P}_u(1,0,1) & \mathbf{P}_u(1,1,1) & \mathbf{P}_{uv}(1,0,1) & \mathbf{P}_{uv}(1,1,1) & \\
& & & & \\
\mathbf{P}_w(0,0,0) & \mathbf{P}_w(0,1,0) & \mathbf{P}_{vw}(0,0,0) & \mathbf{P}_{vw}(0,1,0) & \\
\mathbf{P}_w(1,0,0) & \mathbf{P}_w(1,1,0) & \mathbf{P}_{vw}(1,0,0) & \mathbf{P}_{vw}(1,1,0) & \text{Rate of change} \\
\mathbf{P}_{uw}(0,0,0) & \mathbf{P}_{uw}(0,1,0) & \mathbf{P}_{uvw}(0,0,0) & \mathbf{P}_{uvw}(0,1,0) & \text{at } w=0 \\
\mathbf{P}_{uw}(1,0,0) & \mathbf{P}_{uw}(1,1,0) & \mathbf{P}_{uvw}(1,0,0) & \mathbf{P}_{uvw}(1,1,0) & \\
& & & & \\
\mathbf{P}_w(0,0,1) & \mathbf{P}_w(0,1,1) & \mathbf{P}_{vw}(0,0,1) & \mathbf{P}_{vw}(0,1,1) & \\
\mathbf{P}_w(1,0,1) & \mathbf{P}_w(1,1,1) & \mathbf{P}_{vw}(1,0,1) & \mathbf{P}_{vw}(1,1,1) & \text{Rate of change} \\
\mathbf{P}_{uw}(0,0,1) & \mathbf{P}_{uw}(0,1,1) & \mathbf{P}_{uvw}(0,0,1) & \mathbf{P}_{uvw}(0,1,1) & \text{at } w=1 \\
\mathbf{P}_{uw}(1,0,1) & \mathbf{P}_{uw}(1,1,1) & \mathbf{P}_{uvw}(1,0,1) & \mathbf{P}_{uvw}(1,1,1) &
\end{array}
\tag{6.32}
$$

Blending hyperpatches is done in a manner analogous to that of blending curves to make a spline and that of blending patches to make a continuous surface. Hyperpatch blending requires that the two mating surfaces be coincident and also that both parametric slopes match on the four coincident edges.

In practice, description of a hyperpatch by Eq. (6.30), (6.31), or (6.32) is cumbersome. Available solid modeling systems[42] offer various ways to generate hyperpatches, including (1) rotating or translating a patch or (2) specifying two, three, or four parametrically parallel patches.

Hyperpatches can be constructed from parametric functions other than the parametric cubic. If, for example, linear functions are used, a brick results. Higher than cubic order functions can be used to generate solids with more elaborately sculptured surfaces. The blending functions associated with Bezier and B-spline geometry can also be used in conjunction with 3-D arrays of control points to describe parametric solids.

6.7 CLOSURE

Parametric geometry provides a unified and systematic method for representation of curves, surfaces, and solids. Compared to tessellated surfaces, parametrically described curved surfaces have much higher accuracy with significantly less data storage. In addition to providing the surface normal vectors for light-source shading, these descriptions facilitate the computation of other geometric properties such as surface area.

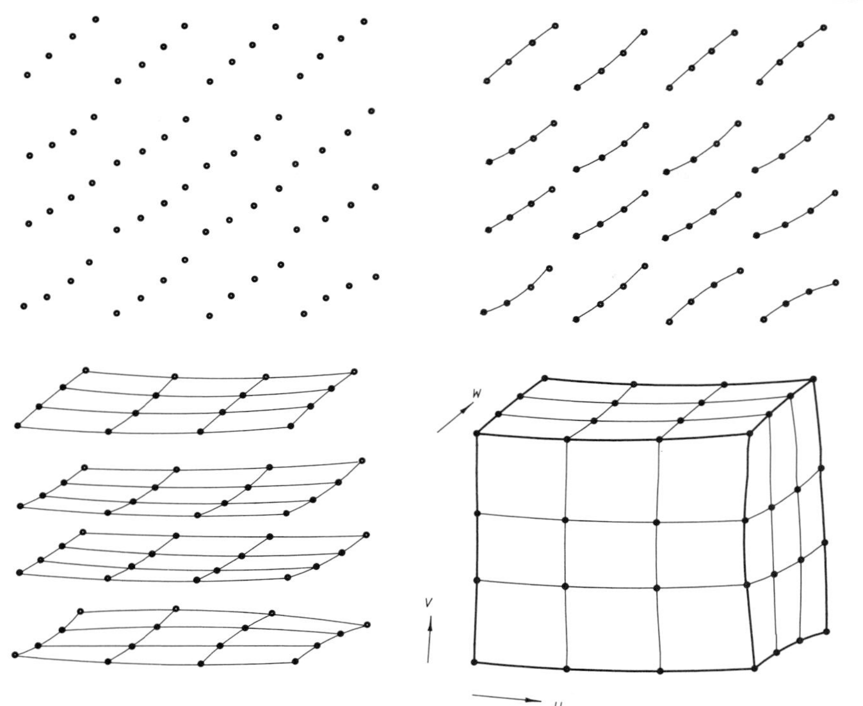

Figure 6.15 Parametric cubic solid as described by 64 points, 16 curves, or 4 surfaces.

Solid objects can be described by their bounding surfaces, although this gives no information about the interior of the object. Such a surface or boundary representation of a solid may be adequate if it is ensured that the solid is homogeneous. However, greater generality is possible with a hyperpatch (parametric solid), which does describe the interior. Direct evaluation of the hyperpatch yields, for example, the volume of a solid. Use of the foregoing descriptions to obtain geometric properties is treated in the next chapter.

PROBLEMS

6.1. A bilinear patch is described by the four points

$$\begin{bmatrix} 0 & 0 & 0 \\ 4 & 0 & 2 \\ -1 & 4 & 2 \\ 5 & 3 & 1 \end{bmatrix}$$

Determine the x-coordinate of the point where $y = 2$ and $z = 1$.

6.2. Determine the parametric equation of the curve which lies on the bilinear patch of Problem 6.1 and has $y = 1 = \text{constant}$.

6.3. For the bilinear patch of Problem 6.1, determine the parametric equation of the curve which lies on the patch and which (a) connects $\mathbf{P}(0, 0)$ to $\mathbf{P}(0, 1)$ and (b) connects $\mathbf{P}(0, 0)$ to $\mathbf{P}(1, 1)$.

6.4. Fit a bilinear patch for the four corner points described in Example 6.3. Determine the percent difference between the radius to the surface described by the bilinear patch and the true spherical surface at the parametric midpoint of the bilinear patch.

6.5. Determine the matrix of geometric coefficients for a parametric cubic F-patch which has the same corners as the patch typical of a four-patch approximation in Example 6.1. Compare the fit of the resulting patch at the parametric midpoint to the surface of the conical frustum.

6.6. Using the geometric coefficient matrix for a parametric cubic patch,

$$[\mathbf{B}_p] = \begin{bmatrix} [0 & 0 & 0] & [0 & 3 & 1] & [2 & 0 & 1] & [2 & 0 & -1] \\ [3 & 0 & 0] & [3 & 3 & 0] & [2 & 0 & 1] & [2 & 0 & -1] \\ [2 & -1 & 0] & [2 & 1 & 0] & [0 & 0 & 0] & [0 & 0 & 0] \\ [2 & 1 & 0] & [2 & -1 & 0] & [0 & 0 & 0] & [0 & 0 & 0] \end{bmatrix}$$

determine the z position of the surface when (a) $x = 1, y = 1$ and (b) $x = 1, y = 2$.

Note: Use of a digital computer or progammable calculator is recommended to solve the resulting equations.

6.7. Modify the geometric coefficient matrix given above to (a) retain the same corners, (b) make the four edges straight lines, and (c) make a symmetrical patch with its parametric midpoint patch passing through the point (1.5, 1.5, 1).

6.8. A straight line connects the points (2, 0, 0) and (1, 2, 0). The line is rotated 180° around the y axis to form a conical surface. Determine the matrix of geometric coefficients for a parametric cubic patch. Compare the accuracy of the resulting patch at the parametric midpoint with the true conical surface.

6.9. Repeat Problem 6.5 except that a cubic Bezier surface patch is used instead of a parametric cubic surface patch.

6.10. Repeat Problem 6.8 using a cubic Bezier surface patch instead of the parametric cubic surface patch.

6.11. Use two blended cubic Bezier patches to approximate the surface of a right circular cylinder of radius r and length h. Make the cylinder symmetric with respect to the y axis such that the patches meet on the $z = 0$ plane. Compare the accuracy of one of the patches at the parametric third point ($u = v = 1/3$) with the true surface.

6.12 A nonperiodic quadratic B-spline surface of four patches is approximately the surface described in Problem 6.8. Compare the accuracy at the parametric quarter point ($u = v = 1/4$) of one of the patches with the true surface.

6.13. Construct a quadratic B-spline surface of eight patches to best approximate the surface of a right circular cylinder with a diameter of 2 m and a height of 9 m. Check the accuracy of the approximation at the parametric center of one of the patches.

6.14. Construct a nonperiodic quadratic B-spline surface with four patches to approximate the surface of half a right circular cylinder. The diameter of the half cylinder is 4 m and its length is 5 m. Check the accuracy of the approximation at the parametric center of one of the patches.

6.15. Write out the matrix of geometric coefficients for a parametric cubic solid which describes a cube bounded by the planes $x = 0$, $y = 0$, $z = 0$, $x = 1$, $y = 1$, and $z = 1$.

6.16. A parametric cubic solid is bounded by two spherical surfaces—one at radius = 10 m, as described in Example 6.3, and the other at radius = 5 m. Both patches subtend the same angles as in the example. Write out the matrix of geometric coefficients.

PROJECTS

NOTE: An available system may be used to execute parts of the following projects. Appropriate images should be produced.

6.1. Generate a perspective view of a church building with its roof formed from two bilinear patches. The corners of the first patch should be placed (coordinates, in feet, are in counterclockwise order looking from the outside, where the y axis is vertical) at the points (0, 0, 0), (100, 0, 30), (100, 30, 0), (0, 30, 0), and the corners of the second patch at (0, 0, 0), (0, 30, 0), (100, 30, 0), (100, 0, −30). The open end and the floor are triangular. The resulting image should have appropriate isoparametric curves to aid in visualization.

6.2. Develop a program which demonstrates how twist vectors control the shape of a parametric cubic patch. The edges of the patch should be straight lines; the patch should be a square when four zero twist vectors are specified. In operation, the user inputs three components for each of the four twist vectors followed by the display of the new patch. The patches should have curves of constant u and v to show shape.

6.3. Generate a model of a hemispherical surface with eight parametric cubic patches, four of which are spherical triangles. Intermediate isoparametric curves should be included in the final image. Is there a problem representing the spherical triangles?

6.4. Duplicate the vaulted roof structure in Example 6.4 with four parametric cubic patches. Include isoparametric curves for visualization. What will be the differences in the shape of the resulting structure?

6.5. Fit a cubic Bezier surface patch to the surface of half a right circular cylinder with a radius of 2 m and a length of 6 m. Include isoparametric curves for visualization, the Bezier control points, and the tessellated surface which bounds the surface patch. How well does the Bezier patch fit at the center?

6.6. Repeat Project 6.5, but fit a quartic Bezier surface patch instead of the cubic one.

6.7. Make a model of a thin-walled section of pipe 1.5 m in diameter and 4 m long with two blended cubic Bezier surface patches. Show appropriate intermediate curves for visualization, control points, and the bounding tessellated surface. Check the accuracy of the approximation at the midpoint of one of the patches.

6.8. Repeat Project 6.7, but use four patches instead of two. Arrange the patches so that each subtends 90° of arc.

6.9. Duplicate the vaulted roof structure of Example 6.4 with nine nonperiodic quadratic B-spline patches. The resulting image should show the control points and the curves where the edges of the patches join.

6.10. Fit eight B-spline patches to approximate a straight section of hollow pipe that is 4 inches in diameter and 10 inches long. Use the same arrangement and type of patches as in Example 6.5.

6.11. Approximate a hemispherical surface with B-spline patches. Discuss any problems that arise.

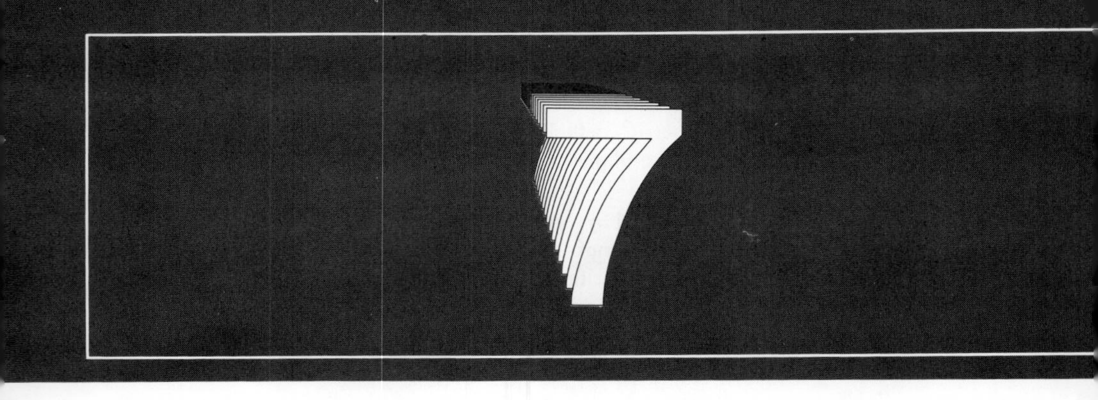

Properties and Relationships

Up to this point, the development has led to the one important result—a picture. In engineering design, quantitative information has even greater importance. A *geometric model* of an object contains mathematical descriptions of curves, lines, solids, and other characteristics. Besides the obvious use for producing pictures, such models may also be used to determine geometric properties, to solve descriptive geometry problems, and to prepare input data for analysis codes.

Geometric properties include curve length, surface area, volume, mass center location, and others. Often, numerical integration is used to compute these properties. Geometric properties supply the design engineer with information on weight, amount of material, size, dynamical characteristics, and so forth. Geometric properties appear in many of the formulations found in dynamics, mechanics of materials, and fluid mechanics.

Descriptive geometry, a well-developed technique using instrumented drafting, addresses spatial relationship problems, such as determining the distance from a point to a plane, and intersection problems such as finding the common line between two planes. Descriptive geometry finds application in mining engineering, structural design, machine design, layout, assembly, and other such work.

Computer codes for stress analysis, computational fluid mechanics, vibration analysis, magnetic field studies, and so on predict performance of proposed designs without requiring the construction and testing of physical models. The savings in time and cost can be significant with the use of such codes. Input and output from these codes also use the computer graphics data base. A brief introduction to use of geometric models in analysis can be found in Chapter 10.

7.1 DEFINITIONS

Geometric properties, often defined as integrals, are treated in textbooks on calculus and engineering mechanics.[34] A summary of the definitions of some of the more commonly used properties follows, where the convention is followed that planar properties are arbitrarily in the x-y plane.

The distance measured along a curve, *arc length,* is

$$L = \int ds \tag{7.1}$$

where ds is the differential of arc length, as shown in Figure 7.1.

The *area* of a surface is

$$A = \int\int dx\, dy \tag{7.2}$$

where dx and dy define the differential area dA in the usual Cartesian coordinates, as shown in Figure 7.2.

The *volume* of a solid is defined by

$$V = \int\int\int dx\, dy\, dz \tag{7.3}$$

where dx, dy, and dz define the incremental volume dV, as shown in Figure 7.3. Multiplication of the integrand of Eq. (7.3) by ρ (density) or γ (specific weight) produces mass or weight.

The *centroid* of an area is the first moment divided by the area. The x coordinate of the centroid is

$$\bar{x} = \frac{\int x\, dA}{\int dA} \tag{7.4}$$

The y coordinate is found similarly. The centroid of a volume is found by integrating over a volume instead of an area. The *mass center* and the *center of gravity* are the same as the centroid for homogeneous solids. For nonhomogeneous solids, the variable density or specific weight, as appropriate, must appear in the integrand.

In mechanics of materials, the formula for bending stress in beams involves the *moment of inertia of area,* a quantity which describes the geometry of the cross section in question. With respect to the x axis,

$$I_x = \int y^2\, dA \tag{7.5}$$

Figure 7.1 Differential arc length.

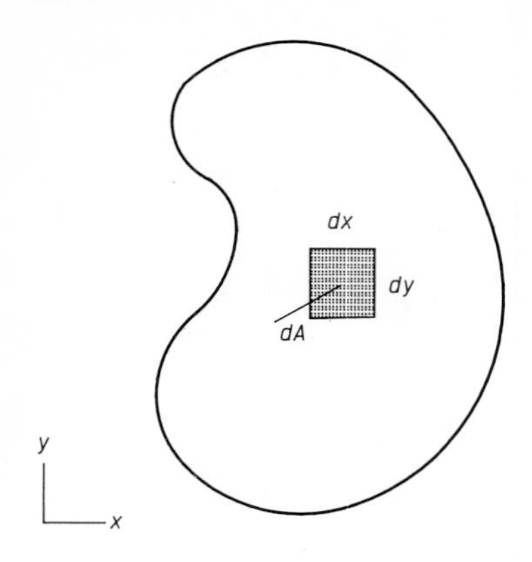

Figure 7.2 Differential area.

The moment with respect to the y axis is found similarly. The polar moment of inertia used in computation of torsional stress is

$$J_0 = \int r^2 \, dA = \int (x^2 + y^2) \, dA \tag{7.6}$$

In asymmetric bending stress problems, the *product of inertia of area* occurs, where

$$I_{xy} = \int xy \, dA \tag{7.7}$$

All of the foregoing may be transformed to other, parallel axes by the transfer-of-axis theorems,

$$\begin{aligned} I_x &= \bar{I}_x + d_x^2 A \\ I_y &= \bar{I}_y + d_y^2 A \\ I_{xy} &= \bar{I}_{xy} + d_x d_y A \end{aligned} \tag{7.8}$$

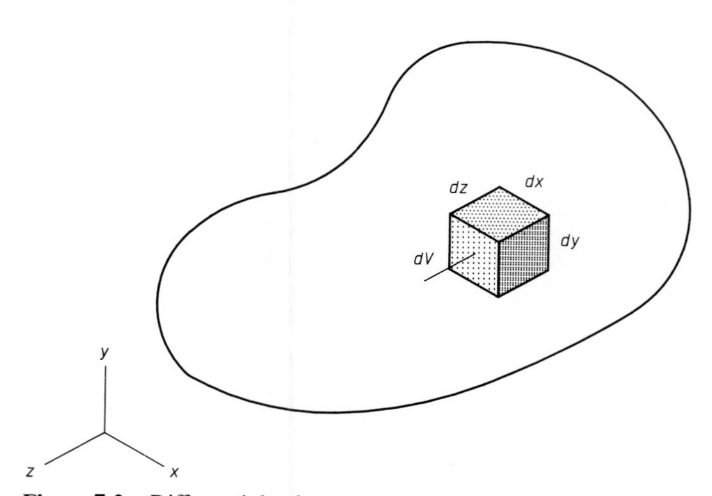

Figure 7.3 Differential volume.

where the bar indicates the quantity with respect to an axis through the centroid and the distances d_x and d_y are measured from the centroid to the x and y axes, respectively.

Problems involving prediction of the dynamic response of systems involve the *moment of inertia of mass*. With respect to the x axis, this quantity is defined by

$$I_{xx} = \int r_x^2 \, dm \tag{7.9}$$

where $dm = \rho \, dV$ and $r_x^2 = y^2 + z^2$, the square of the distance from the x axis. Expressions for the mass moment of inertia are similar for the other two coordinate axes. Besides moments of inertia of mass, use is made in three-dimensional (3-D) dynamics of the *product of inertia of mass,*

$$I_{xy} = \int xy \, dm \tag{7.10}$$

with respect to the x and y axes. Expressions for the moments of inertia with the x and z axes and the y and z axes are similar. As for areas, the moment and product of inertia expressions can be transformed between parallel axes. In this case, the A in Eqs. (7.8) is replaced by m.

Detailed techniques for the closed-form computation and manipulation of these properties as well as tables listing the area and volume properties for elementary shapes are readily available.[34] All of the properties have the same form, namely a single, double, or triple integral

$$\begin{aligned} I_1 &= \int f_1(x) \, dx \\ I_2 &= \int\int f_2(x, y) \, dx \, dy \\ I_3 &= \int\int\int f_3(x, y, z) \, dx \, dy \, dz \end{aligned} \tag{7.11}$$

which admits simple numerical methods for evaluation.

7.2 OVERVIEW OF GAUSSIAN QUADRATURE

Geometric properties of curves, surfaces, and solids are usually defined by integrals. *Gaussian quadrature,* a fast and accurate method of numerical integration, approximates an integral with a small number of sampling points and corresponding weighting functions. The unequally spaced sampling points in Gaussian quadrature[59,60] define approximating polynomials with optimum accuracy. Increasing the number of sampling points improves accuracy. For compact storage, the sampling points and weighting functions match a definite integral with the range arbitrarily set as 0 to 1. For a single integral, the quadrature formula is

$$I_1 = \int_0^1 f_1(u) \, du = \sum_{i=1}^{n} h_i f_1(u_i) \tag{7.12}$$

where the weights h_i and the sampling points u_i are given in Table 7.1. The process for determining the area under a curve with three sampling points is illustrated in Figure 7.4.

Approximation of double and triple integrals makes use of the same weighting functions and sampling points, except that double and triple summations are used. For a double integral,

$$I_2 = \int_0^1 \int_0^1 f_2(u, v)\, du\, dv = \sum_{i=1}^{m} \sum_{j=1}^{n} h_i h_j f_2(u_i, v_j) \tag{7.13}$$

Similarly, a triple integral is

$$I_3 = \int_0^1 \int_0^1 \int_0^1 f_3(u, v, w)\, du\, dv\, dw = \sum_{i=1}^{m} \sum_{j=1}^{n} \sum_{k=1}^{p} h_i h_j h_k f_3(u_i, v_j, w_k) \tag{7.14}$$

The sampling points for double summation can be interpreted as an array of $m \times n$ points spread over a rectangular region. For triple summation, the $m \times n \times p$ array is spread over a boxlike volume.

TABLE 7.1 WEIGHTING COEFFICIENTS AND SAMPLING POINTS FOR GAUSSIAN QUADRATURE BASED ON $(0 \le u \le 1)$

i	h_i	u_i
	$n = 2$	
1	0.500 000 000	0.211 324 865
2	0.500 000 000	0.788 675 135
	$n = 3$	
1	0.277 777 778	0.112 701 665
2	0.444 444 444	0.500 000 000
3	0.277 777 778	0.887 298 335
	$n = 4$	
1	0.173 927 423	0.069 431 844
2	0.326 072 577	0.330 009 478
3	0.326 072 577	0.669 990 522
4	0.173 927 423	0.930 568 156
	$n = 5$	
1	0.118 463 443	0.046 910 077
2	0.239 314 335	0.230 765 345
3	0.284 444 444	0.500 000 000
4	0.239 314 335	0.769 234 655
5	0.118 463 443	0.953 089 923
	$n = 6$	
1	0.085 662 246	0.033 765 243
2	0.180 380 787	0.169 395 307
3	0.233 956 967	0.380 690 407
4	0.233 956 967	0.619 309 593
5	0.180 380 787	0.830 604 693
6	0.085 662 246	0.966 234 757

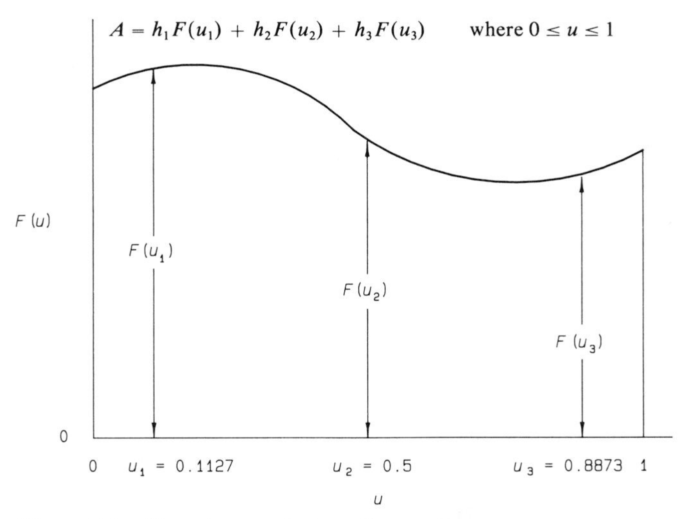

Figure 7.4 Gaussian quadrature for $n = 3$.

EXAMPLE 7.1

Use Gaussian quadrature to determine the polar moment of inertia of a hollow shaft with an inner radius of 2.0 cm and an outer radius of 2.5 cm. Compare the result with the closed-form solution.

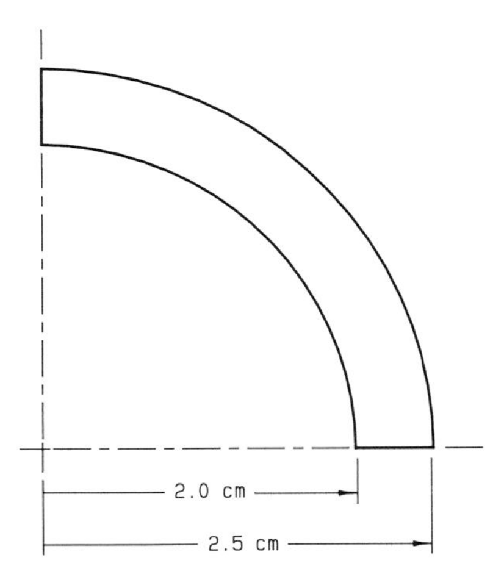

Solution. The integral formulation for J_0 is

$$J_0 = 4 \int_0^{\pi/2} \int_2^{2.5} r^3 \, dr \, d\theta$$

In order to use the weighting coefficients and sampling points of Table 7.1, the limits on the integrals must be 0 and 1. Thus, the substitutions

$$u = 2(r - 2) \quad \text{and} \quad v = \frac{2\theta}{\pi}$$

are made, which lead to

$$r = \frac{u}{2} + 2 \quad \text{or} \quad dr = \tfrac{1}{2}\, du$$

and

$$\theta = \frac{\pi}{2} v \quad \text{or} \quad d\theta = \frac{\pi}{2} dv$$

Thus, the integral is reformulated with the required limits and three quadrature points as

$$\begin{aligned}
J_0 &= 4 \frac{1}{2} \frac{\pi}{2} \int_0^1 \int_0^1 \left(\frac{u}{2} + 2\right)^3 du\, dv \\
&= \frac{\pi}{8} \sum_{i=1}^{3} \sum_{j=1}^{3} h_i h_j (u_i + 4)^3 \\
&= \frac{\pi}{8} [(0.277778)^2(4.112702)^3 + (0.444444)(0.277778)(4.5)^3 \\
&\quad + (0.277778)^2(4.887298)^3 + (0.277778)(0.444444)(4.112702)^3 \\
&\quad + (0.444444)^2(4.5)^3 + (0.277778)(0.444444)(4.887298)^3 \\
&\quad + (0.277778)^2(4.112702)^3 + (0.444444)(0.277778)(4.5)^3 \\
&\quad + (0.277778)^2(4.887298)^3] \\
&= 36.2265
\end{aligned}$$

For comparison, the exact value is

$$J_0 = \frac{\pi}{2}(r_o^4 - r_i^4) = \frac{\pi}{2}(2.5^4 - 2^4)$$

$$= 36.2265$$

which agrees exactly with the six places carried in the approximate calculations.

7.3 POLYGONS AND POLYHEDRONS

In many applications, curved lines are approximated by a series of straight line segments. A planar area bounded by a curve can be approximated by a series of straight lines forming a polygon; sculptured surfaces can be approximated by an array of polygons. Similarly, solids may be represented by polyhedrons. When

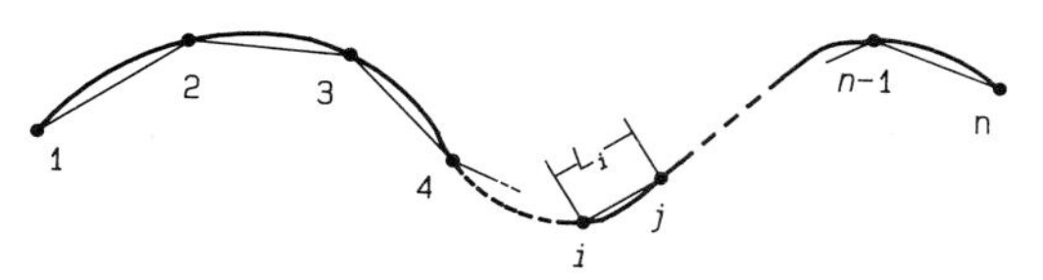

Figure 7.5 Approximation of a curve.

curves, surfaces, and solids are thus approximated, relatively simple formulas can be used to determine geometric properties.

Curve Length

As seen in Figure 7.5, the arc length of a curved line, Eq. 7.1, can be approximated by

$$L = \sum_{i=1}^{n} L_i \tag{7.15}$$

where the lengths of the line segments L_i are computed from

$$L_i = \sqrt{(x_i - x_j)^2 + (y_i - y_j)^2 + (z_i - z_j)^2} \tag{7.16}$$

This tends to underestimate the length of the arc and is somewhat computationally inefficient. A better method for determining the arc length of curves that are described with parametric geometry is given in Section 7.4.

Area Properties

The magnitude of a planar area described by two vectors is related to the familiar formulas for the area of a parallelogram or a triangle. Defined by vectors **a** and **b** as shown in Figure 7.6(a), the area A of a parallelogram is given by the cross product as

$$A = |\mathbf{a} \times \mathbf{b}| \tag{7.17}$$

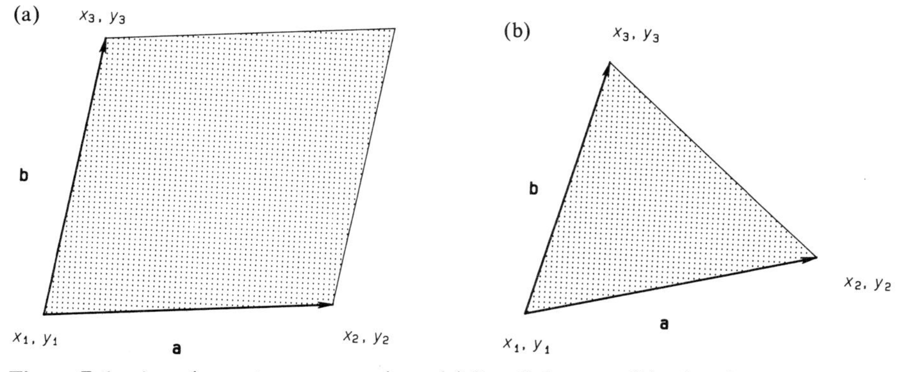

Figure 7.6 Area by vector cross product. (a) Parallelogram; (b) triangle.

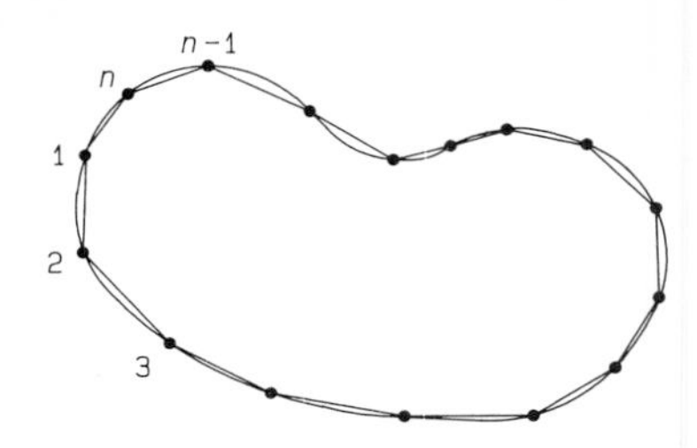

Figure 7.7 Closed polygon approximated by n line segments.

where $\mathbf{a} = (x_2 - x_1)\mathbf{i} + (y_2 - y_1)\mathbf{j}$ and $\mathbf{b} = (x_3 - x_1)\mathbf{i} + (y_3 - y_1)\mathbf{j}$. Similarly, the area of the triangle in Figure 7.6(b) is

$$A = \tfrac{1}{2}|\mathbf{a} \times \mathbf{b}| \tag{7.18}$$

For purposes of evaluation, the magnitude of the cross product in Eq. (7.17) or (7.18) can be written out with the use of Eq. (A.19), Appendix A, as

$$|\mathbf{a} \times \mathbf{b}| = |(x_2 - x_1)(y_3 - y_1) - (x_3 - x_1)(y_2 - y_1)| \tag{7.19}$$

The foregoing development can be extended[12] to determine the area within a 2-D curve enclosed by n line segments, Figure 7.7, which is

$$\begin{aligned} A = \tfrac{1}{2} \big(&(x_1y_2 + x_2y_3 + \cdots + x_{n-1}y_n + x_ny_1) \\ &- (x_2y_1 + x_3y_2 + \cdots + x_ny_{n-1} + x_1y_n)\big) \end{aligned} \tag{7.20}$$

where the polygon closes at node n.

For triangular areas, the integrals needed for Eqs. (7.5) through (7.7) can be shown[60] to be

$$\iint x^2\,dx\,dy = \frac{(x_1^2 + x_2^2 + x_3^2)A}{12} \tag{7.21}$$

and

$$\iint xy\,dx\,dy = \frac{(x_1y_1 + x_2y_2 + x_3y_3)A}{12} \tag{7.22}$$

where A is the area of the triangle, and the x-y coordinate system is located at the centroid of the triangle as defined by the relations

$$\frac{x_1 + x_2 + x_3}{3} = \frac{y_1 + y_2 + y_3}{3} = 0 \tag{7.23}$$

The foregoing relations are useful for the evaluation of properties of planar areas approximated by an array of triangles. In application, the points matrix for each triangle must be translated so that the centroid is at the origin before using Eqs. (7.21) and (7.22).

Volume Properties

The *volume* of a parallelepiped, Figure 7.8(a), is given by the scalar triple product,

$$V = \mathbf{a} \times \mathbf{b} \cdot \mathbf{c} \tag{7.24}$$

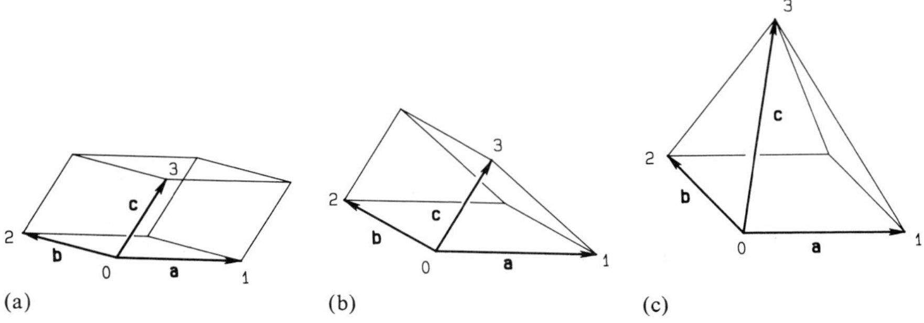

Figure 7.8 Polyhedrons. (a) Parallelepiped; (b) wedge; (c) pyramid.

which with a properly located coordinate system may be written, Eq. (A.23), as the determinant

$$V = \begin{vmatrix} x_1 & y_1 & z_1 \\ x_2 & y_2 & z_2 \\ x_3 & y_3 & z_3 \end{vmatrix} \tag{7.25}$$

where the node 0 is located at the origin, where $x_0 = y_0 = z_0 = 0$. It follows that the volume of the wedge, Figure 7.8(b), is

$$V = \tfrac{1}{2}\,(\mathbf{a} \times \mathbf{b} \cdot \mathbf{c}) \tag{7.26}$$

and the pyramid, Figure 7.8(c), is

$$V = \tfrac{1}{3}\,(\mathbf{a} \times \mathbf{b} \cdot \mathbf{c}) \tag{7.27}$$

Finding the volume of a tetrahedron, Figure 7.9, is simplified with a coordinate system which places the origin at the centroid, defined by

$$\begin{aligned} \frac{x_1 + x_2 + x_3 + x_4}{4} &= 0 \\ \frac{y_1 + y_2 + y_3 + y_4}{4} &= 0 \\ \frac{z_1 + z_2 + z_3 + z_4}{4} &= 0 \end{aligned} \tag{7.28}$$

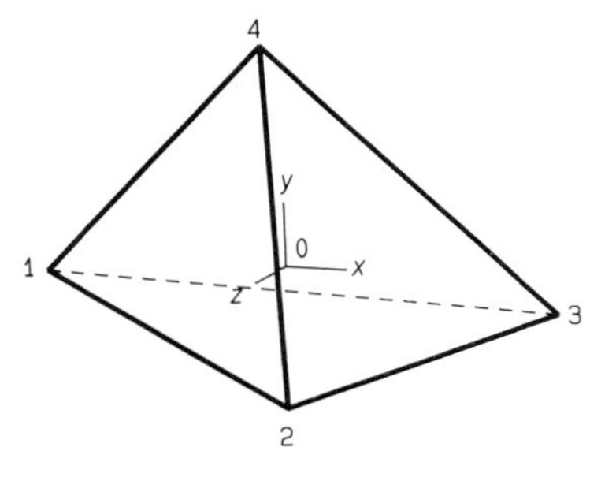

Figure 7.9 Tetrahedron with origin at centroid located by Eq. (7.28).

The volume of a tetrahedron is given by the determinant[60]

$$V = \frac{1}{6} \begin{vmatrix} 1 & 1 & 1 & 1 \\ x_1 & x_2 & x_3 & x_4 \\ y_1 & y_2 & y_3 & y_4 \\ z_1 & z_2 & z_3 & z_4 \end{vmatrix} \tag{7.29}$$

Integrals which involve volume properties of a tetrahedron, for example, Eqs. (7.9) and (7.10), can be expressed by the formulas

$$\begin{aligned} \iiint x\, dx\, dy\, dz &= 0 \\ \iiint x^2\, dx\, dy\, dz &= \frac{(x_1^2 + x_2^2 + x_3^2 + x_4^2)V}{20} \\ \iiint xy\, dx\, dy\, dz &= \frac{(x_1y_1 + x_2y_2 + x_3y_3 + x_4y_4)V}{20} \end{aligned} \tag{7.30}$$

and so forth, where V is given by Eq. (7.29) and the order of numbering is required to be that in Figure 7.9.

The equations in this section are useful for objects constructed from the basic elements used here. *Cell decomposition*[18,36] is a solid modeling strategy which breaks curves, surfaces, and solids into such elements for evaluation of geometric properties. Accuracy, of course, is improved by dividing objects into smaller subdivisions. Geometric properties of other elements, such as spheres or wedges, are available[60] or may be evaluated with numerical techniques.

7.4 PARAMETRIC CURVES, SURFACES, AND SOLIDS

Analytical descriptions of geometric entities with parametric cubic, Bezier, and B-spline representations have been developed in Chapters 5 and 6. These representations for curved lines, surfaces, and solids permit accurate and efficient determination of properties. Although the descriptions that follow use parametric cubic geometry, extension to other parametric systems is routine.

Curve Length

The length of a parametric cubic curve can be computed directly from Eq. (7.1) instead of the piecewise approximation in Eqs. (7.15) and (7.16). Using the parametric derivative defined in Eq. (5.36), the differential arc length ds is

$$ds = \left| \frac{d\mathbf{P}}{du} \right| du = |\mathbf{P}_u| du \tag{7.31}$$

The magnitude of $\mathbf{P}_u$ is given by $\sqrt{\mathbf{P}_u \cdot \mathbf{P}_u}$. After some algebra, the product $\mathbf{P}_u \cdot \mathbf{P}_u$ can be written as

$$\mathbf{P}_u \cdot \mathbf{P}_u = c_0 + c_1u + c_2u^2 + c_3u^3 + c_4u^4 \tag{7.32}$$

where
$$c_0 = \mathbf{a}_2 \cdot \mathbf{a}_2$$
$$c_1 = 4\mathbf{a}_2 \cdot \mathbf{a}_3$$
$$c_2 = 6\mathbf{a}_2 \cdot \mathbf{a}_4 + 4\mathbf{a}_3 \cdot \mathbf{a}_3$$
$$c_3 = 12\mathbf{a}_3 \cdot \mathbf{a}_4$$
$$c_4 = 9\mathbf{a}_4 \cdot \mathbf{a}_4$$

Here, the **a**'s are the algebraic coefficients defined in Eq. (5.33). A similar expression can be obtained in terms of the geometric coefficients. The resulting integral for arc length

$$L = \int_0^1 \sqrt{(c_0 + c_1 u + c_2 u^2 + c_3 u^3 + c_4 u^4)}\, du \tag{7.33}$$

may be evaluated with Gaussian quadrature.

Surface Area

A surface is described parametrically as $\mathbf{P}(u, v)$. The partial derivatives for a parametric cubic surface are given in Eq. (6.19). Using the area subtended by two vectors, Eq. (7.17), the differential area shown in Figure 7.10 is

$$dA = \left| \frac{\partial \mathbf{P}}{\partial u} \times \frac{\partial \mathbf{P}}{\partial v} \right| du\, dv = |\mathbf{P}_u \times \mathbf{P}_v|\, du\, dv \tag{7.34a}$$

which is a polynomial $f_1(u, v)$. Consequently, the area is

$$A = \int_0^1 \int_0^1 f_1(u, v)\, du\, dv \tag{7.34b}$$

Volume

The volume enclosed by a parametric cubic surface $\mathbf{P}(u, v)$, which has the pyramidal differential element of Figure 7.11 is defined by a double integral. Here, Eq. (7.27) gives the differential volume,

$$dV = \tfrac{1}{3}(\mathbf{P} \cdot (\mathbf{P}_u \times \mathbf{P}_v))\, du\, dv = \tfrac{1}{3}((\mathbf{P} \times \mathbf{P}_u) \cdot \mathbf{P}_v)\, du\, dv \tag{7.35}$$

The volume is found from an integral of the form of Eq. (7.34b).

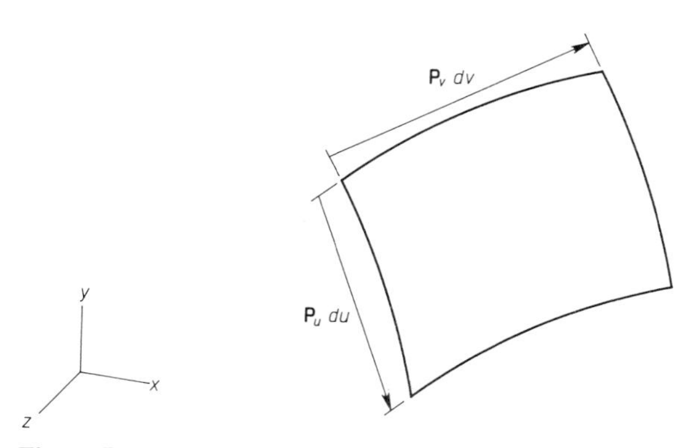

Figure 7.10 Differential surface area.

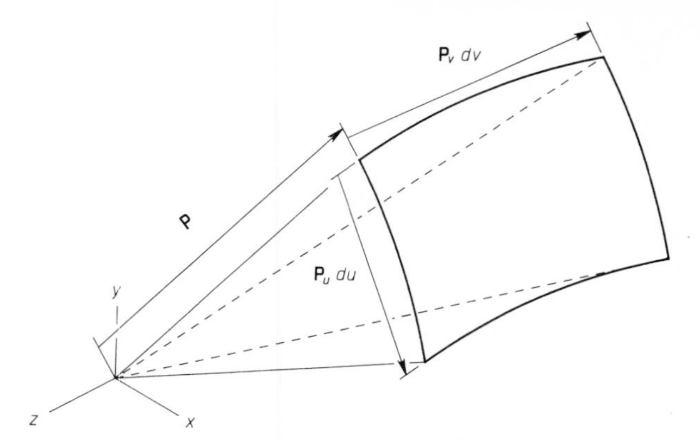

Figure 7.11 Differential volume bounded by surface.

Based on Eq. (7.24), the differential volume of a hyperpatch $\mathbf{P}(u, v, w)$, Eqs. (6.29) through (6.32), is

$$dV = ((\mathbf{P}_u \times \mathbf{P}_v) \cdot \mathbf{P}_w)\, du\, dv\, dw \tag{7.36}$$

The volume is determined by numerical integration of the resulting triple integral. The interested reader should see the references[36,44] for more details on the foregoing topics. The integrals developed in this section are evaluated with numerical integration methods such as Gaussian quadrature. Alternatively, properties of models assembled from an array of simpler elements such as triangles (in the case of surfaces) or tetrahedra (in the case of solids) are determined by combining the properties of each of the elements.

Distances between Points, Curves, and Surfaces

With parametric geometry, it is always possible to determine distances between entities by a trial-and-error process. In some special cases, however, there are more efficient methods.

The most basic problem is finding the minimum distance between two points $\mathbf{P}_1 = [x_1 \quad y_1 \quad z_1]$ and $\mathbf{P}_2 = [x_2 \quad y_2 \quad z_2]$, which from Eq. (7.16) is

$$d_{\min} = \sqrt{(x_2 - x_1)^2 + (y_2 - y_1)^2 + (z_2 - z_1)^2} \tag{7.37}$$

The minimum distance between a point $\mathbf{P}_O$ and a curve $\mathbf{P}(u)$ (for example, Eq. (5.40)) is a line perpendicular to a *tangent* to the curve as shown in Figure 7.12. The problem is to find the value of u which locates $\mathbf{P}_C$, the point on the curve where a line from P_O is perpendicular to the tangent line.

The vector from $\mathbf{P}_O$ to the curve is $(\mathbf{P}_C - \mathbf{P}_O)$. For this vector and the tangent vector to be perpendicular,

$$(\mathbf{P}_C - \mathbf{P}_O) \cdot \mathbf{P}'_C = 0 \tag{7.38}$$

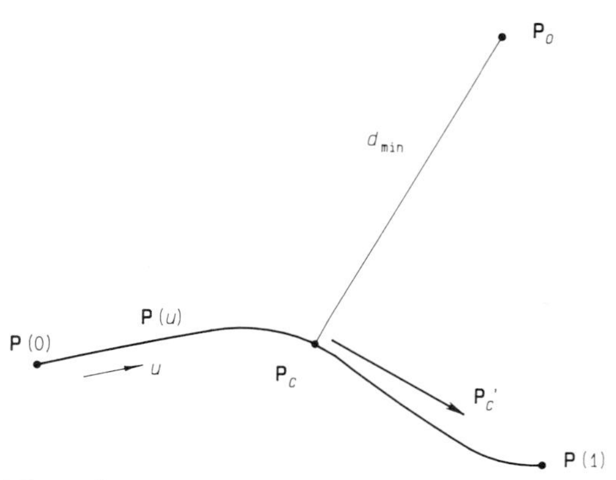

Figure 7.12 Minimum distance between a point and a curve.

In terms of the algebraic coefficients describing $\mathbf{P}_C$ (Eq. (5.33)), Eq. (7.38) may be expanded to give

$$c_0 + c_1 u + c_2 u^2 + c_3 u^3 + c_4 u^4 + c_5 u^5 = 0 \tag{7.39}$$

where
$$\begin{aligned}
c_0 &= \mathbf{a}_1 \cdot \mathbf{a}_2 - \mathbf{P}_O \cdot \mathbf{a}_2 \\
c_1 &= 2\mathbf{a}_1 \cdot \mathbf{a}_3 + \mathbf{a}_2 \cdot \mathbf{a}_2 - 2\mathbf{P}_O \cdot \mathbf{a}_3 \\
c_2 &= 3\mathbf{a}_1 \cdot \mathbf{a}_4 + 3\mathbf{a}_2 \cdot \mathbf{a}_3 - 3\mathbf{P}_O \cdot \mathbf{a}_4 \\
c_3 &= 4\mathbf{a}_2 \cdot \mathbf{a}_4 + 2\mathbf{a}_3 \cdot \mathbf{a}_3 \\
c_4 &= 5\mathbf{a}_3 \cdot \mathbf{a}_4 \\
c_5 &= 3\mathbf{a}_4 \cdot \mathbf{a}_4
\end{aligned}$$

After solution of Eq. (7.39) for one or more real roots, $0 \le u \le 1$, $\mathbf{P}_C$ is evaluated and the distance(s) is determined from Eq. (7.37). Since Eq. (7.39) may produce a maximum distance, the distances from the two end points of the curve should also be checked. In the case that Eq. (7.39) has no roots, the minimum is one of the two end point distances.

When $\mathbf{P}(u)$ is a straight line segment, $\mathbf{P}(u) = \mathbf{a}_1 + \mathbf{a}_2 u = \mathbf{P}(0) + c\mathbf{e}u$, where c is the length of the line and $\mathbf{e}$ is a unit vector along the line, Figure 7.13. Solution of Eq. (7.39) with $\mathbf{a}_3 = \mathbf{a}_4 = 0$ yields a single value for u, which, if in the range 0 to 1, describes the location of $\mathbf{P}_C$. If it is outside this range, the minimum distance is to one of the two end points.

For a line, not necessarily a finite segment, there is another, more common construction for the minimum distance. A unit vector $\mathbf{n}$ normal to the line in the plane of $\mathbf{P}_O$ and $\mathbf{e}$ is

$$\mathbf{n} = \frac{((\mathbf{P}(0) - \mathbf{P}_O) \times \mathbf{e}) \times \mathbf{e}}{|((\mathbf{P}(0) - \mathbf{P}_O) \times \mathbf{e}) \times \mathbf{e}|} \tag{7.40}$$

The minimum distance from the point to the line is thus

$$d_{\min} = |(\mathbf{P}_O - \mathbf{P}(0)) \cdot \mathbf{n}| \tag{7.41}$$

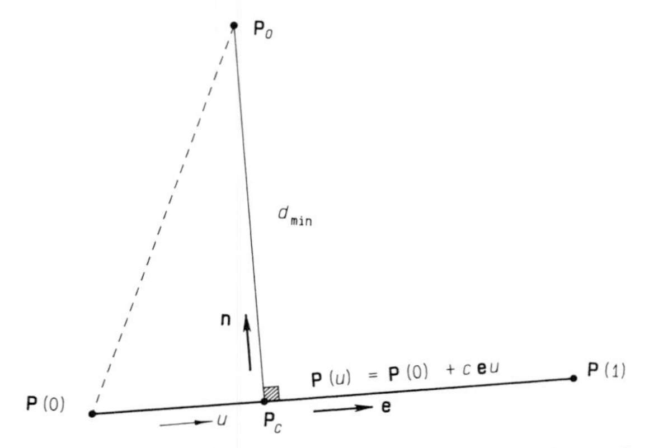

Figure 7.13 Minimum distance between a point and a line.

The minimum distance between two curves in space, $\mathbf{P}(u)$ and $\mathbf{P}(v)$, Figure 7.14, is found by iteration. Repeated application of the distance formula, Eq. (7.37), for a sequence of values of u and v is used to search for the smallest distance. Note that using the *square* of the distance in the search requires less computation.

The distance between two line segments can be determined directly, without iteration. The two lines shown in Figure 7.15 can be represented by the equations

$$\mathbf{P}_1(u) = \mathbf{P}_1(0) + c_1\mathbf{e}_1 u \qquad (0 \leq u \leq 1)$$

and

$$\mathbf{P}_2(v) = \mathbf{P}_2(0) + c_2\mathbf{e}_2 v \qquad (0 \leq v \leq 1) \tag{7.42}$$

where $\mathbf{P}_1(0)$ and $\mathbf{P}_2(0)$ are the points which locate the start of each line, $\mathbf{e}_1$ and $\mathbf{e}_2$ are unit vectors in the directions of the lines, and c_1 and c_2 are constants controlling the length of the line. The minimum distance (unknown) d_{min} has the direction given by the unit vector $\mathbf{e}_0$, which is perpendicular to $\mathbf{e}_1$ and $\mathbf{e}_2$.

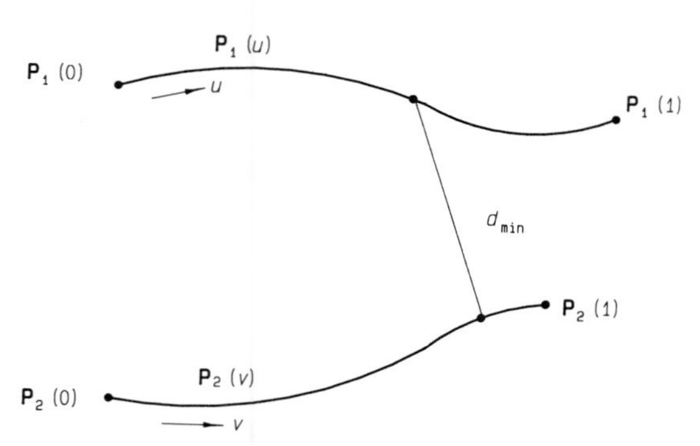

Figure 7.14 Minimum distance between two curves, an iterative process.

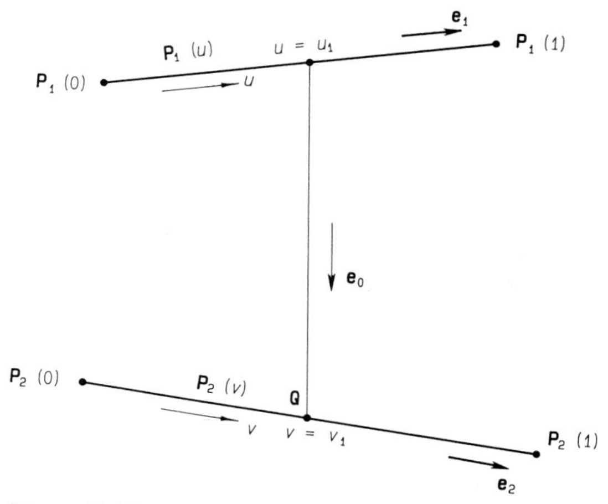

Figure 7.15 Notation for distance between two lines in space.

Two different paths from the origin to the same point Q provide the following relation:

$$\mathbf{P}_1(0) + c_1\mathbf{e}_1 u_1 + d_{\min}\mathbf{e}_0 = \mathbf{P}_2(0) + c_2\mathbf{e}_2 v_1 \tag{7.43}$$

The dot product of $\mathbf{e}_0$ with both sides of this equation yields

$$\mathbf{e}_0 \cdot (\mathbf{P}_1(0) + c_1\mathbf{e}_1 u_1 + d_{\min}\mathbf{e}_0) = \mathbf{e}_0 \cdot (\mathbf{P}_2(0) + c_2\mathbf{e}_2 v_1) \tag{7.44}$$

Since $\mathbf{e}_0$ is perpendicular to $\mathbf{e}_1$ and $\mathbf{e}_2$, $\mathbf{e}_0 \cdot \mathbf{e}_1 = \mathbf{e}_0 \cdot \mathbf{e}_2 = 0$. Thus,

$$d_{\min} = (\mathbf{P}_2(0) - \mathbf{P}_1(0)) \cdot \mathbf{e}_0 \tag{7.45}$$

The direction of $\mathbf{e}_0$ is given by the cross product,

$$\mathbf{e}_0 = \frac{\mathbf{e}_1 \times \mathbf{e}_2}{|\mathbf{e}_1 \times \mathbf{e}_2|} \tag{7.46}$$

Combination of Eqs. (7.45) and (7.46) completes the expression for minimum distance,

$$d_{\min} = \frac{(\mathbf{P}_2(0) - \mathbf{P}_1(0)) \cdot (\mathbf{e}_1 \times \mathbf{e}_2)}{|\mathbf{e}_1 \times \mathbf{e}_2|} \tag{7.47}$$

The $d_{\min}$ thus determined should be substituted into Eq. (7.43), which is a vector equation yielding three scalar equations. Any two of the three are solved for u_1 and v_1. If u_1 or v_1 are not in the range 0 to 1, the minimum distance will be from one of the two end points of the lines.

The minimum distance between a point in space and a curved surface can be determined iteratively. The formula for distance (or the square of the distance) is repeatedly evaluated as u and v for the surface are varied stepwise through the range 0 to 1.

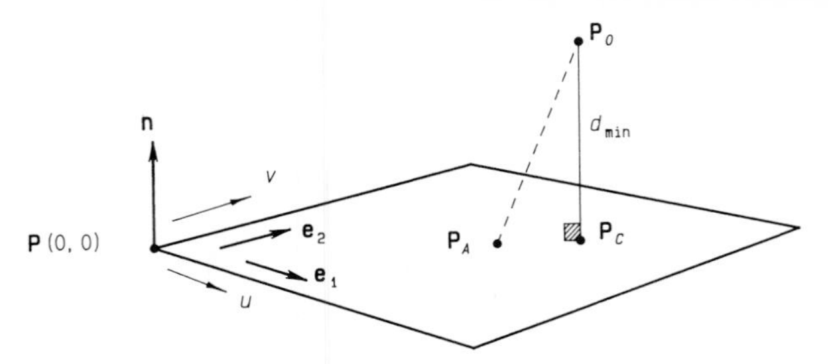

Figure 7.16 Distance from point to a plane.

The problem of determining the minimum distance between a point $\mathbf{P}_O = [x_O \quad y_O \quad z_O]$ and a plane $\mathbf{P}(u, v) = \mathbf{P}(0, 0) + c_1\mathbf{e}_1u + c_2\mathbf{e}_2v$ is shown in Figure 7.16. A vector from any arbitrary point on the plane $\mathbf{P}_A$ to $\mathbf{P}_O$ is $(\mathbf{P}_O - \mathbf{P}_A)$. For a plane with infinite extent,

$$d_{\text{min}} = (\mathbf{P}_O - \mathbf{P}_A) \cdot \mathbf{n} \tag{7.48}$$

where $\mathbf{n} = \mathbf{e}_1 \times \mathbf{e}_2/|\mathbf{e}_1 \times \mathbf{e}_2|$.

If the plane has definite bounds, the distance found from Eq. (7.48) is the solution if the vector which represents the minimum distance intersects the plane within its bounds (which are conventionally $0 \leq u \leq 1$ and $0 \leq v \leq 1$). This is done by solving any two of the scalar equations from

$$\mathbf{P}_O = \mathbf{P}(0, 0) + c_1\mathbf{e}_1u_1 + c_2\mathbf{e}_2v_1 + d_{\text{min}}\mathbf{n} \tag{7.49}$$

for u and v. If there is no solution in the range $0 \leq u \leq 1$ and $0 \leq v \leq 1$, the minimum distance can be determined with an adaption of Eq. (7.39) for each of the edges. If still no perpendicular minimum distance is found in the valid range of u and v, the minimum distance will be that distance to one of the vertices.

More details on methods for other distance problems are available.[36] When maximum generality is desired, iterative algorithms are the common choice.

EXAMPLE 7.2

Determine the distance between the point P_O:(4, 3, 2) and the triangular plane which lies in the first quadrant with its vertices at (1, 0, 0), (0, 1.5, 0), and (0, 0, 2).

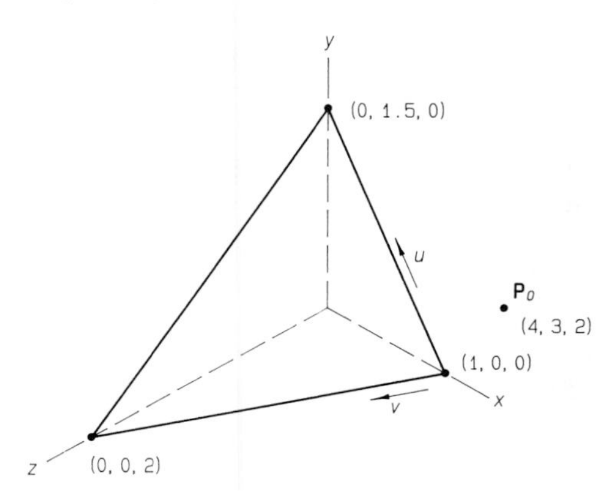

Solution. The triangular plane is described by the vector equation

$$\mathbf{P}(u, v) = \mathbf{P}(0, 0) + c_1\mathbf{e}_1 u + c_2\mathbf{e}_2 v$$
$$= \mathbf{i} + (-\mathbf{i} + 1.5\mathbf{j})u + (-\mathbf{i} + 2\mathbf{k})v$$
$$0 \le u \le (1 - v), \quad 0 \le v \le (1 - u)$$

The unit normal to the plane is

$$\mathbf{n} = \frac{(c_1\mathbf{e}_1 \times c_2\mathbf{e}_2)}{|(c_1\mathbf{e}_1 \times c_2\mathbf{e}_2)|} = \frac{(-\mathbf{i} + 1.5\mathbf{j}) \times (-\mathbf{i} + 2\mathbf{k})}{|(-\mathbf{i} + 1.5\mathbf{j}) \times (-\mathbf{i} + 2\mathbf{k})|}$$
$$= 0.768\mathbf{i} + 0.512\mathbf{j} + 0.384\mathbf{k}$$

From Eq. (7.48) the distance from $\mathbf{P}_O$ to an *unbounded* plane is

$$d_{\min} = (\mathbf{P}_O - \mathbf{P}_A) \cdot \mathbf{n}$$
$$= (3\mathbf{i} + 3\mathbf{j} + 2\mathbf{k}) \cdot (0.768\mathbf{i} + 0.512\mathbf{j} + 0.384\mathbf{k})$$
$$= 4.608$$

where $\mathbf{P}_A = \mathbf{P}(0, 0)$.

Next, the $d_{\min}$ just found is tested to determine whether it is the minimum distance to the *bounded* plane. Using Eq. (7.49),

$$\mathbf{P}_O = d_{\min}\mathbf{n} + \mathbf{P}(0, 0) + c_1\mathbf{e}_1 u_1 + c_2\mathbf{e}_2 v_1$$

or

$$(4\mathbf{i} + 3\mathbf{j} + 2\mathbf{k}) = 4.608(0.768\mathbf{i} + 0.512\mathbf{j} + 0.384\mathbf{k}) + \mathbf{i}$$
$$+ (-\mathbf{i} + 1.5\mathbf{j})u_1 + (-\mathbf{i} + 2\mathbf{k})v_1$$

Two scalar equations are conveniently formed from the coefficients of $\mathbf{j}$ and $\mathbf{k}$,

$$3 = 2.359 + 1.5u_1 \quad \text{and} \quad 2 = 1.769 + 2v_1$$

which have the solutions $u_1 = 0.427$ and $v_1 = 0.115$. These solutions fall within the triangular region, since $u_1 > 0$, $v_1 > 0$, and $u_1 + v_1 < 1$. Thus, $d_{\min} = 4.608$.

Intersections

Intersections of curves and planes are used in applications to construct surfaces and to check for interference. While cases involving curved lines and surfaces often require iterative solutions, intersections involving straight lines and flat planes can be found directly.

Two straight lines may intersect at one point or may be parallel, collinear, or skew. With the equations for the two lines given by Eq. (7.42), simultaneous linear equations may be formed to define the intersection point $\mathbf{P}_i$,

$$\mathbf{P}_i = \mathbf{P}_1(0) + c_1\mathbf{e}_1 u = \mathbf{P}_2(0) + c_2\mathbf{e}_2 v \tag{7.50}$$

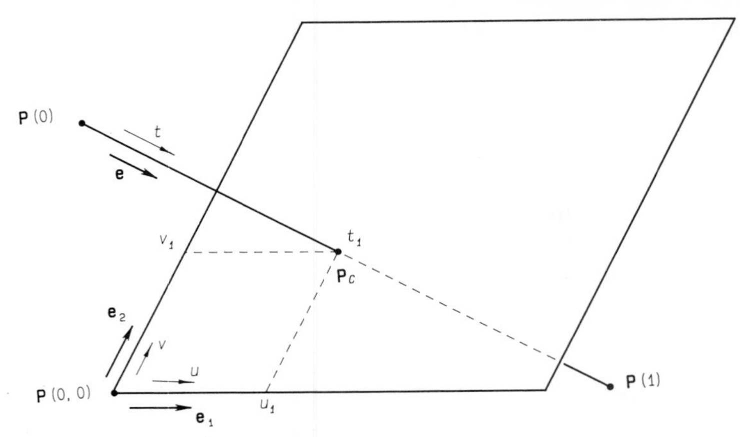

Figure 7.17 Intersection of line and plane.

There are two unknowns, u and v, and three equations available—one for each of the coordinate directions. The extra equation can be disregarded, or it can be used to check the solution.

The intersection of a straight line with a plane is shown in Figure 7.17. The equation of the plane is $\mathbf{P}(u, v) = \mathbf{P}(0, 0) + c_1\mathbf{e}_1 u + c_2\mathbf{e}_2 v$ and the equation of the line is $\mathbf{P}(t) = \mathbf{P}(0) + c\mathbf{e}t$, where the parameters u, v, and t have the usual range 0 to 1. The system of equations which describes the point of intersection is

$$\mathbf{P}_C = \mathbf{P}(0, 0) + c_1\mathbf{e}_1 u_1 + c_2\mathbf{e}_2 v_1 = \mathbf{P}(0) + c\mathbf{e}t_1 \tag{7.51}$$

Here, the three unknowns u_1, v_1, and t_1 are uniquely determined by the three available equations in the coordinate directions. If the solutions are out of the

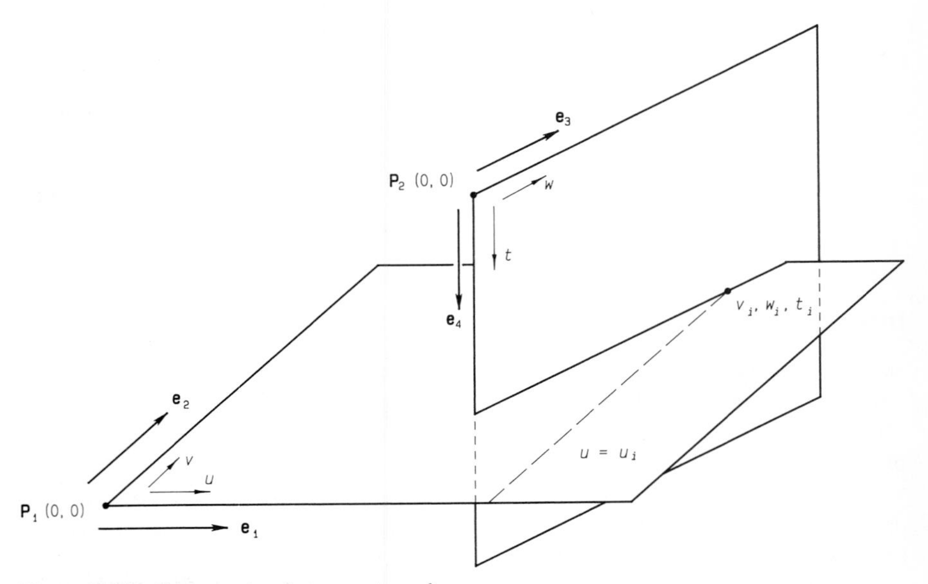

Figure 7.18 Intersection between two planes.

range 0 to 1, the intersection lies on the extended plane (or line). If the line is parallel to the plane or if the line lies in the plane, there is no unique solution.

Two flat planes, Figure 7.18, intersect in a straight line unless they are parallel or coincident. The equations of the two planes are $\mathbf{P}_1(u, v) = \mathbf{P}_1(0, 0) + c_1\mathbf{e}_1 u + c_2\mathbf{e}_2 v$ and $\mathbf{P}_2(w, t) = \mathbf{P}_2(0, 0) + c_3\mathbf{e}_3 w + c_4\mathbf{e}_4 t$. The resulting system of linear equations has four unknowns. If one of the unknowns takes on some assumed value, for example, $u = u_1$, the available three equations can be used to determine the corresponding values of the remaining three unknowns. Clearly, a series of points which lie on the intersection can be determined. Again, any solutions for the parameters which are outside the range 0 to 1 indicate an intersection on the extended plane(s).

Details of intersection problems involving curves and curved surfaces can be found in the literature.[10,36] In general these are iterative processes.

EXAMPLE 7.3

Determine the line of intersection between the triangular plane $\mathbf{P}(u, v)$ of Example 7.2 and the rectangular plane described by the equation

$$\mathbf{P}(w, t) = 3\mathbf{k}w + (2\mathbf{i} + \mathbf{j})t \qquad 0 \le w \le 1, \quad 0 \le t \le 1$$

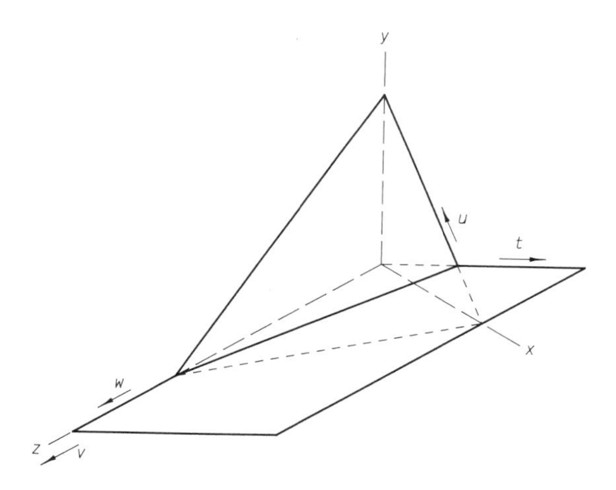

Solution. The components of $\mathbf{P}(u, v)$ and $\mathbf{P}(w, t)$ corresponding to $\mathbf{i}$, $\mathbf{j}$, and $\mathbf{k}$ yield the three equations

$$1 - u - v = 2t$$

$$1.5u = t$$

$$2v = 3w$$

Because both planes are in the first quadrant, the line of intersection could be assumed to have its ends at $w = 0$ and $t = 0$. For the end $w = 0$, the system of three equations yields the solutions $u = 1/4$, $t = 3/8$, and $v = 0$. For the end $t = 0$, the solutions are $u = 0$, $v = 1$, and $w = 2/3$. (The fact that $v = 0$ at one end of the line and $u = 0$ at the other demonstrates that the line of intersection starts and ends at the edges of both planes.) Substitution of the given values into the equation of either plane yields the end points $(\frac{3}{4}\mathbf{i} + \frac{3}{8}\mathbf{j})$ and $2\mathbf{k}$. With this information, the equation of the line of intersection can be written in parametric form as

$$\mathbf{P}(s) = (\tfrac{3}{4}\mathbf{i} + \tfrac{3}{8}\mathbf{j}) + (-\tfrac{3}{4}\mathbf{i} - \tfrac{3}{8}\mathbf{j} + 2\mathbf{k})s \qquad (0 \le s \le 1)$$

Angular and Directional Properties

Engineering problems in descriptive geometry also deal with determination of directions of lines and planes. While the direction of a line is completely specified by $\mathbf{e} = e_x\mathbf{i} + e_y\mathbf{j} + e_z\mathbf{k}$, a unit vector along the line, and the direction of a plane is completely specified by $\mathbf{n} = n_x\mathbf{i} + n_y\mathbf{j} + n_z\mathbf{k}$, a unit vector normal to the plane, it is useful to relate these to the terms found in engineering practice.

In Figure 7.19, the unit vectors are abitrarily arranged so that **i** is east, **j** is north, and **k** is up. The *bearing* of a line described by the unit vector **e** is

$$\text{Bearing} = \{N \text{ or } S\}\theta\{E \text{ or } W\} \tag{7.52}$$

where N is used if $e_y \ge 0$ or S is used if $e_y < 0$, $\theta = |\tan^{-1}(e_x/e_y)|$, and E is used if $e_x \ge 0$ or W is used if $e_x < 0$.

The bearing is undefined if $e_z = 1$. Examples of use are the bearings $S\ 30°\ E$ and $N\ 5°\ W$.

The slope ϕ of the line is defined as the angle the line makes with the horizontal. With reference to Figure 7.19, the slope is

$$\text{Slope} = \phi = 90° - \cos^{-1}(\mathbf{e} \cdot \mathbf{k}) \tag{7.53}$$

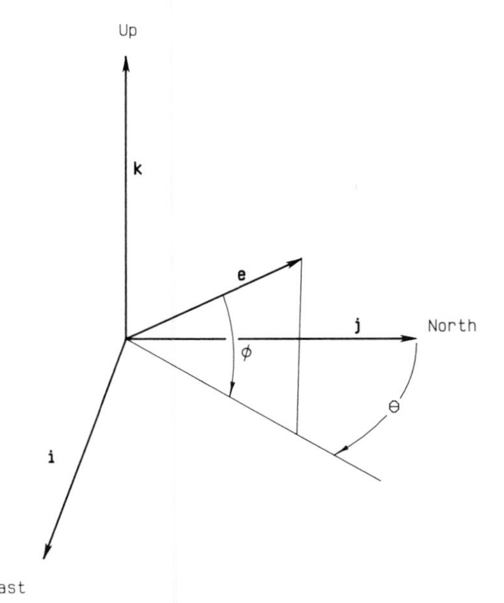

Figure 7.19 Bearing and slope of a line.

where ϕ denotes the slope angle in degrees. The value of $\tan \phi$ converted to percent is known as *grade*. The positive or negative sense applied to grade denotes "uphill" or "downhill," respectively.

Geological and mining engineers describe the directions of strata of ore bodies by *strike* and *dip*. To define strike and dip of a plane, an *upward* unit normal **n** is required, that is, $n_z > 0$. A *downward* normal unit vector can be corrected by setting $\mathbf{n} \equiv -\mathbf{n}$.

The *strike* of a plane is defined as the bearing of a horizontal line in the plane, which is $\mathbf{k} \times \mathbf{n} = (-n_y\mathbf{i} + n_x\mathbf{j})$. A horizontal line in the opposite direction results from $\mathbf{n} \times \mathbf{k} = (n_y\mathbf{i} - n_x\mathbf{j})$. Thus, there are two possible ways to write the same strike, since a horizontal line in the plane is bidirectional. It follows that strike need only be written with respect to north, as

$$\text{Strike} = N\theta\{E \text{ or } W\} \tag{7.54}$$

where $\theta = |\tan^{-1}(n_y/n_x)|$ and E is used if $n_x n_y \le 0$ or W is used if $n_x n_y > 0$.

Examples of use are N 23° W, N 15° E, and S 15° W, where the latter two have the same meaning.

The steepest line down a plane is called the *fall line,* so named because of the direction in which objects roll down the plane. The angle the fall line makes with the horizontal is the *dip angle,* or *dip,* which also is the angle between the unit normal **n** to the plane and the **k** (upward) direction. The general direction of the plane is added to the complete specification of dip,

$$\text{Dip} = \phi\{N \text{ or } S\}\{E \text{ or } W\} \tag{7.55}$$

where $\phi = \cos^{-1}(\mathbf{n} \cdot \mathbf{k})$, N is used if $n_y \ge 0$ or S is used if $n_y < 0$, and E is used if $n_x \ge 0$ or W is used if $n_x < 0$.

The dip angle ϕ ranges from 0° for a plane with **n** vertical to 90° for a plane with **n** horizontal. An example of use is the specification 30° SE. As shown in Figure 7.20, strike can be visualized as the direction of the top of a roof and dip as the slope of a rafter.

The *dihedral angle* between two planes is

$$\phi = \cos^{-1}(\mathbf{n}_1 \cdot \mathbf{n}_2) \tag{7.56}$$

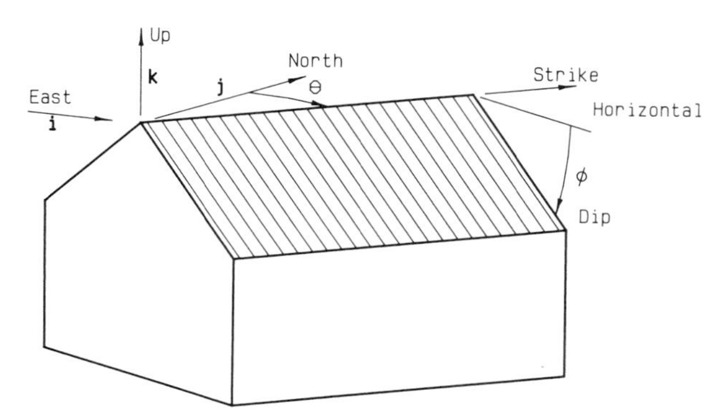

Figure 7.20 Illustration of strike and dip.

where $\mathbf{n}_1$ and $\mathbf{n}_2$ are the unit normals to the two planes. Note that the dip is the dihedral angle between a given plane and a horizontal plane.

Angular and directional properties may be ascribed to points on curved lines and surfaces described by parametric geometry. The geometric slope at the point $\mathbf{P}(u_i)$ on a curve is found from Eq. (5.43). The unit normal at a point $\mathbf{P}(u_i, v_i)$ on a curved surface is given by

$$\mathbf{n} = \frac{\mathbf{P}_u(u_i, v_i) \times \mathbf{P}_v(u_i, v_i)}{|\mathbf{P}_u(u_i, v_i) \times \mathbf{P}_v(u_i, v_i)|} \tag{7.57}$$

where the parametric derivatives are defined in Eq. (6.19). Similar results may be found for curves and surfaces described by Bezier and B-spline geometry.

EXAMPLE 7.4

Core samples show that the top of a stratum of interest is located at the three given points A:(0, 0, −100), B:(150, 0, −120), and C:(100, 120, −80), where the coordinate system is located such that **i** is east, **j** is north, and **k** is up. Determine (a) the bearing and grade of the line from A to C, and (b) the strike and dip of the plane defined by A, B, and C.

Solution. **(a)** A unit vector along the line AC is

$$\mathbf{e} = \frac{\mathbf{P}_C - \mathbf{P}_A}{|\mathbf{P}_C - \mathbf{P}_A|} = \frac{100\mathbf{i} + 120\mathbf{j} + 20\mathbf{k}}{\sqrt{100^2 + 120^2 + 20^2}} = 0.635\mathbf{i} + 0.762\mathbf{j} + 0.127\mathbf{k}$$

The angle θ is determined from Eq. (7.52) as $|\tan^{-1}(0.635/0.762)| = 39.8°$. Since e_y and e_x are both positive,

$$\text{Bearing} = N\ 39.8°\ E$$

The slope, from Eq. (7.53), is

$$\phi = 90° - \cos^{-1}(e_z) = 90° - \cos^{-1}(0.127) = 7.30°$$

Since tan (7.30°) = 0.128,

$$\text{Grade} = +12.8\%$$

(b) A unit normal to this plane is formed with the use of the cross product between $(\mathbf{P}_B - \mathbf{P}_A)$ and $(\mathbf{P}_C - \mathbf{P}_A)$,

$$\mathbf{n} = \frac{(150\mathbf{i} - 20\mathbf{k}) \times (100\mathbf{i} + 120\mathbf{j} + 20\mathbf{k})}{|(150\mathbf{i} - 20\mathbf{k}) \times (100\mathbf{i} + 120\mathbf{j} + 20\mathbf{k})|}$$

$$= 0.127\mathbf{i} - 0.265\mathbf{j} + 0.956\mathbf{k}$$

which is an upward normal as required. With Eq. (7.54) the strike is found from $|\tan^{-1}(-0.265/0.127)| = 64.4°$ as

$$\text{Strike} = N\ 64.4°\ E$$

and dip is found with Eq. (7.55) with $\phi = \cos^{-1}(0.956) = 17.1°$ as

$$\text{Dip} = 17.1°\ SE$$

7.5 CLOSURE

The foregoing procedures show some of the ways in which information in a computer graphics data base can be used to evaluate geometric properties and solve problems involving distances, angles, and intersections. Features common to most procedures include numerical integration and iteration, both of which are well adapted to computing.

Another strategy used for geometric property determination is cell decomposition, which involves subdividing the object into smaller constitutive parts. As shown in Chapter 10, cell decomposition is the basis for finite element analysis, where complex objects are represented by an assembly of simpler objects—finite elements—that have known properties. A special case of cell decomposition, *spatial occupancy enumeration,* uses cells that are cubical and located in a regular 3-D grid. The resolution of the model is determined by the cell size. Volume is determined by counting the cells; other geometric properties are computed by replacing the integral process with a summation process.

Evaluation of geometric properties and solution of descriptive geometry problems are features expected in solid modeling systems, a still-emerging technology.

PROBLEMS

7.1. Verify that Eq. (7.20) reduces (a) to Eq. (7.18) in the case of a triangle and (b) to the correct result in the case of a parallelogram.

7.2. Rewrite the integrals in Eqs. (7.21) and (7.22) where the limitations on the coordinate system do not apply.

7.3. Rewrite Eq. (7.25) where the special coordinate system with node 0 at $x = y = z = 0$ is replaced by one where node 0 is at an arbitrary location, x_0, y_0, z_0.

7.4. Rewrite Eq. (7.32) so that the geometric coefficients of parametric cubic geometry are used instead of algebraic coefficients.

7.5. A 40-mm-diameter circle is centered on the origin. Check the percentage of difference between the arc length predicted by Eq. (7.15) and the exact solution where the number of chords is (a) 16, (b) 64, and (c) 256.

7.6. Develop a parametric cubic representation for the portion of the circle of Problem 7.5 in the first quadrant. Apply Eq. (7.33) with three-point Gaussian quadrature to approximate the arc length of the quarter circle. Compare the result to the exact value.

7.7. An ellipse is described by the equation

$$x^2/16 + y^2/9 = 1$$

In the first quadrant, define four chords with end points of equally spaced values of x. Taking advantage of symmetry by working in the first quadrant, use Eq. (7.15) to determine the arc length of the entire ellipse.

7.8. Use Eq. (7.20) with the ellipse and chord points of Problem 7.7 to determine the area enclosed. Work in the first quadrant and apply symmetry.

7.9. Determine the area enclosed under the ellipse of Problem 7.7 using four-point Gaussian quadrature. Use symmetry properties.

7.10. Determine the product of inertia of area with respect to the origin of the first quadrant portion of the ellipse described in Problem 7.7. Use four-point Gaussian quadrature and symmetry properties.

7.11. A triangle has vertices at (0, 0), (3, 0), and (3, 6). With the aid of Eq. (7.21), determine the polar moment of inertia of this triangle with respect to its centroid.

7.12. A regular hexagon has sides of length b. By decomposing the figure into triangles and using symmetry properties, use Eq. (7.21) to compute the area moment of inertia of this figure with respect to a symmetrical axis that connects two vertices.

7.13. A tetrahedron has vertices at (0, 0, 0), (0, 1, 0), (2, 2, 1), and (0, 3, −2). Determine the enclosed volume.

7.14. Determine the moment of inertia of mass with respect to the x axis of the tetrahedron in Problem 7.13. Assume that the density ρ is a constant, so that $dm = \rho\, dx\, dy\, dz$. Leave your answer in terms of ρ.

7.15. Divide a right circular cylinder of diameter d and height h into wedges and make use of Eq. (7.26) to determine its volume. Compare the exact solution with (a) 8 wedges, (b) 16 wedges.

7.16. A curve which passes through the four points in the x-y plane (0, 0), (1, 2), (2, 1.5), and (3, 0.5) is translated 4 units in the z direction to form a surface. Estimate the curve length and surface area with Eq. (7.15).

7.17. Repeat Problem 7.16, but use a parametric cubic approximation of the curve and surface. Integrate with three-point Gaussian quadrature.

7.18. Determine the arc length of a planar cubic Bezier curve with the control points (0, 0), (1, 2), (3, 2), and (4, 1). Use the parametric derivative of Eq. (5.63) in Eqs. (7.32) and (7.33) appropriately modified for the Bezier form.

7.19. Repeat Problem 7.18 for a Bezier curve which has the control points (−1, 1), (1, 2), (3, 2), and (0, 4).

7.20. A planar parametric cubic curve passes through the points (0, 0), (1, 1), (2, 1), and (4, 0). Determine the minimum distance between this curve and the point (3, 3). *Note:* Use of a digital computer or programmable calculator is recommended to iteratively find the real roots of the resulting polynomial.

7.21. A planar cubic Bezier curve is defined by the control points (0, 0), (2, 2), (5, 3), and (5, 0). Determine the minimum distance between this curve and the point (4, 3). See Note, Problem 7.20.

7.22. Determine the minimum distance between the line from (0, −2, 2) to (4, 2, 4) and the point (3, 0, 0).

7.23. Determine the minimum distance between the line from the origin to (2, 3.75, 0) and the point (2.5, 0, 1).

7.24. A line segment AB connects the points (0, 0, 0) and (2, 1, 2). Line segment CD connects the points (−1, 1, 4) and (3, 2, 0). Determine the minimum distance between these two line segments.

7.25. Determine the minimum distance between the two line segments EF and GH. EF has length 5 units with the direction cosines (0.6667, −0.6667, 0.3333) and starts at the origin. GH connects the points (−1.0, 2.5, −2.0) and (4.0, 2.0, 3.0).

7.26. A triangular plane has vertices on the coordinate axes at $x = 5$, $y = 6$, and $z = 9$. Determine the minimum distance from the origin to the plane.

7.27. Determine the minimum distance from (6, 6, 9) to the plane of Problem 7.26.

7.28. An unbounded plane with the normal $\mathbf{n} = 0.346\mathbf{i} - 0.773\mathbf{j} + 0.532\mathbf{k}$ intersects the y axis at -2. Determine the minimum distance between this plane and the point $(2, 0, -1)$.

7.29. A line segment connects the points $(-1, -2, -1)$ and $(2, 3, 2)$. Where does this line intersect the y axis? If it does not intersect, determine the minimum distance to the y axis.

7.30. Determine the intersection, if any, between the two line segments OA and BC, where the points are defined by O:(0, 0, 0), A:(3, 6, 6), B:(3, −2, 0), and C:(−1, 6, 4).

7.31. A line segment connects the points (2, 3, 2) and (−6, 1, 3). Where does this line segment intersect the y-z plane?

7.32. An unbounded line is parallel to the y axis with $x = 3$ and $z = -2$. Determine the point of intersection between this line and an unbounded plane that cuts the coordinate axes at $x = 6$, $y = 2$, and $z = 3$.

7.33. The direction cosines of a line of length 10 with one end at the origin are (0.280, 0.576, 0.768). Determine where this line intersects the unbounded plane which cuts the coordinate axes at $x = 2$, $y = 5$, and $z = 3$.

7.34 Determine the equation of the line of intersection between the plane of Problem 7.26 with an unbounded plane which is parallel to and 2 units above the x-z plane.

7.35. Determine the equation of the line of intersection between an unbounded plane which is parallel to the x-y plane and which intersects the z axis at -1, and an unbounded plane with the normal $\mathbf{n} = 0.3845\mathbf{i} + 0.9231\mathbf{j}$.

7.36. Determine the angle between the line with the direction cosines $(0.2308\mathbf{i} - 0.9231\mathbf{j} + 0.3077\mathbf{k})$ and the unbounded plane which has the normal $\mathbf{n} = (0.3411\mathbf{i} + 0.4571\mathbf{j} - 0.8214\mathbf{k})$.

7.37. Determine the bearing and slope of the line in Problem 7.36, assuming that **i** is east and **j** is north.

7.38. Determine the angle between the line and plane described in Problem 7.33.

7.39. Determine the strike and dip of the plane of Problem 7.26. Assume that **i** is east and **j** is north.

7.40. Three coordinates of the top surface of an ore body are (0, 0, −2000), (1000, 0, −2450), and (0, 500, −2700). Determine the strike and dip of this surface.

7.41. Determine the angle between two unbounded planes: One is the plane which passes through the points (0, 0, 0), (1, 2, 2) and (−2, 1, 2) and the other is the plane $y = 1$.

7.42. Determine the dihedral angle between two unbounded planes: Plane 1 cuts the coordinate axes at $x = -1$, $y = 2$, and $z = 1$. Plane 2 cuts the coordinate axes at $x = 1$, $y = -1$, and $z = 3$.

PROJECTS

For the projects that do not require the development of a program, an available solid modeling system may be used.

7.1. Develop a program which digitizes plane areas and computes the surface area enclosed. As a first step, two points of known distance apart should be digitized and used as a scaling factor. Then the user should digitize any arbitrary number of points around a closed contour to approximate the enclosed area. The program should have the ability to subtract "holes." The computed area should be plotted and filled (if convenient). Check your program on known problems.

7.2. Add the ability to Project 7.1 to determine the location of the centroid of the areas.

7.3. Develop a program which plots parametic cubic curves as seen from any selected viewpoint and computes the length using Eq. (7.15). The input should be four arbitrary points. Devise tests of accuracy of your program.

7.4. Same as Project 7.3, except that the length is computed with Eq. (7.33).

7.5. Develop a program which fits a parametric cubic spline through an arbitrary number of given points and computes its length. The resulting spline should be capable of being viewed from any selected viewpoint.

7.6. Same as Project 7.5, except the program fits a cubic B-spline.

7.7. Design a water tank to contain 80,000 gallons of water which provides a head of 100 feet when it is 80% full and a head of 80 feet when it is 20% full. Work to minimize material by minimizing surface area of the tank. Illustrate your design with a support structure included.

7.8. Design a helium-filled dirigible which has a cylindrical shape 10 meters in diameter, with a length of 80 meters. The ends should have curved, aerodynamic tapers. Assume the skin and structure taken together have an average mass (based on surface area) of 8 kg/m^2. Determine the payload of your design, assuming that the density of helium is 28% that of air. Provide an illustration of your design.

7.9. Design a rocket. Determine the surface area of the rocket and its fins, and the volume enclosed. Provide an illustration.

7.10. Obtain a topographical map of a site suitable for construction of a dam. Develop and demonstrate a methodology for estimating the volume of water (acre-feet) impounded for various appropriate heights of the dam.

7.11. Develop a program which computes the minimum distance between two separate 3-D parametric cubic curves which are specified in geometric form. The program will require iteration. Test the program on some simple problems with known solutions, including ones with straight lines. The input curves and the line representing the distance should be presented graphically.

7.12. Same as Project 7.11, except that B-spline curves are used.

7.13. Same as Project 7.11, except that Bezier curves are used.

7.14. Same as Project 7.11, except that the program should compute and show the minimum distance between a parametric cubic curve and a parametric cubic patch.

7.15. Same as Project 7.11, except that the program should compute and show the minimum distance between two parametric cubic patches.

7.16. Develop a program which determines the intersection between a parametric cubic patch and a parametric cubic curve, both specified in geometric form. The program should provide an illustration.

7.17. Develop a program which determines the intersection between two parametric cubic patches, both in geometric form. The two patches and the intersection line should be displayed graphically.

7.18. Obtain data for core samples, and reconstruct an illustration of an ore body. Estimate the volume of the ore body.

4

Applications

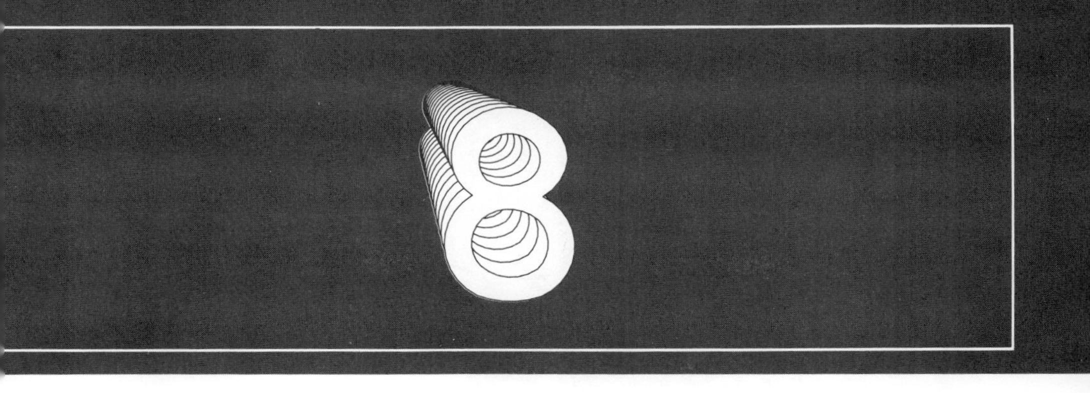

Data Presentation Graphics

A common engineering use of computer graphics is the presentation of quantitative data. By condensing and organizing information, graphics enhances interpretation and understanding. Comparison of numerical data is aided because the relative magnitudes of the data items are shown pictorially rather than as numbers. Furthermore, presentation graphics can be used to compare spatial data relationships such as locations on maps and flowcharts.

In engineering, the most widely used data presentation graphic is the *line graph* or *two-dimensional data plot*. One such plot can convey a large amount of information, since several sets of data can share the same axes. Commonly, 2-D data plots use Cartesian coordinates, as in the example of Figure 8.1. In this type of presentation, it is conventional that the horizontal axis (x axis) displays the *independent* variable and the vertical axis (y axis) displays the *dependent* variable. The quality and visual impact of 2-D data plots may be enhanced by using curve smoothing and fitting techniques such as those described in Sections 8.4 and 8.5. In cases where the independent variable is an angular or directional quantity, the 2-D data plot may use polar coordinates, as in the example of Figure 8.2.

A *pie chart* compares the proportional distribution of a single set of data, as in the example in Figure 8.3. This type of presentation graphic is effective in applications which show how a budget is divided, compare the proportion of employees by division in a company, and so forth. A pie chart should be used only when a small number of data are to be compared, say six or less.

Related to the pie chart is the *Venn diagram* which finds use in relational analysis, constructive solid geometry modeling, statistics, and logic. Venn diagrams employ circles or ellipses to indicate what classes of data are included or excluded and how the classes overlap. The example of Figure 8.4 shows interrelationships among certain classes of software capabilities in a particular

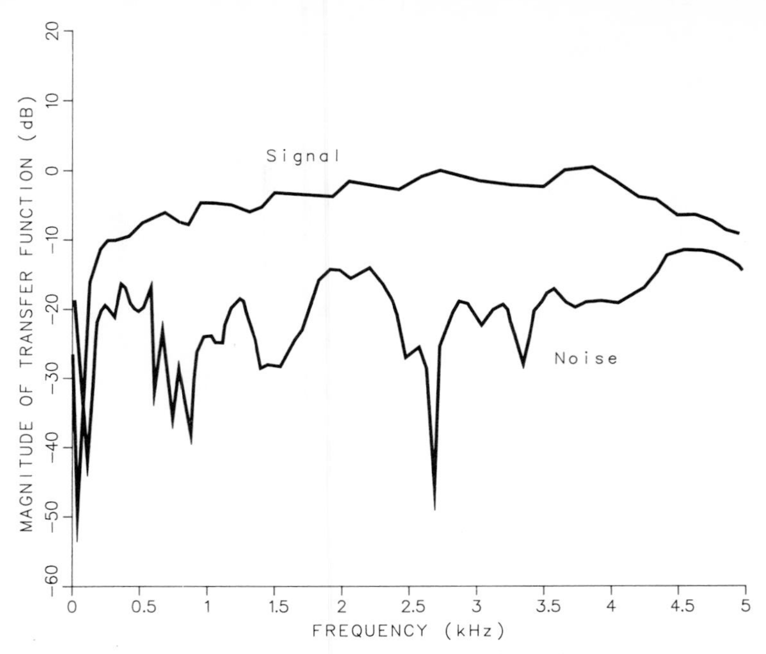

Figure 8.1 Two dependent variables in a 2-D data plot.

CAE system. The type of shading or color in the circles can be used to distinguish between classes of sets. The Boolean operations of union, difference, and intersection can be illustrated with Venn diagrams.

A *bar graph* or column chart is suited for dealing with a small number of data. This type of presentation effectively compares relationships between discrete groups of data, as in Figure 8.5. Bar graphs are particularly appropriate when one of the data axes is nonnumeric. Classes of data may be differentiated or emphasized by the use of different shading or coloration of the bars.

The *histogram* or step chart is similar to a bar chart, except there are no spaces between the bars and both axes have numerical values. The width of the bars, sometimes referred to as "buckets," represents the size of interval covered by the data. The height of the bars represents the number of occurrences captured in that particular bucket. As shown in Figure 8.6, a histogram shows trends in probabilistic situations. Histograms should be employed with care: Possibly very different histograms can be constructed with minor changes in the locations and sizes of the buckets.

A histogram in polar coordinates is called a *rose diagram,* Figure 8.7, which is useful for comparing frequency distributions of quantities which are directional. This type of plot is very similar to a polar 2-D data plot, except that the data are in buckets. Applications are found in electrical engineering, atmospheric science, oceanography, oil and gas production, and so forth.

A *contour plot,* such as the one in Figure 8.8, is used to show 3-D data relationships in geotechnical and other engineering work. The axes of a contour

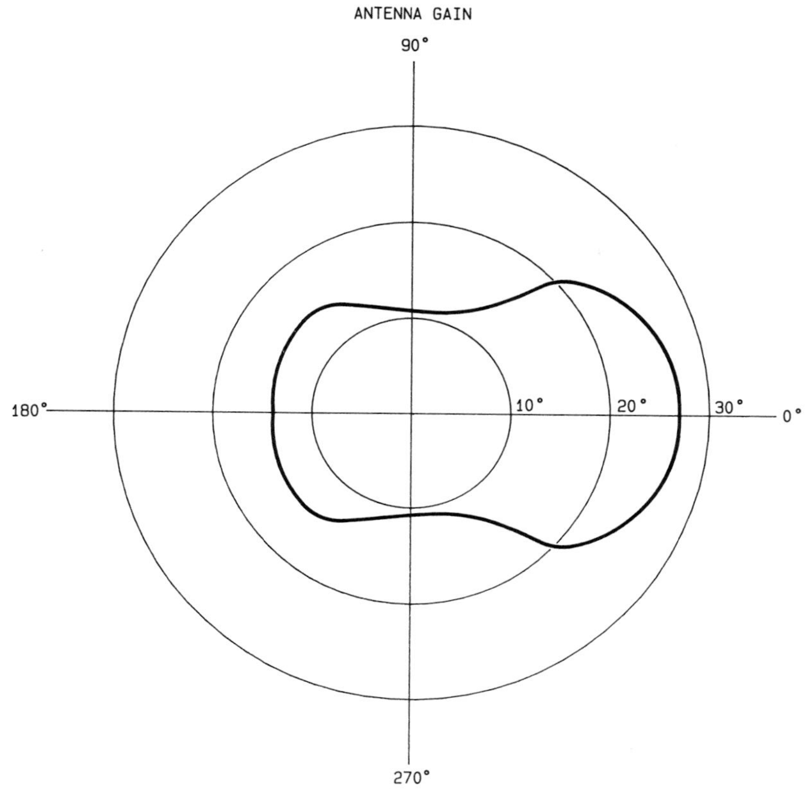

Figure 8.2 Two-dimensional data plot in polar coordinates, showing the variation of a continuous quantity as a function of direction.

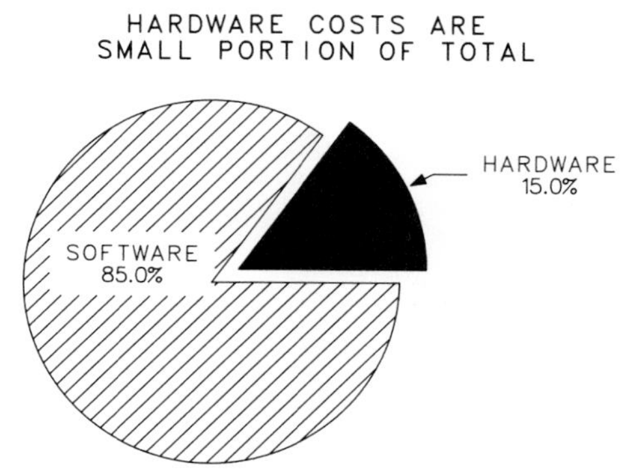

Figure 8.3 Pie chart that focuses attention on a single detail.

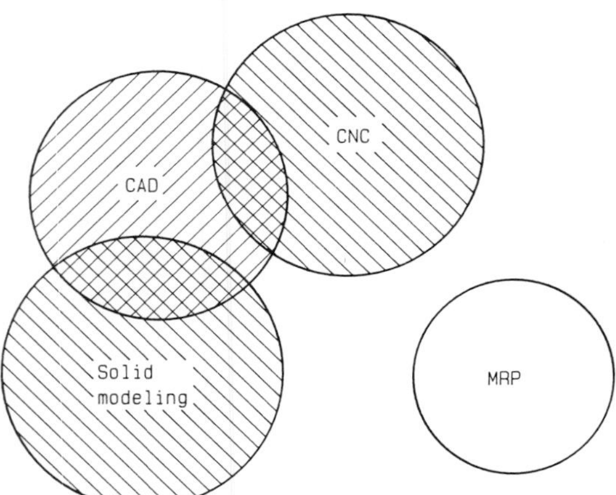

Figure 8.4 Venn diagram showing software availability in a particular CAE system and the extent of overlapping capabilities.

plot are the two independent variables; constant values of the dependent variable are plotted as contour lines. These plots are useful for presenting data in surveying, stress analysis, heat transfer, hydrology, magnetic fields, weather, and other such applications. High-quality contour plots often require some kind of curve smoothing. More on contour plots can be found in Section 8.6. Color can be

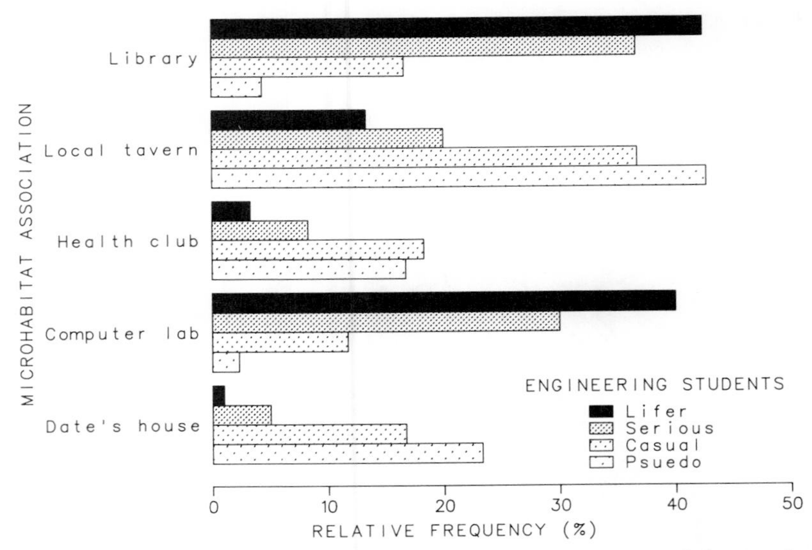

Figure 8.5 Bar graph comparing similar quantities, where one of the axes is not necessarily a numerical quantity.

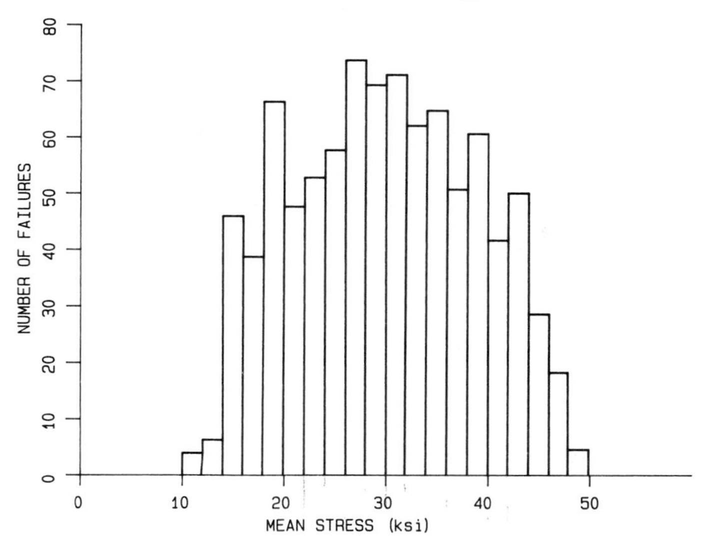

Figure 8.6 Histogram showing a frequency distribution.

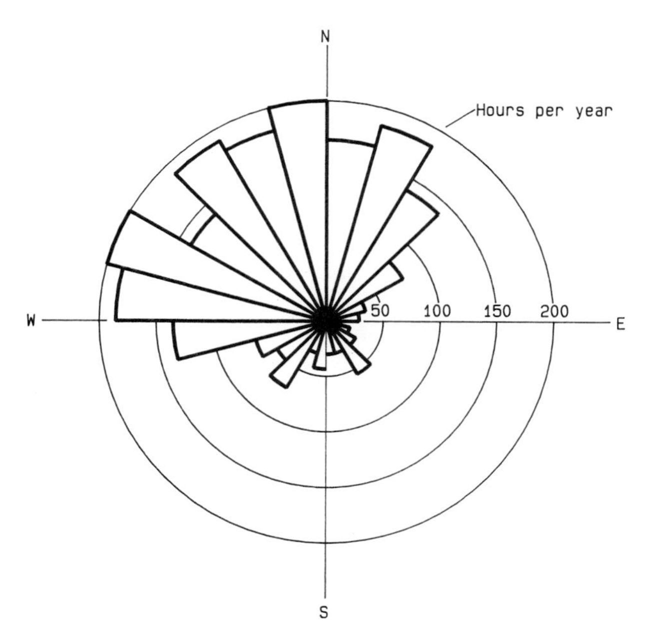

Figure 8.7 Rose showing the frequency distribution of directional quantities.

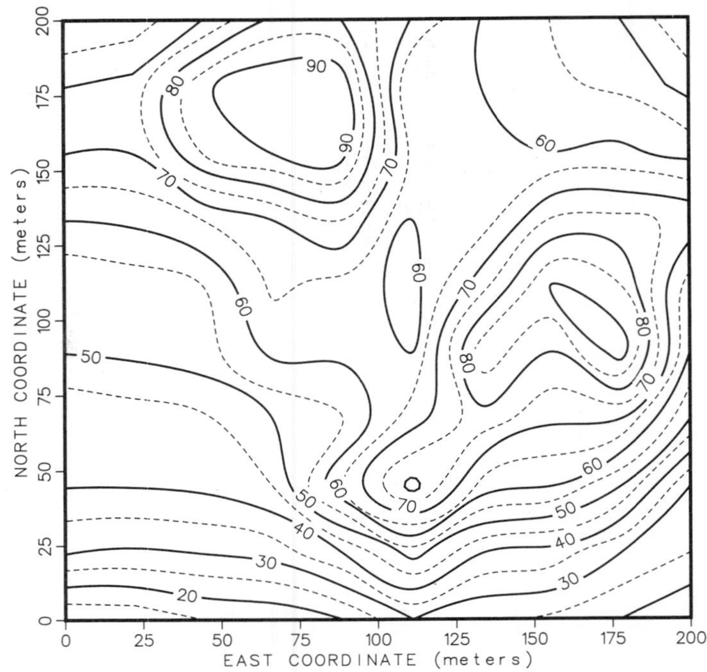

Figure 8.8 Contour plot showing a function of x and y.

used effectively in contour plots, as in Plates 10.1 and 10.2, which shows a temperature distribution ranging from cool (blue) to hot (red).

The 3-D extension of a line graph, the *surface plot* or *carpet plot,* presents data as a function of two variables. Hidden-surface removal and 3-D viewing transformations are used to manipulate the surface plot to most clearly show the data. Carpet plots are usually too cluttered if more than one surface set is displayed on the same axes. The example shown in Figure 8.9 makes use of a feature that prepares a perspective projection of a contour plot directly under the surface plot. More on surface plots may be found in Section 8.7.

Contour plots and surface plots can be used to show the same information in different forms. If numerical values of the dependent variable are to be read from the graph for particular values of the two independent variables, the contour plot is obviously superior. On the other hand, surface plots more clearly show data trends such as the locations of maximum and minimum values. A study reported in Section 8.8 confirms that surface plots are generally more easily interpreted than contour plots.

A *vector plot* or *field plot* is effective for comparing the magnitudes and directions of vector quantities in a selected plane. Here, the length of the plotted vector is proportional to magnitude and the arrow clearly indicates the direction. The arrangement of plotted vectors in rows and columns matches the tabular

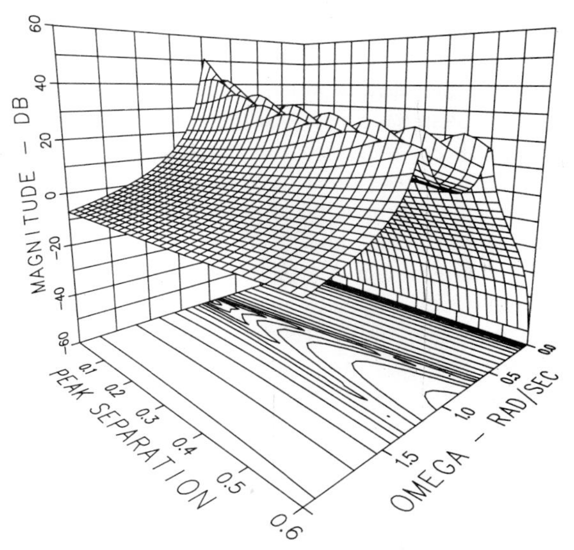

Figure 8.9 Surface plot showing a function of two variables. The function is also displayed as a contour plot.

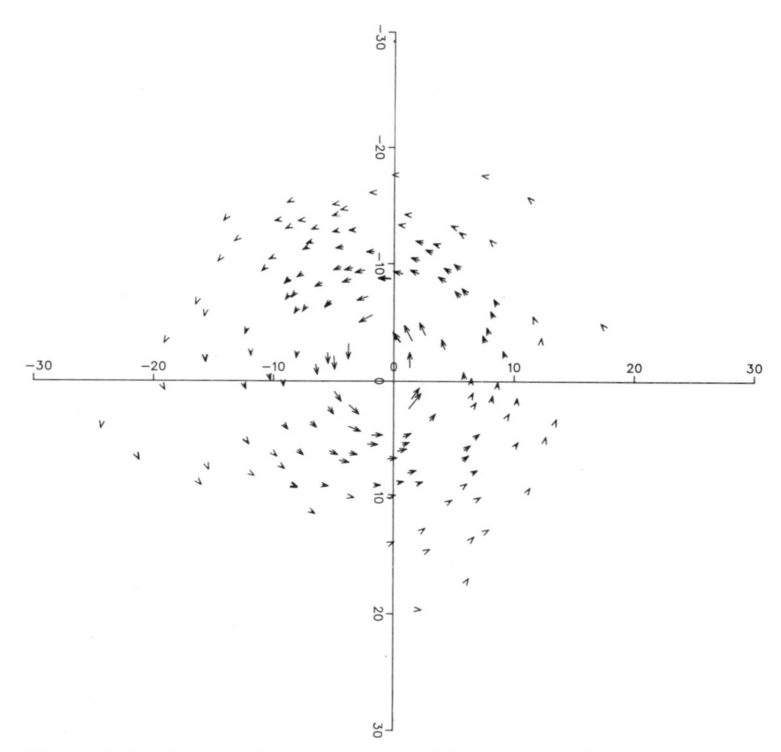

Figure 8.10 Vector plot showing velocities at selected points in a flow field.

output common in many analysis programs. Vector plots are useful in stress analysis, where the magnitudes and directions of the principal stresses can be illustrated. In aerodynamics and fluid flow analyses, the velocities at points in a grid can be illustrated, as in Figure 8.10. For vibration analysis, vector plots can be used to show amplitudes, velocities, and accelerations.

A *flowchart* or network, for example, Figure 8.11, finds wide use in engineering presentations. Sequencing of processes, electrical circuit and logical diagrams, computer programs, and interconnection of equipment are typical uses. Like the Venn diagram, a flowchart shows relationships among logical elements but does not necessarily show quantitative information.

A *map* aids in understanding relationships in the context of geographic data. Maps support operations management and decision making. Computer-produced maps such as the one in Figure 8.12 are important to land surveying, land-use planning, oil and gas exploration, power line or pipeline networking, mining, construction, transportation, and other related applications. In these and other applications, interactive computer-aided drafting (CAD) systems are useful in capturing, modifying, and displaying maps. In addition to the basic features described in Chapter 9, CAD specific to cartography[9] and surveying contains special symbols, scaling conventions, contouring algorithms, and mapping data base features.

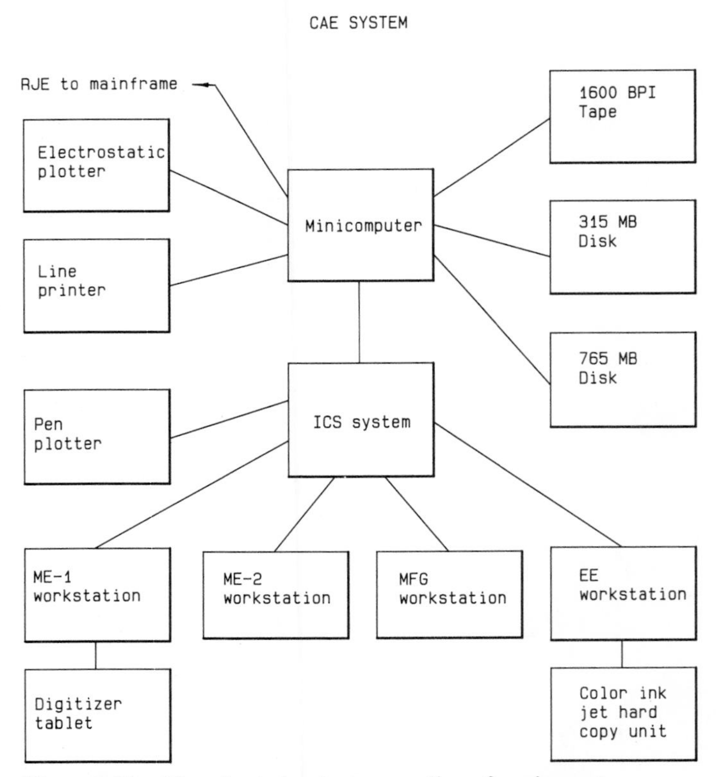

Figure 8.11 Flowchart showing connection of equipment.

HAZARDOUS MATERIAL FLOWS ON THE U.S. RAILROAD SYSTEM
ICC 1977 1% WAYBILL SAMPLE

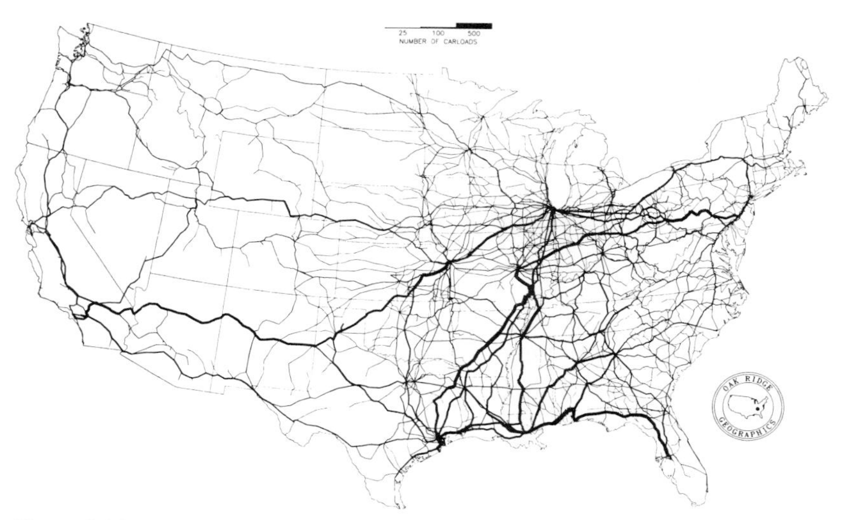

Figure 8.12 Example of a map generated from an existing geographic data base. Courtesy of Computing and Telecommunications Division at Oak Ridge National Laboratory, operated by Martin Marietta Energy Systems.

The computer is well suited for generation of data presentation graphics. A typical strategy involves keeping numerical data in a computer file in tabular format, perhaps as a spreadsheet. Routinely, 2-D data plots, contour plots, and other appropriate displays can be generated from the data file. The examples here should help in the selection of presentation graphics to support better communication of ideas and information for a wide variety of situations.

8.1 EFFECTIVE PRESENTATION TECHNIQUES

The most important factor in effective data presentation graphics is concentration on the data.[13,57] Presentations should be tailored to the audience, with the necessary information presented in a clear, unambiguous way. It is well known that graphics are a valuable enhancement to any presentation, oral or written. To have the most impact, presentation graphics must be in such a form and present the data in such a way that they are easily understood. The following guidelines will help make better presentation graphics:

- Focus attention on the *data*. Shaded patterns, decorative frames, and other nondata features which detract from clarity should be avoided. Keep it simple.
- Use scales on the axes which are easy to read and use for interpolation of numerical values. Logarithmic axes should be used only for technical audiences.

- Convey data truthfully. Axes with discontinuities and axes not starting at zero are deceptive. If there must be a discontinuity or an axis does not start at zero, attention should be drawn to this fact.
- Make titles and axis labels as descriptive as possible. Units should be given. Titles should not convey the same information as the axis labels.
- Use color to enhance data presentation, but take care not to overuse color. On a white background axes and titles should be black, while yellow lines and pastel colors should be reserved for nondata lines.
- Use simple (sans serif) fonts. Do not mix fonts. Keep labels horizontal when possible.

The examples in Figures 8.13 through 8.19 highlight some applications of the foregoing strategies to improve the quality of presentations. More detailed descriptions and examples are available.[13,57]

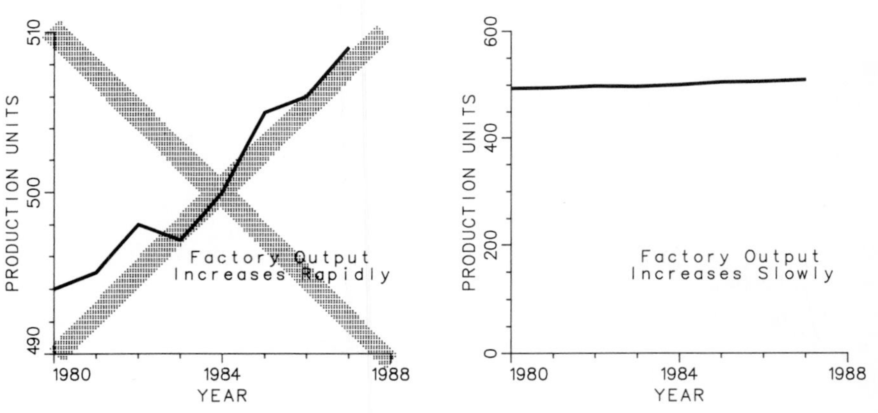

Figure 8.13 Accuracy in data comprehension is enhanced by the inclusion of zero in plots which compare amounts, levels, totals, and trends.

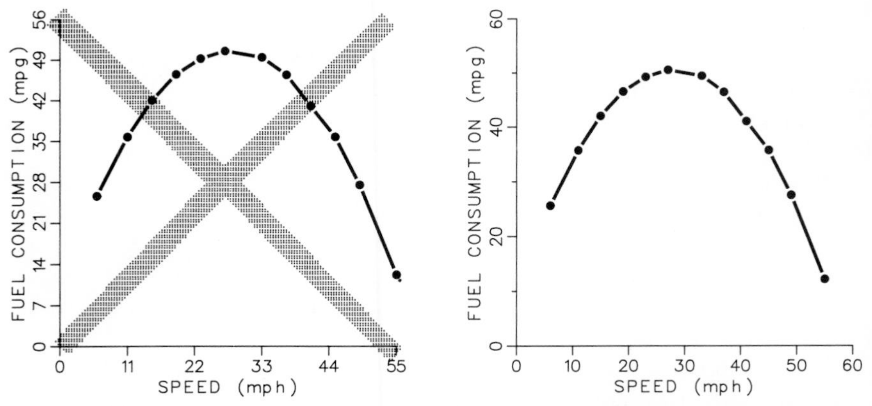

Figure 8.14 Interpolation and ease of reading are helped by using axes with tick marks and scales in a multiple of 1, 2, or 5.

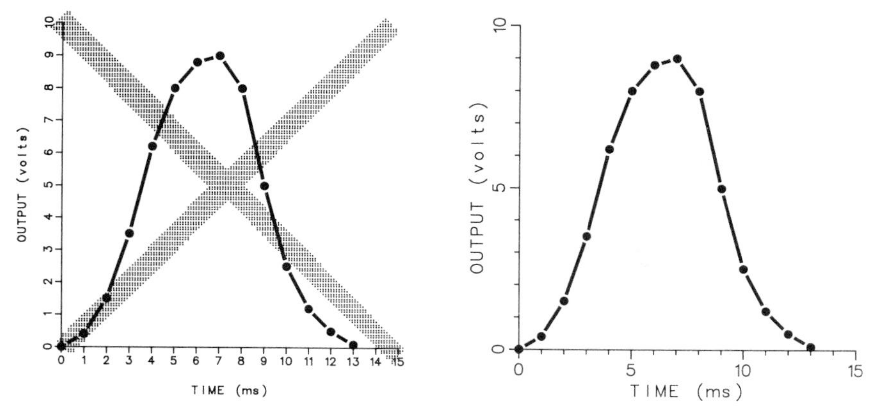

Figure 8.15 Ease of reading is aided by proper-sized text. A common problem is using typewritten characters (about $\frac{1}{8}$ inch high) on transparencies for overhead projection.

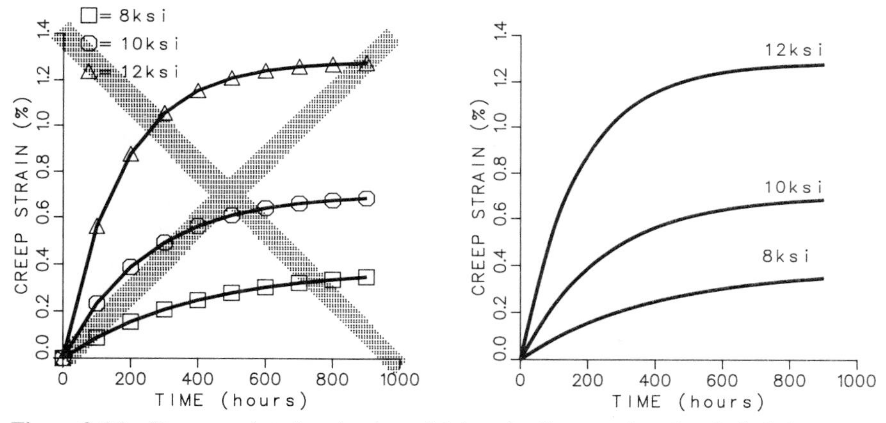

Figure 8.16 Focus on data by plotting with heavier lines, or in color. Label the curves if there is room. Do not use large data markers. Do not use data markers at all if the plotted data points are computed from a formula.

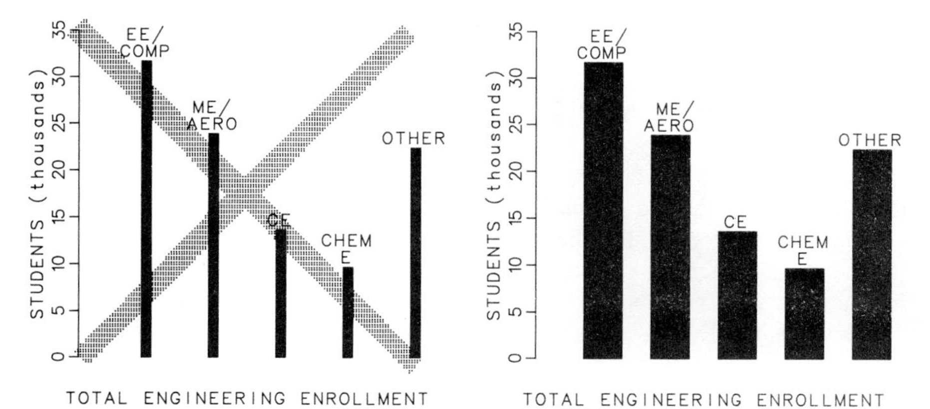

Figure 8.17 Emphasize data with bars wider than the intervening space.

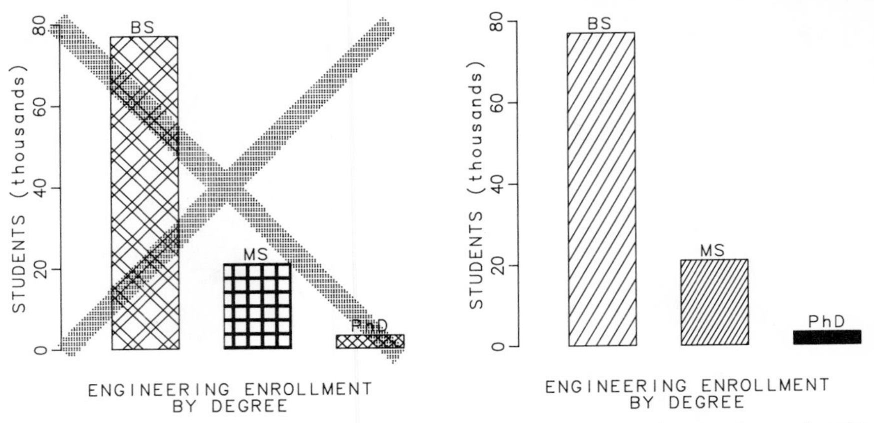

Figure 8.18 "Loud" hatch patterns detract from the data. Use harmonious colors if available, or subtle patterns instead.

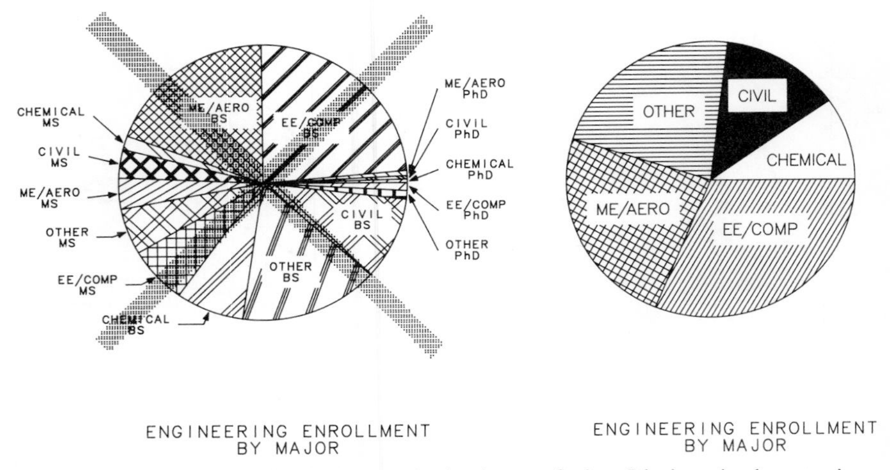

Figure 8.19 Too many details in one plot lead to confusion. Limit a pie chart to six or fewer sectors. Label the sectors internally if possible.

8.2 PIE CHARTS AND BAR GRAPHS

If a small number of data are to be presented, the *pie chart* effectively shows the distribution of data, and the *bar graph* communicates quantitative relationships among different groups. Pie charts and bar graphs can be constructed without an independent numerical variable. In contrast, the familiar 2-D data plot shows how a dependent numerical variable varies as a function of an independent numerical variable.

Most presentation graphics packages have the capability to generate pie charts and bar graphs. Excellent charts can also be executed with a CAD system.

COMPOSITION OF TYPE II PORTLAND CEMENT
Typical low alkali specification
Tricalcium silicate
45%
Gypsum 5%
Tricalcium
aluminate
8%
Dicalcium silicate
30%
Tetracalcium
aluminoferrite
12%

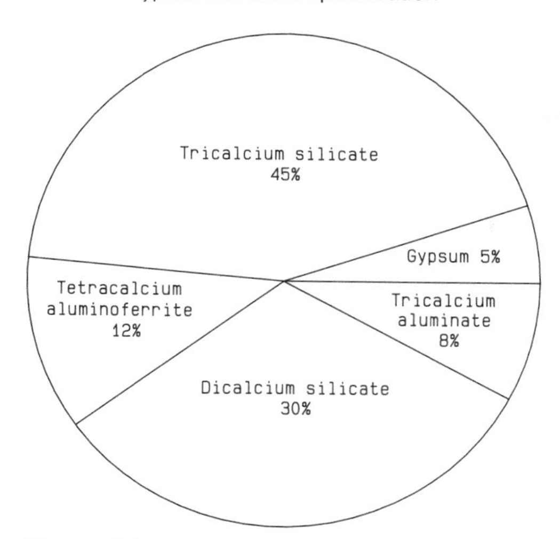

Figure 8.20 Example of an engineering pie chart showing chemical composition.

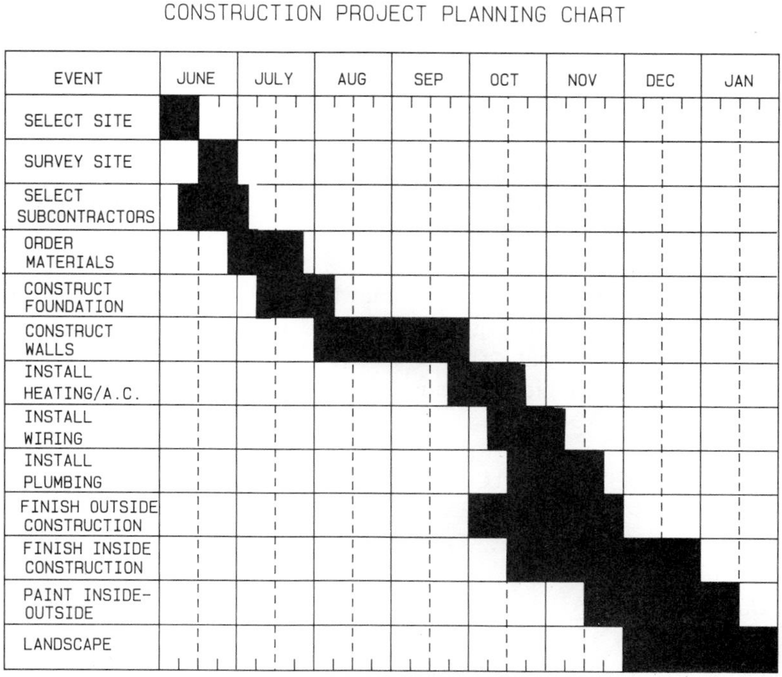

CONSTRUCTION PROJECT PLANNING CHART
EVENT
JUNE
JULY
AUG
SEP
OCT
NOV
DEC
JAN
SELECT SITE
SURVEY SITE
SELECT
SUBCONTRACTORS
ORDER
MATERIALS
CONSTRUCT
FOUNDATION
CONSTRUCT
WALLS
INSTALL
HEATING/A.C.
INSTALL
WIRING
INSTALL
PLUMBING
FINISH OUTSIDE
CONSTRUCTION
FINISH INSIDE
CONSTRUCTION
PAINT INSIDE-
OUTSIDE
LANDSCAPE

Figure 8.21 Gantt chart for planning construction project.

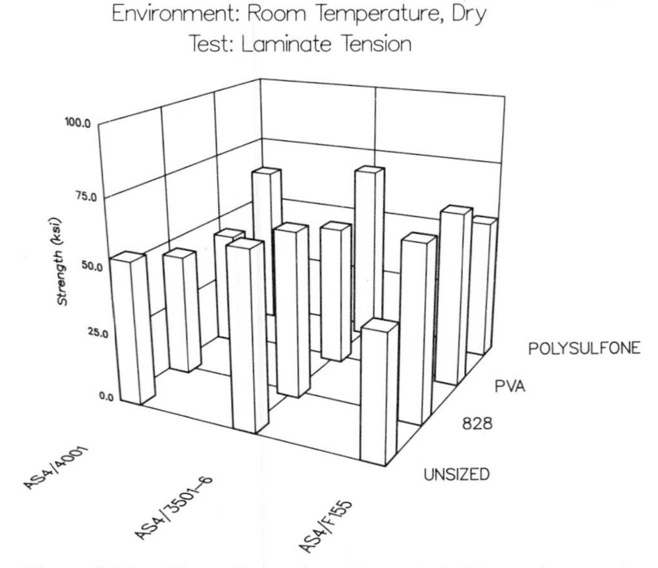

Figure 8.22 Three-dimensional bar graph illustrating results of a laminated composite material under tensile test as a function of resin type and fiber treatment. Since the axes of the two independent variables are nonnumeric, a bar chart is more appropriate than a 3-D surface plot.

The examples elsewhere in this chapter should be consulted by users contemplating these forms of presentation graphics.

Pie charts and bar graphs are sometimes classified as " business graphics" because of their wide use in financial reporting. However, there are some excellent engineering uses, besides the obvious one of showing budgets. Pie charts can show the breakdown of constituents, such as the composition of an alloy or the distribution of time for different phases of a project. An engineering pie chart is shown in Figure 8.20.

One special form of bar graph known as the Gantt chart is useful for scheduling and is shown in Figure 8.21. A "3-D" bar graph, Figure 8.22, is effective for presenting data that are functions of nonnumeric independent variables which can be logically arranged in rows and columns.

8.3 TWO-DIMENSIONAL DATA PLOTS

Generation of the ubiquitous 2-D data plots on the computer saves time and improves quality.

Conventional practice dictates that the horizontal axis, also known as the x axis or *abscissa,* is used for the independent variable. The vertical axis, also known as the y axis or *ordinate,* is used for the dependent variable. Multiple data sets which are all functions of the same independent variable are conveniently

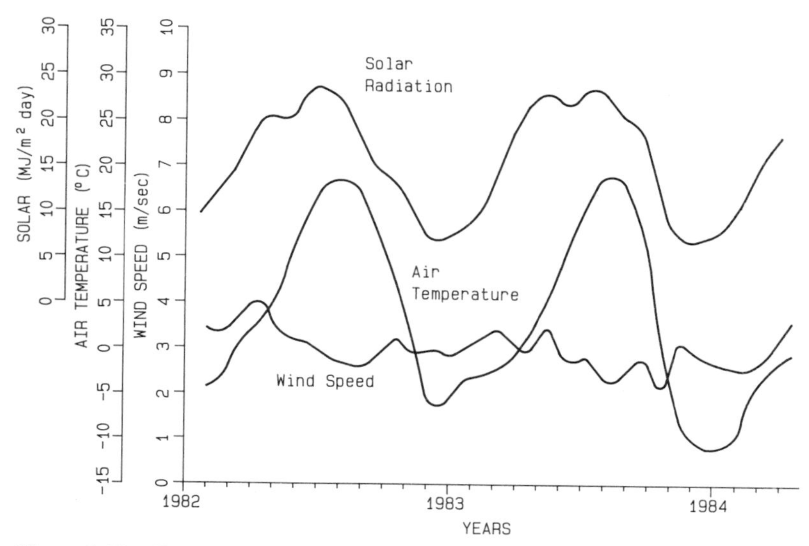

Figure 8.23 Example of multiple axes for dependent variables.

graphed with 2-D data plots, such as the example in Figure 8.23. In this example, a separate *y* axis is provided for each dependent variable being graphed.

It is good practice to name the variables and include units on the axis labels. Scales in scientific notation should be labeled as shown in Figure 8.24(a). However, it is better to use prefixes on the units if possible, as in Figure 8.24(b).

Sometimes, nonlinear axes may be used in 2-D data plots. When the data span a wide range, one or two of the axes may be logarithmic. A *semilogarithmic* or *semilog* plot has one axis with a linear scale and one with a logarithmic scale, as in Figure 8.25. A *full logarithmic* or *log-log* plot has both axes with

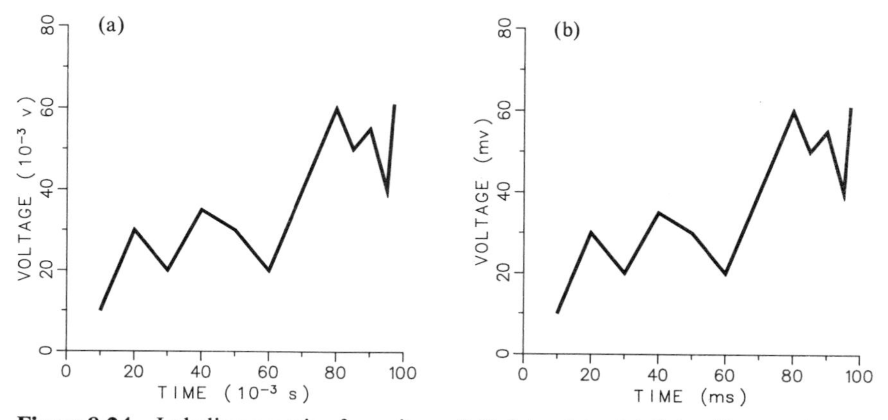

Figure 8.24 Labeling practice for units on 2-D data plots. (a) Scientific notation; (b) prefixes (preferred).

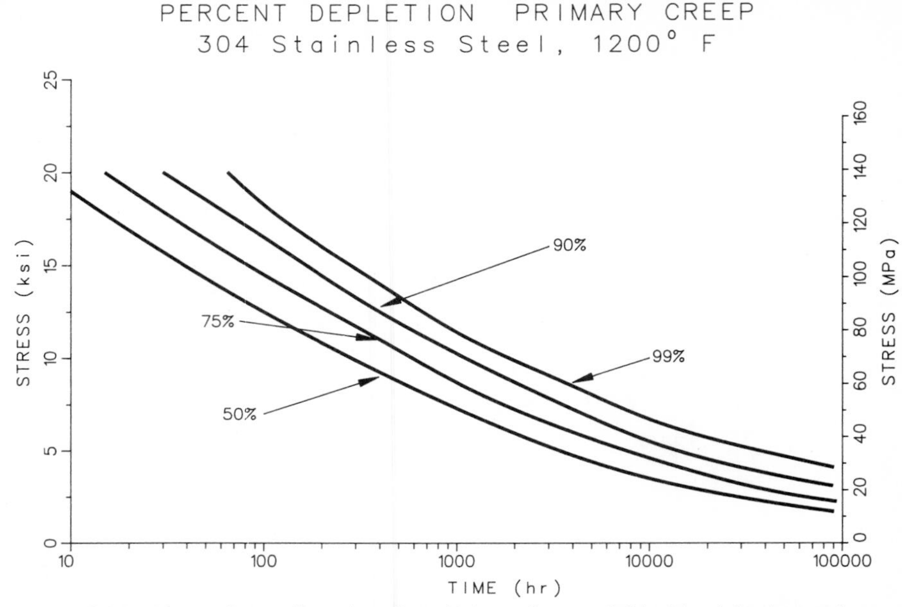

Figure 8.25 Example semilog plot where independent variable (time) is observed over an extensive range. If these data were plotted on a linear abscissa, the variations for the smaller values of time would be obscured.

logarithmic scales, as in Figure 8.26. A *polar* plot, Figure 8.2, is actually linear, except that the axes show radius and angle information. When possible, linear Cartesian axes should be used, because the rate of change and relative magnitudes can easily be inferred. A common practice for indicating both U.S. customary and SI units on a graph is shown in Figures 8.25 and 8.26.

8.4 CURVE SMOOTHING

If only a few points are available for plotting, drawing straight line segments between the points may not be entirely satisfactory. Interpolation can be used for *curve smoothing* on the data to replace the straight line segments with curved lines that pass exactly through the data points. Alternatively, noisy data, or data that are known to have a certain form, can be processed by *curve fitting,* as explained in Section 8.5. Three popular interpolation methods are presented here. Many useful interpolation algorithms can be found in the literature.[16,59]

Point-to-Point Polynomials

As one of the simplest and least computationally intensive methods of curve smoothing, piecewise polynomials work well in many applications. Because of symmetry, third- or fifth-order polynomials are better for smoothing than are

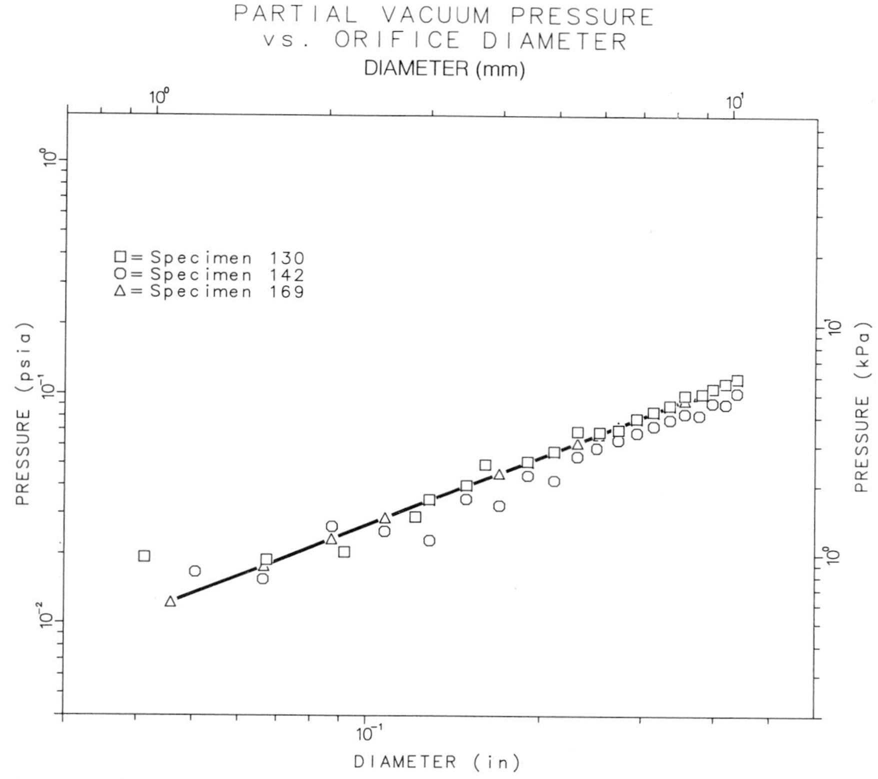

Figure 8.26 Example plot to support a theoretical prediction which plots as a straight line on log-log axes.

second- or fourth-order polynomials. A conceptual example of using this type of smoothing is given in Figure 8.27.

A choice can be made as to whether the smoothing is done with nonparametric or parametric polynomials. Smoothing with nonparametric polynomials is described first.

The equation of a nonparametric cubic polynomial is

$$y = C_1 + C_2x + C_3x^2 + C_4x^3 \tag{8.1}$$

where the coefficients C_1, C_2, C_3, and C_4 are determined such that the curve passes through the data points, or knots, (x_i, y_i), (x_{i+1}, y_{i+1}), (x_{i+2}, y_{i+2}), and (x_{i+3}, y_{i+3}). This is done with the system of equations

$$\begin{bmatrix} 1 & x_i & x_i^2 & x_i^3 \\ 1 & x_{i+1} & x_{i+1}^2 & x_{i+1}^3 \\ 1 & x_{i+2} & x_{i+2}^2 & x_{i+2}^3 \\ 1 & x_{i+3} & x_{i+3}^2 & x_{i+3}^3 \end{bmatrix} \begin{bmatrix} C_1 \\ C_2 \\ C_3 \\ C_4 \end{bmatrix}_i = \begin{bmatrix} y_i \\ y_{i+1} \\ y_{i+2} \\ y_{i+3} \end{bmatrix} \tag{8.2}$$

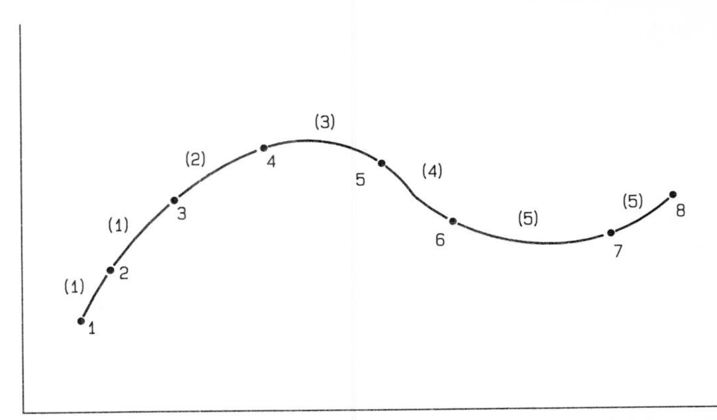

Figure 8.27 Example of smoothing point to point with cubic polynomials. The numbers of the five polynomials are in parentheses. Note that a single cubic polynomial passes through four points and that, in general, only the middle part of the cubic polynomial is used.

which is solved numerically for $[\mathbf{C}]_i$, where the subscript on the matrix refers to the ith polynomial. If n points are to be smoothed, there will be $n - 3$ sets of coefficients. Knots $i + 1$ and $i + 2$ are connected by a curve computed from the ith polynomial, as shown in Figure 8.28. The exceptions are the ends where the first polynomial interpolates between the first three points and the last polynomial interpolates between the last three points.

Substitution of 5 to 10 evenly spaced values of x, such that $x_{i+1} \leq x \leq x_{i+2}$, into Eq. (8.1) for each segment is usually sufficient to produce a smooth curve. If two data points have the same x, part of the curve is vertical and Eq. (8.2) is singular. In this case, use of a parametric polynomial is indicated.

A parametric cubic polynomial adapted from Eq. (5.33) for a planar plot is

$$[x \quad y] = [1 \quad u \quad u^2 \quad u^3] \begin{bmatrix} a_{1x} & a_{1y} \\ a_{2x} & a_{2y} \\ a_{3x} & a_{3y} \\ a_{4x} & a_{4y} \end{bmatrix} = [1 \quad u \quad u^2 \quad u^3][\mathbf{A}]_i \qquad (8.3)$$

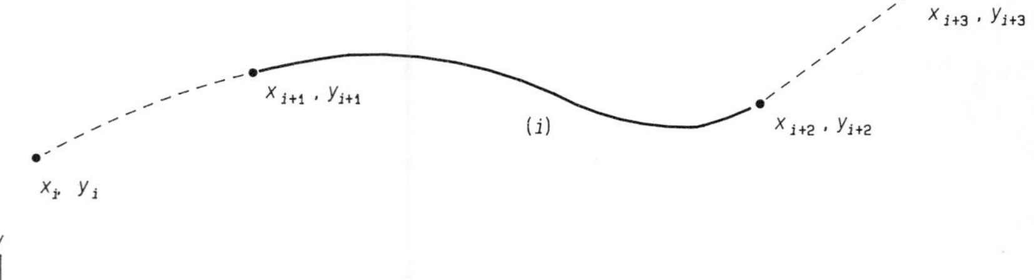

Figure 8.28 Cubic polynomial (i) interpolating between the second and third of the four defining knots.

which can be parametrized over any arbitrary range of u, customarily $0 \leq u \leq 1$. The matrix $[\mathbf{A}]_i$ is distinct for each of the i polynomials. Fitting the four given points through the third points in u, that is, $u = 0, 1/3, 2/3, 1$, permits direct use of Eq. (5.35) for the solution of $[\mathbf{A}]_i$, or

$$\begin{bmatrix} a_{1x} & a_{1y} \\ a_{2x} & a_{2y} \\ a_{3x} & a_{3y} \\ a_{4x} & a_{4y} \end{bmatrix}_i = \begin{bmatrix} 1 & 0 & 0 & 0 \\ -11/2 & 9 & -9/2 & 1 \\ 9 & -45/2 & 18 & -9/2 \\ -9/2 & 27/2 & -27/2 & 9/2 \end{bmatrix} \begin{bmatrix} x_i & y_i \\ x_{i+1} & y_{i+1} \\ x_{i+2} & y_{i+2} \\ x_{i+3} & y_{i+3} \end{bmatrix} \tag{8.4}$$

In plotting the ith polynomial for the smoothed curve using the $(n - 3)$ sets of coefficients derived from Eq. (8.4), the parameter u is varied on the range 1/3 to 2/3. The first polynomial is plotted with $0 \leq u \leq 2/3$ and the last polynomial is plotted with $1/3 \leq u \leq 1$. Again, 5 to 10 line segments between knots should be enough for a smooth plot.

The foregoing methods can be extended to a polynomial of any degree. The fifth-order polynomial, which fits through six knots, leaves two sections on each side. Its use will produce a slightly smoother curve than the cubic polynomial.

"Smoothness" is measured by continuity of the curve and its derivatives. Piecewise polynomials have slopes at all the knots (except the first two and last two) which are theoretically discontinuous. Such discontinuities may be noticeable with some data sets.

Cubic Spline

Continuity of both the first and second derivatives at the n knots is provided with the *cubic spline,* discussed in Section 5.3. A cubic spline can be nonparametric or parametric and has some similarity to the piecewise polynomial of Figure 8.27. Instead of the $n - 3$ sets of 4×4 linear equations of Eq. (8.2), the cubic spline requires the solution of an $n \times n$ (or slightly smaller, depending on the end conditions) system of linear equations. For smoothing a relatively large number of data, the cubic spline can become computationally intensive.

The commonly used nonparametric or "natural" cubic spline fits a cubic polynomial, Eq. (8.1), between each knot. For convenience, the notation used in Eq. (8.1) is changed so that the segment which connects knot i to knot $i + 1$ is given by

$$y = a_i + b_i(x - x_i) + c_i(x - x_i)^2 + d_i(x - x_i)^3 \qquad (i = 1, 2, \ldots, n - 1) \tag{8.5}$$

The coefficients a_i, b_i, c_i, and d_i are determined by the conditions that the curve must pass through each knot and that the first and second derivatives of y must be continuous at the knots. The details of the algebra may be found in Yakowitz and Szidarovszky[59] and other numerical analysis references. It can be demonstrated that the coefficients a_i, b_i, and d_i can be written in terms of c_i and the given data points as

$$a_i = y_i \tag{8.6}$$

$$b_i = \frac{y_{i+1} - y_i}{x_{i+1} - x_i} - \frac{x_{i+1} - x_i}{3}(2c_i + c_{i+1}) \tag{8.7}$$

and

$$d_i = \frac{c_{i+1} - c_i}{3(x_{i+1} - x_i)} \qquad (i = 1, 2, \ldots, n-1) \tag{8.8}$$

where the coefficients c_i are determined from the system of equations

$$\begin{bmatrix} 2(x_3 - x_1) & (x_3 - x_2) & 0 & \cdots & 0 \\ (x_3 - x_2) & 2(x_4 - x_2) & (x_4 - x_3) & \cdots & 0 \\ 0 & (x_4 - x_3) & 2(x_5 - x_3) & \cdots & 0 \\ & \cdots & & & \\ 0 & \cdots & & (x_{n-1} - x_{n-2}) & 2(x_n - x_{n-2}) \end{bmatrix} \begin{bmatrix} c_2 \\ c_3 \\ \vdots \\ c_{n-1} \end{bmatrix}$$

$$= \begin{bmatrix} 3\dfrac{y_3 - y_2}{x_3 - x_2} - 3\dfrac{y_2 - y_1}{x_2 - x_1} \\ 3\dfrac{y_4 - y_3}{x_4 - x_3} - 3\dfrac{y_3 - y_2}{x_3 - x_2} \\ 3\dfrac{y_5 - y_4}{x_5 - x_4} - 3\dfrac{y_4 - y_3}{x_4 - x_3} \\ \vdots \\ 3\dfrac{y_n - y_{n-1}}{x_n - x_{n-1}} - 3\dfrac{y_{n-1} - y_{n-2}}{x_{n-1} - x_{n-2}} \end{bmatrix} \tag{8.9}$$

At the first and the nth knot,

$$c_1 = c_n = 0 \tag{8.10}$$

The natural cubic spline, Eqs. (8.5) through (8.10), produces a somewhat lengthy computer program, where a tridiagonal solution routine is recommended for Eq. (8.9). A potential problem arises because the nonparametric cubic spline is undefined if the curve is vertical. (i.e., $x_i = x_{i+1}$) For this and other reasons, a parametric cubic spline is more versatile.

The parametric cubic spline with relaxed ends (the conventional condition) is given in Eqs. (5.52) and (5.60). Like the natural cubic spline, the parametric cubic spline provides continuity through the second derivative at the knots. To describe a plane curve which connects knots i and $i + 1$, Eq. (5.52) can be written as

$$\begin{aligned} x(u) &= x_i + x_i' u + (3x_{i+1} - 3x_i - 2x_i' - x_{i+1}')u^2 \\ &\quad + (2x_i - 2x_{i+1} + x_i' + x_{i+1}')u^3 \\ y(u) &= y_i + y_i' u + (3y_{i+1} - 3y_i - 2y_i' - y_{i+1}')u^2 \\ &\quad + (2y_i - 2y_{i+1} + y_i' + y_{i+1}')u^3 \end{aligned} \tag{8.11}$$

where $0 \le u \le 1$ provides interpolation between knots. The quantities $x' \equiv dx/du$ and $y' \equiv dy/du$ are the x and y components of the parametric slopes at the knots. These are the unknowns, found from the system of linear equations in Eq. (5.60),

$$\begin{bmatrix} 2 & 1 & 0 & & 0 & \cdots & 0 \\ 1 & 4 & 1 & & 0 & \cdots & 0 \\ 0 & 1 & 4 & & 1 & \cdots & 0 \\ \cdots & \cdots & \cdots & \cdots & \cdots & \cdots & \cdots \\ 0 & \cdots & 1 & & 4 & 1 & 0 \\ 0 & \cdots & 0 & & 1 & 4 & 1 \\ 0 & \cdots & 0 & & 0 & 1 & 2 \end{bmatrix} \begin{bmatrix} x_1' & y_1' \\ x_2' & y_2' \\ x_3' & y_3' \\ \vdots & \vdots \\ x_{n-2}' & y_{n-2}' \\ x_{n-1}' & y_{n-1}' \\ x_n' & y_n' \end{bmatrix}$$

$$= \begin{bmatrix} 3x_2 - 3x_1 & 3y_2 - 3y_1 \\ 3x_3 - 3x_1 & 3y_3 - 3y_1 \\ 3x_4 - 3x_2 & 3y_4 - 3y_2 \\ \vdots & \vdots \\ 3x_{n-1} - 3x_{n-3} & 3y_{n-1} - 3y_{n-3} \\ 3x_n - 3x_{n-2} & 3y_n - 3y_{n-2} \\ 3x_n - 3x_{n-1} & 3y_n - 3y_{n-1} \end{bmatrix} \tag{8.12}$$

for the condition of relaxed ends.

The systems of linear equations in Eqs. (8.9) and (8.12) can most efficiently be handled with a tridiagonal solution routine.[59] The square matrix for the parametric cubic spline, which contains predefined numbers, can be generated more simply than the square matrix for the nonparametric cubic spline, which contains algebraic expressions. On the other hand, the parametric formulation requires solution for two sets of unknowns. Computationally, this does not involve too much more work if the reduced square matrix is saved and used for back substitution. The additional computation required for a parametric cubic spline is justified by the increased versatility of handling vertical slopes. A simple straightening algorithm, described next, can also be applied to the parametric cubic spline.

Straightened Cubic Spline

Cubic splines, both parametric and nonparametric, maintain continuity through the second derivative at the knots but sometimes produce undesirable oscillations, such as that shown in Figure 8.29. This problem can be corrected with a "cubic spline under tension." Schweikert[52] used a mechanics of materials solution for a beam with an axial load to develop a nonparametric cubic spline under tension. The resulting algorithm, which requires that the curve be drawn with hyperbolic functions, is somewhat lengthy and is not presented here.

The cubic spline under tension has been incorporated in some presentation graphics software. The user specifies a parameter, analogous to the axial load in a beam, that pulls out the oscillations any arbitrary amount. Typically, the parameter, R, may be varied from 0 to 1.0. At $R = 0$ there is no tension and the cubic spline is unchanged. At $R = 1.0$ the tension is "infinite" and the segments between the knots are straight lines. A satisfactory value of R may be determined by experimentation.

In Figure 5.5 it can be observed that changing the magnitudes of the parametric slope terms in the geometric form of a parametric cubic curve affects the curve shape. But, as long as the ratio x_i'/y_i' is constant, the geometric slope of the curve at knot i is unchanged. As the curve connecting knot i to knot $i + 1$ becomes straight, the magnitudes of parametric slopes at each end point approach zero as shown in Figure 5.5. Thus, a parametric cubic spline can be straightened by using the tension parameter R such that a modified set of parametric slopes is given by

$$x_i'^* = x_i'(1 - R) \quad \text{and} \quad y_i'^* = y_i'(1 - R) \tag{8.13}$$

where $0 \leq R \leq 1$. In application, all that is necessary is that the parametric slope terms be modified before the curve is generated by Eq. (8.11). When this modification is made, the resulting curve has continuous first derivatives but

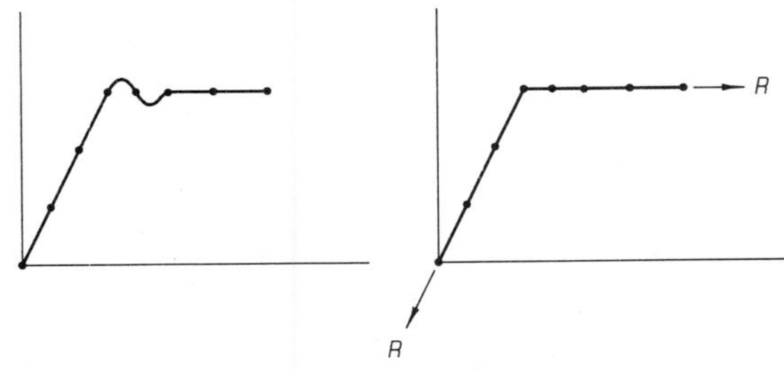

Figure 8.29 Illustration of cubic spline under tension. Oscillations in the curve are removed, dependent on the value of R.

discontinuous second derivatives at the knots. The visual effect of this loss in continuity is generally not a problem in presentation graphics. Use of this type of curve straightening is illustrated in the following example.

EXAMPLE 8.1

Develop a parametric cubic spline which interpolates between the data given below. Demonstrate how the curve changes with the tension parameters (a) $R = 0$, (b) $R = 0.4$, and (c) $R = 1.0$.

i	x	y
1	0	3
2	1	4
3	2	2
4	3	3
5	3	1
6	4	3

Solution. The system of linear equations generated by substitution of the given data into Eq. (8.12) is

$$\begin{bmatrix} 2 & 1 & 0 & 0 & 0 & 0 \\ 1 & 4 & 1 & 0 & 0 & 0 \\ 0 & 1 & 4 & 1 & 0 & 0 \\ 0 & 0 & 1 & 4 & 1 & 0 \\ 0 & 0 & 0 & 1 & 4 & 1 \\ 0 & 0 & 0 & 0 & 1 & 2 \end{bmatrix} \begin{bmatrix} x'_1 & y'_1 \\ x'_2 & y'_2 \\ x'_3 & y'_3 \\ x'_4 & y'_4 \\ x'_5 & y'_5 \\ x'_6 & y'_6 \end{bmatrix} = \begin{bmatrix} 3 & 3 \\ 6 & -3 \\ 6 & -3 \\ 3 & -3 \\ 3 & 0 \\ 3 & 6 \end{bmatrix}$$

The solutions are found to be

$$\begin{bmatrix} x'_1 & y'_1 \\ x'_2 & y'_2 \\ x'_3 & y'_3 \\ x'_4 & y'_4 \\ x'_5 & y'_5 \\ x'_6 & y'_6 \end{bmatrix} = \begin{bmatrix} 1.02392 & 2.09569 \\ 0.95215 & -1.19139 \\ 1.16746 & -0.33014 \\ 0.37799 & -0.48804 \\ 0.32057 & -0.71770 \\ 1.33971 & 3.35885 \end{bmatrix}$$

which are used with Eq. (8.11) to generate the curve marked $R = 0$ shown on the top of the next page.

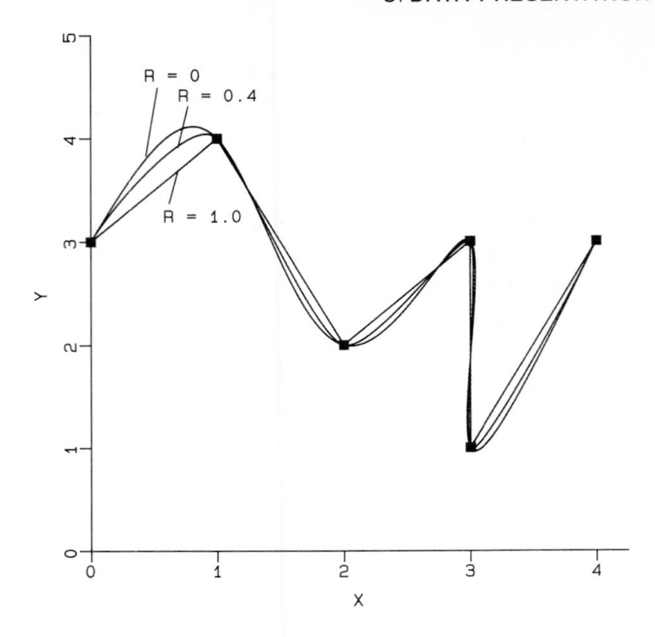

Application of Eq. (8.13) with tension parameter $R = 0.4$ yields the corrected parametric slopes

$$\begin{bmatrix} x_1'^* & y_1'^* \\ x_2'^* & y_2'^* \\ x_3'^* & y_3'^* \\ x_4'^* & y_4'^* \\ x_5'^* & y_5'^* \\ x_6'^* & y_6'^* \end{bmatrix} = \begin{bmatrix} 0.61435 & 1.25741 \\ 0.57129 & -0.71483 \\ 0.70048 & -0.19808 \\ 0.26979 & -0.29282 \\ 0.19234 & -0.43062 \\ 0.80383 & 2.01531 \end{bmatrix}$$

and $R = 1.0$,

$$\begin{bmatrix} x_1'^* & y_1'^* \\ x_2'^* & y_2'^* \\ x_3'^* & y_3'^* \\ x_4'^* & y_4'^* \\ x_5'^* & y_5'^* \\ x_6'^* & y_6'^* \end{bmatrix} = \begin{bmatrix} 0 & 0 \\ 0 & 0 \\ 0 & 0 \\ 0 & 0 \\ 0 & 0 \\ 0 & 0 \end{bmatrix}$$

Curves generated with Eq. (8.11) using the above parametric slopes are shown for comparison.

8.5 CURVE FITTING

Curve fitting is the process of adapting an approximating function to match a given data set. This process is particularly useful for obtaining a continuous mathematical function to represent experimentally determined data. With curve fitting, the inevitable scatter in experimental data can be filtered and represented in a mean sense. As with the data of Figure 8.26, the physical law governing the phenomenon usually leads to the best approximating function. When such a governing equation is not available, straight lines and low-order polynomials are common choices.

The statistical procedure known as *regression analysis* deals with the systematic adjustment of approximating functions to best fit data. The method known as *linear least squares regression* seeks to determine approximating functions such that the squares of the residuals are minimized. Here, the term residual means the difference between the actual data and the value predicted by the model; linear refers to the unknown coefficients in the approximating functions.

Generalized Least Squares Procedure

In the regression process the observed data are first tabulated, where conventionally x is the independent and y is the dependent variable. It is convenient to write these data as column matrices,

$$[\mathbf{X}] = \begin{bmatrix} x_1 \\ x_2 \\ x_3 \\ \cdot \\ \cdot \\ \cdot \\ x_n \end{bmatrix} \qquad [\mathbf{Y}] = \begin{bmatrix} y_1 \\ y_2 \\ y_3 \\ \cdot \\ \cdot \\ \cdot \\ y_n \end{bmatrix} \tag{8.14}$$

Previewing the data with a "scatter plot" on linear axes is easily done with most general-purpose presentation graphics software. Such a plot, Figure 8.30, helps to visualize the possible form of the approximating function. Once chosen, the unknown coefficients of the approximating functions are adjusted to minimize the sum of the squares of the residuals.

In linear least squares fitting, the m linear coefficients $c_1, c_2, c_3, \ldots, c_m$ are used in an approximating function with the form

$$\tilde{y} = c_1 f_1(x) + c_2 f_2(x) + c_3 f_3(x) + \cdots + c_i f_i(x) + \cdots + c_m f_m(x) \tag{8.15}$$

which is an explicit nonparametric function with x as the independent variable and y as the dependent variable. It should be noted that, in general, interchanging the dependent and independent variables produces a different approximating

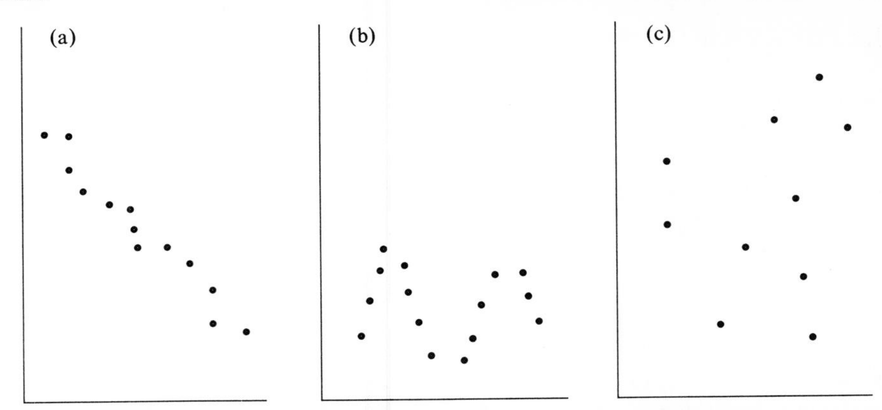

Figure 8.30 Scatter plots preliminary to performing curve fitting routines. (a) Data suggesting a straight line; (b) data suggesting a periodic function (trigonometric series); (c) data suggesting no relationship.

curve. In Eq. (8.15), the $f_i(x)$ can be any arbitrary functions of x, such as x^2, $\sin \cdot x + 2 \cos x$, $\log x$, and even x^0.

Upon selection of a model to fit the data, a matrix of functions can be formed with rows of $f(x)$ terms from Eq. (8.15),

$$[\mathbf{F}] = \begin{bmatrix} f_1(x_1) & f_2(x_1) & \cdots & f_m(x_1) \\ f_1(x_2) & f_2(x_2) & \cdots & f_m(x_2) \\ f_1(x_3) & & \cdots & f_m(x_3) \\ \cdots & \cdots & \cdots & \cdots \\ & & \cdots & f_m(x_{n-1}) \\ f_1(x_n) & f_2(x_n) & \cdots & f_m(x_n) \end{bmatrix} \tag{8.16}$$

The matrix has n rows, corresponding to the n given data, and m columns, corresponding to the m unknown coefficients. The array of linear coefficients which are to be determined is

$$[\mathbf{C}] = \begin{bmatrix} c_1 \\ c_2 \\ \vdots \\ c_m \end{bmatrix} \tag{8.17}$$

The ith residual, r_i, is defined by

$$r_i = \tilde{y}_i - y_i$$

or

$$r_i = c_1 f_1(x_i) + c_2 f_2(x_i) + \cdots + c_m f_m(x_i) - y_i \tag{8.18}$$

with the use of Eq. (8.15). A positive residual means that the predicted curve fit gives a higher-than-actual value. The entire set of residuals can be written as a column matrix,

$$[\mathbf{R}] = \begin{bmatrix} r_1 \\ r_2 \\ r_3 \\ \cdot \\ \cdot \\ \cdot \\ r_n \end{bmatrix} \tag{8.19}$$

In matrix notation, the complete set of residuals determined by using Eq. (8.18) for each is

$$[\mathbf{R}] = [\mathbf{F}][\mathbf{C}] - [\mathbf{Y}] \tag{8.20}$$

In the least squares algorithm, the coefficients [**C**] are chosen so that the sum of the squares of the residuals is minimized. This procedure tends to give outlying points great emphasis while presenting no difficulty with the algebraic signs of the residuals. Stated in mathematical terms, the criterion is

$$\sum_{i=1}^{n} r_i^2 = \min \tag{8.21}$$

By taking partial derivatives of the r_i with respect to the unknown coefficients c_k, the above expression may be minimized,[19]

$$\sum_{i=1}^{n} \frac{\partial r_i}{\partial c_k} r_i = 0 \qquad k = 1, 2, \cdot \cdot \cdot , m \tag{8.22}$$

When the partial derivatives computed from Eq. (8.18) are substituted into Eq. (8.22) and are written in matrix form the result is

$$[\mathbf{F}]^T[\mathbf{R}] = [0] \tag{8.23}$$

Substitution of [**R**] as defined in Eq. (8.20) into Eq. (8.23) gives

$$[\mathbf{F}]^T[\mathbf{F}][\mathbf{C}] = [\mathbf{F}]^T[\mathbf{Y}] \tag{8.24}$$

which is immediately recognized as the standard form for linear algebraic equations

$$[\mathbf{A}][\mathbf{C}] = [\mathbf{B}] \tag{8.25}$$

where $[\mathbf{A}] = [\mathbf{F}]^T[\mathbf{F}]$ and $[\mathbf{B}] = [\mathbf{F}]^T[\mathbf{Y}]$. A check of the matrix multiplications performed will verify that there are m equations and m unknown coefficients. Since the number of equations is usually not large, almost any of the recognized methods of linear equation solution[32,59] may be used to determine [**C**]. Subsequent substitution of [**C**] into Eq. (8.15) completes the regression process.

Next, it is desirable to check how well the chosen approximating function fits (regresses to) the data. Two simple tests are given here. More complete information can be found in statistics texts.[19]

R-Squared Test

Since the immediate objective of least squares fitting is to minimize the sum of squares of the residuals (SSR), where

$$\text{SSR} = \sum_{i=1}^{n} r_i^2 = \sum_{i=1}^{n} (\tilde{y}_i - y_i)^2 \tag{8.26}$$

The magnitude of SSR is a logical choice to compare goodness of fit. However, SSR itself is not particularly useful since its magnitude depends on the relative magnitude of the data being fit. For example, data with units in millimeters produce a million times the SSR produced by the same data with units in meters.

The widely used R-squared test contains a normalized measure of SSR. The starting point is determination of the corrected total sum of squares (CTSS), which measures the squares of the differences between the mean $\bar{y}$ and the observed values y_i,

$$\text{CTSS} = \sum_{i=1}^{n} (y_i - \bar{y})^2 \tag{8.27}$$

where

$$\bar{y} = \frac{1}{n} \sum_{i=1}^{n} y_i \tag{8.28}$$

An alternative expression for CTSS which is easier to compute is

$$\text{CTSS} = \sum_{i=1}^{n} (y_i)^2 - \frac{1}{n}\left(\sum_{i=1}^{n} y_i\right)^2 \tag{8.29}$$

The difference between CTSS and SSR is known as the sum of squares due to regression. This difference is also called the "explained variation" because the regression curve is compared with the mean value of the data. It happens that SSR is called "unexplained variation" because the regression curve is compared with the data values themselves. The two types of variation are shown in Figure 8.31.

The R-squared test measures the ratio of the explained variation to the corrected total sum of squares, or

$$R^2 = \frac{\text{CTSS} - \text{SSR}}{\text{CTSS}} = 1 - \frac{\text{SSR}}{\text{CTSS}} \tag{8.30}$$

It can be observed that $R^2 = 1$ when SSR $= 0$, the case where the regression curve passes exactly through all the data. On the other hand, as SSR approaches CTSS in magnitude, R approaches zero, which shows that the approximating function is not a better predictor of the data than the arithmetic mean.

Standard Error of Estimate

Another measure of how well the regression fits the data is the standard error of the estimate (SEE), which is analogous to the standard deviation. First, a

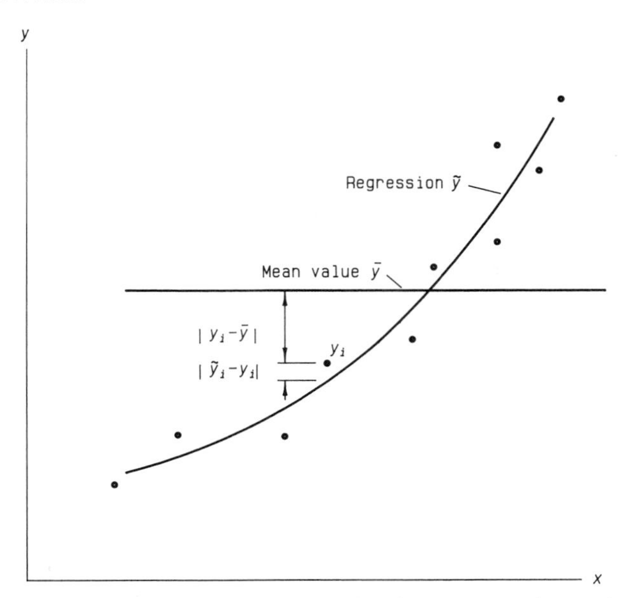

Figure 8.31 Contributions to the data-mean value and data-regression curve variations.

measure of variation can be formed by dividing the SSR by the number of degrees of freedom, which is the difference between the number of data and the number of coefficients in the model. Hence

$$\mathrm{VAR} = \mathrm{SSR}/(n - m) \tag{8.31}$$

where SSR is defined by Eq. (8.26). The SEE is defined as the square root of the variance, which gives the same units as those of the independent variable. Thus,

$$\mathrm{SEE} = \sqrt{\mathrm{SSR}/(n - m)} \tag{8.32}$$

A physical interpretation of Eq. (8.32) is shown in Figure 8.32. In a manner analogous to the standard deviation, for a typical value x_i selected out of a large sample size n, about 68% of the y_i values of the data are within a tolerance of ± 1 SEE of the $\tilde{y}_i$ predicted by the model, about 95% of the y values are within ± 2 SEE, and so forth.

Some other comments concerning R^2 and SEE: If a straight line is fit to just two data points ($m = n = 2$), $R^2 = 1$ and SEE is not defined. Obviously, the line fits the existing data perfectly, but there are not enough data to make any statistical inference about the quality of the fit. Hence, it is recommended that $n \gg m$, which means that the unknown coefficients are based on a relatively large number of data.

A plot of the data and the resulting equation checks for spurious peaks and other artifacts, which is sometimes a problem when data are clustered at just a few values of x, as shown in Figure 8.33. The result should be continuous, permit interpolation between data, and support some degree of extrapolation. In some applications, for example, determining velocities from displacement data, it is critical that the function have valid derivatives.

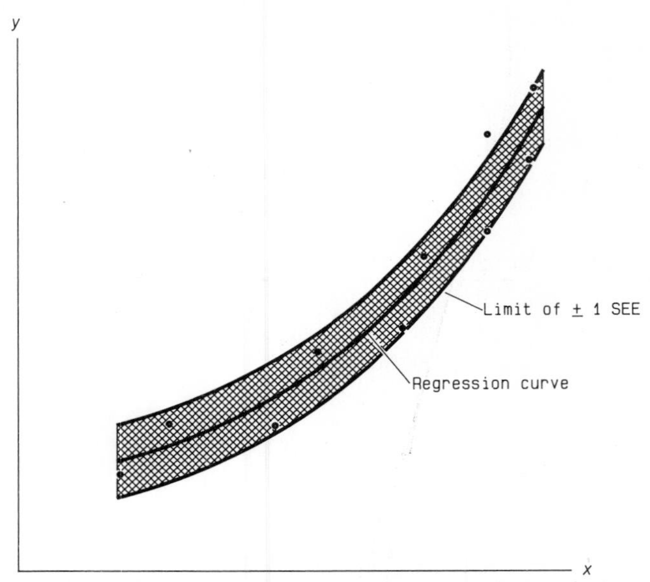

Figure 8.32 Scatter band containing ±1 * SEE, which is predicted to contain approximately 68% of normally distributed data.

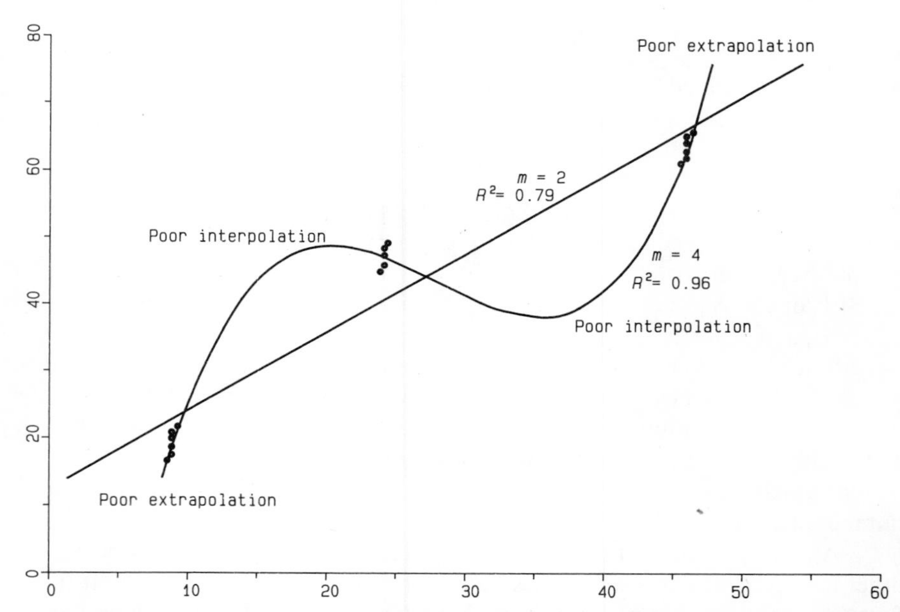

Figure 8.33 A higher-order polynomial does not necessarily fit data better. In this example, a straight line ($m = 2$) and a cubic polynomial ($m = 4$) are fit to 24 data. While R^2 indicates a better fit for the cubic polynomial than the straight line, the former does very poorly for interpolation and extrapolation.

Whenever possible, physical equations which relate the phenomenon being modeled should be used for the regression. Even though some other function might produce better goodness-of-fit statistics, the physical equation will be more satisfactory for extrapolation, derivatives, and so on. The simple straight line is often a good choice. Other functions that work well with physical systems include Fourier series, logarithmic and exponential functions, and polynomials. In any event, the choice should tend toward low values of m.

Linear regression software is widely available both in stand-alone programs and in subroutine libraries. If such software is not available, the programming effort to implement the routines given here is minimal when matrix multiplication, matrix transpose, and simultaneous linear equation solution subroutines (Appendix B) are used.

EXAMPLE 8.2

The tabulated data are observed in a test of the response of a large dashpot subjected to a step load. Determine A, B, and k to best fit the model

$$\tilde{u} = A + \sqrt{Bt} + k \ln t$$

to the data and comment on quality of the resulting fit.

Time t (s)	Speed u (m/s)
1.0	1.40
2.0	2.80
4.0	3.55
10.0	3.05
12.0	2.85

Solution. Because the method requires *linear* coefficients, the unknowns are replaced by $c_1 = A$, $c_2 = \sqrt{B}$, and $c_3 = k$. There are three coefficients ($m = 3$) to be determined with the five given observations ($n = 5$). The matrix [F] from Eq. (8.16) is

$$[\mathbf{F}] = \begin{bmatrix} 1 & 1.000 & 0.000 \\ 1 & 1.414 & 0.693 \\ 1 & 2.000 & 1.386 \\ 1 & 3.162 & 2.303 \\ 1 & 3.464 & 2.485 \end{bmatrix}$$

(The numbers here are rounded off to enhance readability.) Multiplying the matrices $[\mathbf{F}]^T[\mathbf{F}]$ and $[\mathbf{F}]^T[\mathbf{U}]$ followed by substitution into Eq. (8.24) yields the system of three linear equations

$$\begin{bmatrix} 5.000 & 11.040 & 6.867 \\ 11.040 & 29.000 & 19.642 \\ 6.867 & 19.642 & 13.880 \end{bmatrix} \begin{bmatrix} c_1 \\ c_2 \\ c_3 \end{bmatrix} = \begin{bmatrix} 13.650 \\ 31.976 \\ 20.967 \end{bmatrix}$$

Solution of this system for [**C**] yields $c_1 = 5.033$, $c_2 = 3.621$, and $c_3 = 4.145$. Thus,

$$\tilde{u} = 5.033 - \sqrt{13.11t} + 4.145 \ln t$$

which is plotted below.

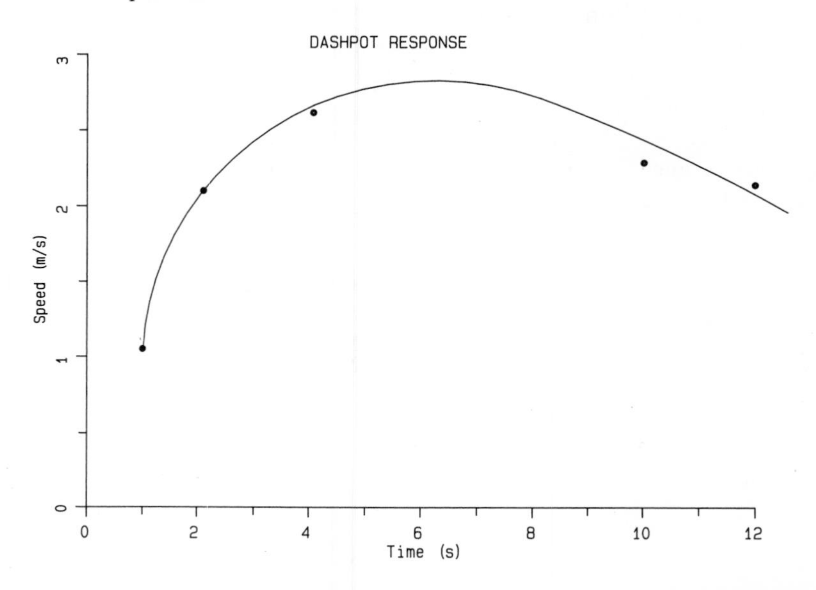

Investigation of the quality of fit begins with construction of a table of residuals:

$t(s)$	u(m/s)	$\tilde{u}$(m/s)	$r = \tilde{u} - u$(m/s)
1.0	1.40	1.41154	0.01154
2.0	2.80	2.78493	−0.01507
4.0	3.55	3.53546	−0.01454
10.0	3.05	3.12864	0.07864
12.0	2.85	2.78943	−0.06057

From the last column, the sum of squares of the residuals is approximately

$$\text{SSR} = 0.01043 \quad \text{m/s}$$

Use of Eq. (8.29) with the data in the second column yields

$$\text{CTSS} = \Sigma u_i^2 - \tfrac{1}{5}(\Sigma u_i)^2 = 2.563 \quad (\text{m/s})^2$$

Thus, from Eq. (8-30) the R-squared test yields

$$R^2 = \frac{\text{CTSS} - \text{SSR}}{\text{CTSS}} = 0.9959$$

and from Eq. (8.32), the standard error of the estimate is

$$\text{SEE} = \sqrt{\text{SSR}/(5-3)} = 0.07221 \text{ m/s}$$

These statistics show R^2 very close to unity, which indicates an excellent fit. The standard error of the estimate, 0.07221 m/s, is reasonably small when dealing with data in the range of 1.4 to 3.5 m/s. Regression models should be used with caution outside the range of the input data, particularly if the function is not a valid physical relation. In the present example, a small value of t yields a large negative velocity ($\tilde{u} = -52.24$ m/s for $t = 1$ μs), which is not reasonable in view of the sign and magnitude of the other data.

8.6 CONTOUR PLOTS

A *contour plot* is a method for presenting a surface which is a function of two variables, for instance, the surface $z = f(x, y)$. The contours are curves of constant z which may be visualized as the intersections between the surface and cutting planes that are parallel to the x-y plane. Contour plots are useful in technical work because numerical values can be obtained by interpolation between contours. Applications include survey maps, where the contour lines show constant elevation, and field theory solutions such as temperature distributions, where the contours show constant values of the field variable.

Contour plots can be generated most easily from data which are gridded. Data derived from a continuous mathematical representation $z = f(x, y)$ are easily gridded by using equally spaced values of x and y. However, when the z's are discrete points arbitrarily located in the x-y plane, interpolation is needed to grid the data.

Grid Interpolation

One simple algorithm for interpolation between arbitrarily spaced input data given by $z_k = f(x_k, y_k)$ to the grid $z_{ij} = f(x_i, y_j)$ is

$$z_{ij} = \frac{\displaystyle\sum_{k=1}^{n} \frac{z_k}{[(x_i - x_k)^2 + (y_j - y_k)^2]^w}}{\displaystyle\sum_{k=1}^{n} \frac{1}{[(x_i - x_k)^2 + (y_j - y_k)^2]^w}} \tag{8.33}$$

where n is the number of arbitrarily spaced input points being used in the interpolation, w is an arbitrary weighting factor applied to the distance between each input point and the derived grid point, and i and j are the row and column location of the grid point. Ordinarily, the grid points are equally spaced in the x and y directions.

The effect of the weighting factor w on the shape of the cross section of a surface interpolated between two points is shown in Figure 8.34. As shown here, $w = 0.5$ gives straight line interpolation between the points. However, the value $w = 1$ spreads the contours out a bit and is a good default value, because evaluation of Eq. (8.33) requires somewhat less computation. The user may want to experiment with different values of w to produce best results. When a very

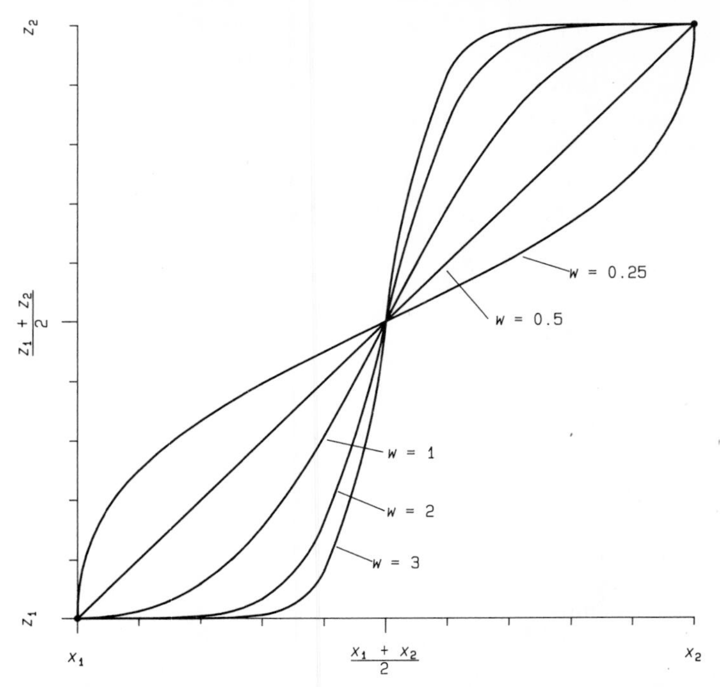

Figure 8.34 Cross sections of surfaces $z = f(x, y)$ for y = constant shown the effect of w in Eq. (8.33) on the interpolation between the two points (x_1, z_1) and (x_2, z_2). It can be observed that some arbitrary contour z_c, where $z_c < (z_1 + z_2)/2$, becomes closer to x_1 as w becomes smaller.

large number of points is to be contoured, it is advantageous to use only the n points nearest to x_i and y_j in Eq. (8.33) to compute the values of z_{ij}. A fine grid and a large number of data both contribute to improving the quality of contour plots.

Construction

The gridded data define an array of cells or elements, which may consist of either rectangles or right triangles. Contour plots can also be constructed on planar finite element meshes, which are arrays of three- or four-sided polygons. The simplest scheme involves processing one cell at a time, with the contour curve being represented by straight line segments crossing each cell.

An arbitrary triangular cell is shown in the example in Figure 8.35. The constant values of z_c to be contoured must first be specified; often this is user input. One cell at a time, the z values at the vertices are checked in the program for the possibility that the specified z_c lies between them. If a contour enters a cell by crossing one edge, then it must leave the cell by crossing one other edge. The x and y coordinates of the intersections between the contour lines and the edges are

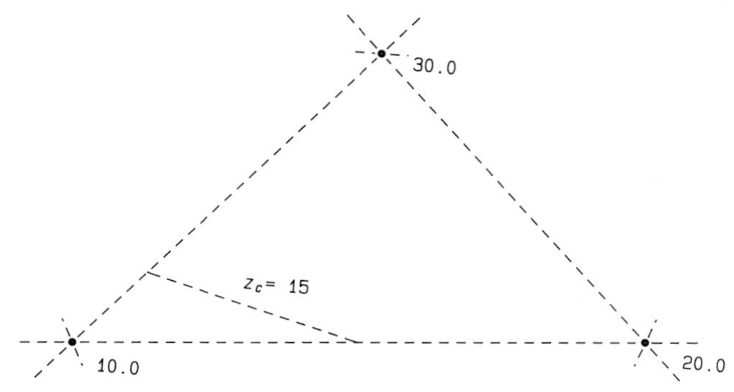

Figure 8.35 Example cell where the contour line $z = 15$ is drawn between two points determined by linear interpolation on the edges of the cell.

determined by interpolation and are connected with a line segment. In this algorithm, triangular cells work better than four-sided cells, because there is no possibility that *four* intersections of z_c can be found on *three* edges.

A smoothing function, such as a parametric cubic spline, will improve the appearance of a contour plot. Programming such smoothing is somewhat complicated, because the points describing z_c must be connected in proper order. The possibility that there may be more than one contour line with the same z_c must also be accommodated.

EXAMPLE 8.3

Compare the contour plots for the data given with a grid of 20×20 for $w = 1$ and $w = 2$. The region in the x-y plane to be contoured is $0 \le x \le 5$ and $0 \le y \le 5$, and contour lines are to have values of 10, 15, 20, and 25.

x	y	z
0.85	0.34	19
0.34	3.00	21
0.65	2.70	23
0.69	3.40	24
0.33	4.42	27
1.55	1.50	18
2.00	1.03	16
2.05	2.40	18
2.00	4.42	24
3.00	0.70	11
2.95	3.25	19
3.00	4.10	21
3.90	4.80	22
4.40	0.66	8
4.50	2.05	13
4.85	3.10	16
4.80	3.92	18

Solution. This example uses a 20 × 20 grid which is subsequently divided into triangles. Because the number of input data are small, all are used in the interpolation. Had there been more input data, the two plots would be more similar. The computer program and the resulting plots follow, where it is noted that the values of contour are inserted under user control. The program also provides for the optional use of markers to distinguish the contours.

```
C
C      PROGRAM CONTUR
C
       COMMON /GEOM/ X0,Y0,DX,DY,Z(21,21),ZC(4),M,N,NCON
       COMMON /CTL/ IFLAG
C
C    X0,Y0 - LOCATE ORIGIN
C    DX,DY - SPECIFY GRID SPACING
C    Z(I,J) - Z(X,Y) AS INTERPOLATED TO GRID POINTS
C    ZC(K) - THE NCON SELECTED CONTOUR VALUES
C    IFLAG - SET TRUE WHEN A CONTOUR LINE IS STARTED
C    M,N - THE DIMENSION OF THE GRID
C
C    INITIALIZE GRAPHICS (SEE APPENDIX B)
C
       CALL GRINIT(4107,7550,1)
       CALL WINDOW(-1.0,8.1,-1.0,6.0)
C
C    INPUT ORIGINAL DATA AND INTERPOLATE TO GRID
C
       CALL GRID
C
C    DRAW FRAME WITH AXES
C
       XE = X0+M*DX
       YE = Y0+N*DY
       CALL AXES(X0,Y0,'X',-1,XE,0,X0,1.0,XE,0,5,0)
       CALL AXES(X0,Y0,'Y', 1,YE,90,Y0,1.0,YE,0,5,1)
       CALL MOVE(XE,Y0)
       CALL DRAW(XE,YE)
       CALL DRAW(X0,YE)
C
C    DIVIDE EACH CELL INTO TRIANGLES    I,JJ  ____ II,JJ
C    TEST EACH EDGE FOR POSSIBLE              | /|
C    INTERSECTION                             |/ |
C    WITH CONTOUR LINE ZC(K)           I,J   ---- II,J
C
```

```
100    DO 200 K=1,NCON
          DO 180 I=1,N
             II = I+1
          DO 180 J=1,M
             JJ = J+1
C
C      TEST LOWER RIGHT TRIANGLE
C
          IFLAG  =  0
          IF(Z(I,J).LT.ZC(K)  .AND.  ZC(K).LT.Z(II,J)
     &          .OR. Z(I,J).GT.ZC(K)  .AND.  ZC(K).GT.Z(II,J))
     &       CALL CONLIN(I,J,II,J,K)
          IF(Z(II,J).LT.ZC(K)  .AND.  ZC(K).LT.Z(II,JJ)
     &          .OR. Z(II,J).GT.ZC(K)  .AND.  ZC(K).GT.Z(II,JJ))
     &       CALL CONLIN(II,J,II,JJ,K)
          IF(Z(II,JJ).LT.ZC(K)  .AND.  ZC(K).LT.Z(I,J)
     &          .OR. Z(II,JJ).GT.ZC(K)  .AND.  ZC(K).GT.Z(I,J))
     &       CALL CONLIN(II,JJ,I,J,K)
C
C    TEST UPPER LEFT TRIANGLE
C
          IFLAG = 0
          IF(Z(I,J).LT.ZC(K)  .AND.  ZC(K).LT.Z(II,JJ)
     &          .OR. Z(I,J).GT.ZC(K)  .AND.  ZC(K).GT.Z(II,JJ))
     &       CALL CONLIN(I,J,II,JJ,K)
          IF(Z(II,JJ).LT.ZC(K)  .AND.  ZC(K).LT.Z(I,JJ)
     &          .OR. Z(II,JJ).GT.ZC(K)  .AND.  ZC(K).GT.Z(I,JJ))
     &       CALL CONLIN(II,JJ,I,JJ,K)
          IF(Z(I,JJ).LT.ZC(K)  .AND.  ZC(K).LT.Z(I,J)
     &          .OR. Z(I,JJ).GT.ZC(K)  .AND.  ZC(K).GT.Z(I,J))
     &       CALL CONLIN(I,JJ,I,J,K)
180    CONTINUE
200    CONTINUE
C
C    PUT ON LABELS
C
300    WRITE(1,*) 'ADD LABEL? (Y/N)'
       READ(1,10) IANS
10     FORMAT(A1)
11     IF(IANS.EQ.'Y') THEN
          CALL LABEL
          GO TO 300
       END IF
       CALL GRSTOP
       STOP
       END
C
C-------------------------------------------------------------
```

```
C
      SUBROUTINE GRID
C
C   ROUTINE FOR INPUT OF RAW DATA ARRAYS XX, YY, AND ZZ
C   AND INTERPOLATION TO GRID FORM
C
      COMMON /GEOM/ X0,Y0,DX,DY,Z(21,21),ZC(4),M,N,NCON
      DIMENSION XX(17),YY(17), ZZ(17)
      DATA XX/0.85,0.34,0.65,0.69,0.33,1.55,2.00,2.05,
     &        2.00,3.00,2.95,3.00,3.90,4.40,4.50,4.85,4.80/
      DATA YY/0.34,3.00,2.70,3.40,4.42,1.50,1.03,2.40,
     &        4.42,0.70,3.25,4.10,4.80,0.66,2.05,3.10,3.92/
      DATA ZZ/19.0,21.0,23.0,24.0,27.0,18.0,16.0,18.0,
     &        24.0,11.0,19.0,21.0,22.0, 8.0,13.0,16.0,18.0/
      DATA X0,Y0,DX,DY/0.0,0.0,0.25,0.25/
      DATA M,N,NP,NCON/20,20,17,4/,ZC/10.0,15.0,20.0,25.0/,EPS/1.0E-10/
C
C   PLOT THE X,Y LOCATIONS OF THE INPUT POINTS
C
      DO 100 I=1,N
         CALL MARKER(XX(I),YY(I),9)
100   CONTINUE
C
C   SELECT THE WEIGHTING FUNCTION W
C
      W = 2.0
C
C     GENERATE A GRID OF N X M RECTANGLES OR N+1 X M+1 POINTS
C
      DO 200 I=1,N+1
      DO 200 J=1,M+1
         X = X0 + (I-1)*DX
         Y = Y0 + (J-1)*DY
         SUMT = 0.0
         SUMB = 0.0
            DO 180 K=1,NP
               DENOM = ((X-XX(K))**2 + (Y-YY(K))**2)**W
C
C   GO OUT OF LOOP IF INPUT POINT IS MATCHED EXACTLY
C
               IF(DENOM.LT.EPS) THEN
                  Z(I,J) = ZZ(K)
                  GO TO 200
                END IF
                SUMT = SUMT + ZZ(K)/DENOM
                SUMB = SUMB + 1.0/DENOM
   180          CONTINUE
            Z(I,J) = SUMT/SUMB
   200   CONTINUE
         RETURN
         END
```

```
C
C-------------------------------------------------------------------
C
      SUBROUTINE CONLIN(I1,J1,I2,J2,K)
C
C   CALLED WHEN Z1 <= ZC(K) <= Z2
C   USES LINEAR INTERPOLATION TO LOCATE END POINTS
C   OF CONTOUR LINE SEGMENT WHICH CROSSES TRIANGLE
C
      COMMON /GEOM/ X0,Y0,DX,DY,Z(21,21),ZC(4),M,N,NCON
      COMMON /CTL/ IFLAG
C
C   DETERMINE COORDINATES OF GRID POINTS
C
      X1 = X0 + (I1-1)*DX
      Y1 = Y0 + (J1-1)*DY
      X2 = X0 + (I2-1)*DX
      Y2 = Y0 + (J2-1)*DY
      Z1 = Z(I1,J1)
      Z2 = Z(I2,J2)
      X = X1 + (X2-X1) * (ZC(K)-Z1) / (Z2-Z1)
      Y = Y1 + (Y2-Y1) * (ZC(K)-Z1) / (Z2-Z1)
      IF (IFLAG.EQ.0) THEN
         CALL MOVE(X,Y)
         IFLAG = 1
C
C   REPLACE THE PRECEDING WITH THE FOLLOWING FOR MARKERS
C        CALL MARKER (X,Y,K)
C
      ELSE
         CALL DRAW(X,Y)
      END IF
      RETURN
      END
C
C-------------------------------------------------------------------
C
      SUBROUTINE LABEL
C
C   PLACES UP TO FOUR CHARACTERS AS LOCATED BY USER
C
      CALL LCATE(X,Y,IDAT)
      WRITE(1,*) 'TYPE UP TO 4 CHARACTERS'
      READ(1,10) INSTR
10    FORMAT(A4)
      CALL SYMBOL(X,Y,1,INSTR,0,4)
      RETURN
      END
```

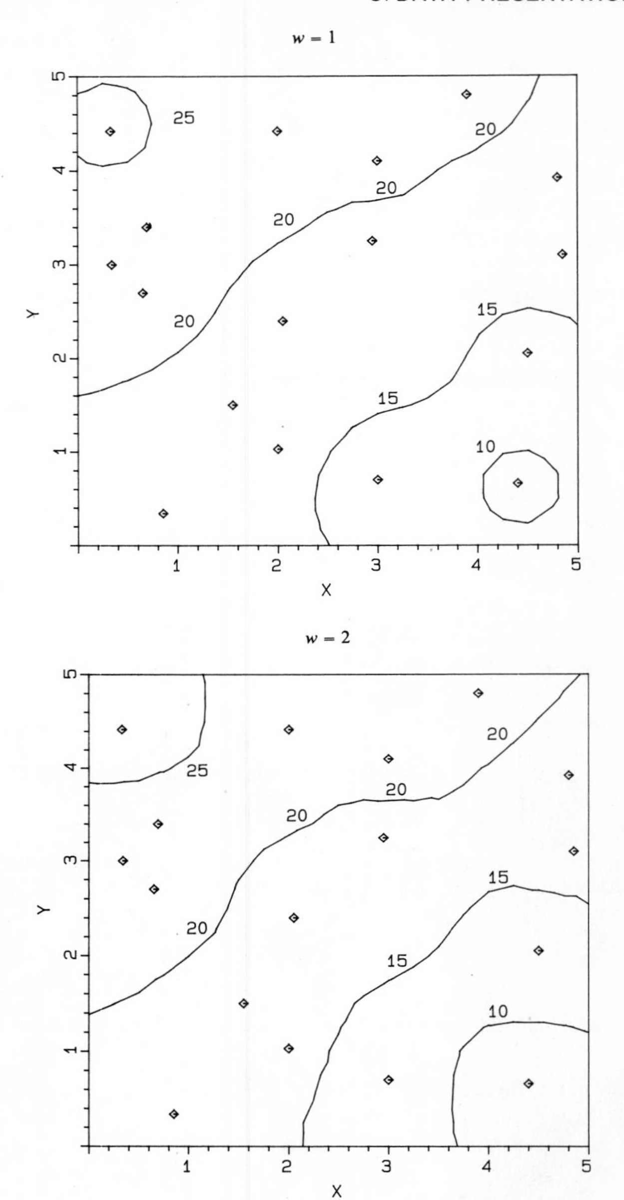

8.7 THREE-DIMENSIONAL SURFACE PLOTS

Another method for displaying functions of the form $z = f(x, y)$ is the 3-D *surface plot* or *carpet plot*. The grid lines on the surface are most conveniently generated by lines of constant x and lines of constant y. An example surface plot appears in Figure 8.9.

Different views of surface plots are produced using the various transformations in Chapter 4. Thus, it is possible to produce surfaces in axonometric, oblique, or perspective projection from any arbitrary viewpoint. Removal of hidden lines is most conveniently done with the masking algorithm of Section 2.5 and Example 2.2.

Just as with contour plotting, a gridded array of data is needed to produce a surface plot. An array with approximately 10 to 50 values on each side produces surface plots of good quality with straight line segments between the z values. For nongridded data, interpolation as in Eq. (8.33) may be used to convert the data to grid form. If the data must be smoothed, multiple regression[19] can be applied, where the surfaces of Chapter 6 are good candidates for approximating functions. The gridded data are then generated from the function resulting from the regression analysis. In some cases, it may be simpler to construct a surface plot with lines of either constant x or constant y.

In the example which follows, no interpolation is needed because the mathematical equation for the surface can be evaluated directly at the grid points.

EXAMPLE 8.4

The function

$$z = \sin\sqrt{x^2 + y^2} \qquad -5.0 \le x \le 5.0, \quad -5.0 \le y \le 5.0$$

is to be displayed as a surface plot. Develop a program to compute values of the function at equally spaced grid points, to produce an axonometric projection of the resulting points, and to render the image with the aid of the masking hidden-line algorithm.

Solution. In the FORTRAN-77 program which follows, $Z(I, J)$ is computed in the code between statements 100 and 150. The necessary input data are hard-coded in the DATA statements for compactness. The surface plotting is done by SUBROUTINE CARPET, which calls graphics subroutines described in Appendix B and the hidden-line subroutines (MASK, VISIBL, ORDER, and DRAWLN) given in Example 2.2.

The grid is 21 × 21. The rotation angles are $\theta_z = 40°$ (the z axis is vertical) and $\theta_x = -70°$ (the x axis is horizontal), which places the origin of the coordinate system in front and gives a top view of the surface.

```
C
C     PROGRAM SURFACE
C
      COMMON /AMASK/ YMAX(131), YMIN(131), XL, XH, NSTOT, LACC
      COMMON /GEOM/ THETAX,THETAZ,X0,Y0,X1,Y1,IFLAG0,IFLAG1
      DIMENSION P(1,3),X(21),Y(21),Z(21,21)
      DATA YMAX/131*-1.0E20/, YMIN/131*1.0E20/, NSTOT/130/
     1 ,IFLAG0/1/, N/21/,LACC/8/,XL,XH,YL,YH/-10.0,10.0,-8.0,8.0/
      DATA THETAZ,THETAX/40.0,-70.0/
C
```

```
C    YMAX AND YMIN - MAX AND MIN VALUES TO DEFINE MASK
C    NSTOT - NUMBER OF STRIPS IN MASK
C    LACC -  NUMBER OF INTERPOLATIONS TO DRAW PARTIALLY HIDDEN LINE
C    XL,XH,YL,YH - GRAPHICS WINDOW IN WORLD COORDINATES
C    X,Y - WORLD COORDINATES OF GRID (INDEPENDENT VARIABLES)
C    Z - DEPENDENT VARIABLE TO DEFINE SURFACE Z = Z(X,Y)
C    THETAZ - VIEWING ROTATION AROUND VERTICAL AXIS  0<THETAZ<90
C    THETAX - VIEWING ROTATION AROUND HORIZONTAL AXIS  0>THETAX>-90
C    P - TEMPORARY STORAGE FOR POINT BEING PROCESSED
C    X0,Y0,X1,Y1 - TEMPORARY END POINTS OF LINE BEING PROCESSED
C    IFLAG - VISIBILITY FLAG ->  0 = NOT VISIBLE,  1 = VISIBLE
C          IFLAG0 - FLAG FOR X0,Y0
C          IFLAG1 - FLAG FOR X1,Y1
C
C    INITIALIZE GRAPHICS (SEE APPENDIX B FOR GRAPHICS SUBROUTINES)
C
      CALL GRINIT (4107,7550,1)
      CALL WINDOW (XL,XH,YL,YH)
      CALL NEWPAG
C
C    GENERATE EQUALLY SPACED X, Y AND Z(X,Y)
C
       DO 100 I=1,N
          X(I) = (I-11.0)/2.0
          Y(I) = (I-11.0)/2.0
100   CONTINUE
      DO 150 I=1,N
      DO 150 J=1,N
         ARG = SQRT(X(I)**2 + Y(J)**2)
         Z(I,J) = SIN(ARG)
150   CONTINUE
C
C    GENERATE CARPET FOR LOWER TRIANGLE
C
      DO 200 I=2,N
         CALL CARPET(X(1),Y(I),Z(1,I),0)
         I1 = I-1
         DO 180 J=1,I1
            K = I-J
            DO 160 M=1,2
               CALL CARPET(X(J+M-1),Y(K),Z(J+M-1,K),1)
160         CONTINUE
180      CONTINUE
200   CONTINUE
C
C    GENERATE CARPET FOR UPPER TRIANGLE
```

```
C
      N1 = N-1
      DO 300 I=1,N1
         CALL CARPET(X(I),Y(N),Z(I,N),0)
         K = N
         I1 = I + 1
         DO 280 J=I1,N
            DO 260 M=1,2
               CALL CARPET(X(J),Y(K),Z(J,K),1)
               K = K-1
260            CONTINUE
            K = K+1
280      CONTINUE
300   CONTINUE
      CALL GRSTOP
      STOP
      END
C
C------------------------------------------------------------------
C
      SUBROUTINE CARPET(X,Y,Z,INTR)
C
C   X,Y,Z ARE COORDINATES OF POINT TO BE PLOTTED
C   INTR = 0 FOR POINT ON EDGE OF SURFACE, = 1 FOR INTERIOR POINT
C
      COMMON /AMASK/ YMAX(131), YMIN(131), XL, XH, NSTOT, LACC
      COMMON /GEOM/ THETAX,THETAZ,X0,Y0,X1,Y1,IFLAG0,IFLAG1
      DIMENSION P(1,3)
C
C   PLACE IN SINGLE ROW POINTS MATRIX
C
      P(1,1) = X
      P(1,2) = Y
      P(1,3) = Z
C
C   ROTATE THIS POINT BEFORE DRAWING
C
      CALL ROTZ(P,1,THETAZ,0.0,0.0)
      CALL ROTX(P,1,THETAX,0.0,0.0)
C
C   EXTRACT SCREEN PROJECTION OF NEW POINT
C
      X1 = P(1,1)
      Y1 = P(1,2)
C
C   POINT ON EDGE IS ALWAYS VISIBLE
```

```
C
      IF (INTR.EQ.0) THEN
         CALL MOVE(X1,Y1)
         IFLAGO = 1
         GO TO 200
      END IF
C
C     TEST INTERIOR POINTS AND DRAW VISIBLE LINE SEGMENTS
C
      CALL VISIBL(X1,Y1,IFLAG1)
      IF (IFLAGO.EQ.1 .AND. IFLAG1.EQ.1) THEN
         CALL DRAW(X1,Y1)
         CALL MASK(X0,Y0,X1,Y1)
      ELSE IF (IFLAGO.EQ.1 .OR. IFLAG1.EQ.1) THEN
         CALL DRAWLN(X0,Y0,X1,Y1, LACC,IFLAGO,IFLAG1)
         CALL MASK(X0,Y0,X1,Y1)
      END IF
C
C     USE END OF LINE SEGMENT TO BEGIN THE NEXT LINE SEGMENT
C
200         X0 = X1
            Y0 = Y1
            IFLAGO = IFLAG1
300      RETURN
         END
C
C     SUBROUTINES MASK, VISIBL, ORDER, AND DRAWLN CAN BE FOUND IN EXAMPLE 2
C
```

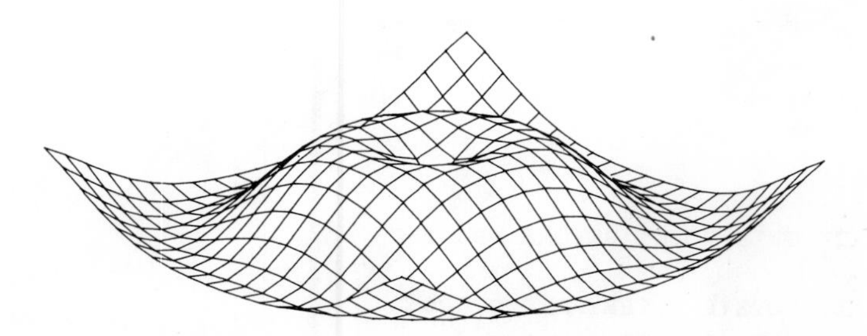

The resulting plot is shown above.

The program can be simplified and the limitation of having a square grid can be removed if lines of constant x or constant y are drawn instead of the grid. If the interpolation—DRAWLN—which draw the partial lines between a visible and an invisible point is eliminated, the quality of the result is not degraded appreciably if the number of grid points is relatively large.

The quality of presentation is improved if axes and other visualization lines are added to the plot, as in Figure 8.9. The additional code can be lengthy, depending on how the axis labels are handled.

8.8 CLOSURE

Presentation graphics greatly aids the interpretation and comprehension of numerical information. Besides the monochromatic presentation graphics examples shown in this chapter, examples of the use of color shading for data presentation are shown in the plates of Chapter 10. These examples show *color-shaded contour plots* where the color-coded temperature replaces labeled contour lines and *color-shaded surface plots* where color coding is superimposed on a surface that is rendered with light-source shading.

Are "fancy" data presentation graphics worthwhile? According to Belie's experiment[5] the answer is yes. Here, presentation graphics depicting eight different temperature distributions in a rectangular field described by 10,000 nodes were prepared in the following forms:

Sorted printouts. Several pages of output listed x, y locations and their corresponding temperatures. Analysis programs typically supply results in this form.

Numerical plots. The temperature at every third node was printed in corresponding row and column position to map the distribution. The 33 × 33 array of three-digit numbers is a practical limit to the amount of printed content on a single piece of paper.

Contour plots. Single-color contour plots from two different software packages were used for the test. The contour lines of constant temperature were labeled with appropriate numerical values.

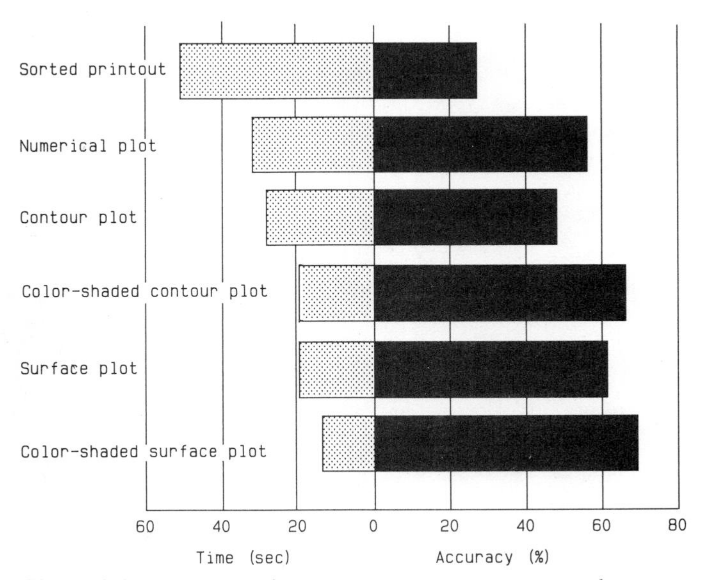

Figure 8.36 Average time to generate a response and average accuracy of the responses for various presentation formats.

Surface plots. Single-color surface plots were rendered for each of the eight temperature distributions.

Color-shaded contour plots. Color-coded temperature distributions were rendered with two different packages. The hottest region was coded with red and the coolest with blue.

Color-shaded surface plots. The color-shaded contour plots above were rendered on a 3-D surface which was light-source shaded to show form. Production of these plots is somewhat more expensive in terms of the computation and hardware needed.

Twenty-five people of various backgrounds and education above high school were asked to identify the number and shape of the areas of maximum and minimum temperature. The time to arrive at a response and the correctness of the response were recorded for each of the eight temperature distributions. The average of the results for the eight test items is shown graphically (of course) in Figure 8.36. Graphical presentation doubles the accuracy of interpretation. Faster response and higher accuracy go together. And, in general, surface plots appear to be more effective communicators than contour plots.

PROBLEMS

8.1. Comment on the practice of labeling axes in scientific notation as Volts $\times 10^3$. Are the plotted quantities kilovolts or millivolts?

8.2. Sketch the plot that is a straight line in Figure 8.26 on (a) linear axes and (b) linear x axis and logarithmic y axis.

8.3. Why is the parametric cubic polynomial in Eq. (8.3) plotted over the range $1/3 \leq u \leq 2/3$, whereas that in Eq. (8.11) is plotted over the range $0 \leq u \leq 1$?

8.4. Develop a set of equations in a form similar to Eqs. (8.1) through (8.4) for a fifth-degree polynomial. *Note:* Use of a digital computer or programmable calculator is recommended to invert the resulting matrix.

8.5. Reformulate Eqs. (8.12) for the condition where the *parametric* slope at the beginning of the spline is specified as (x_1', y_1') and the other end is free. If the *geometric* slope was specified instead of the parametric slope, how would this equation be used?

8.6. Rewrite the equations for a 3-D quadratic B-spline curve, Eqs. (5.66) through (5.68), in 2-D (x-y) form. Comment on the use of these equations for smoothing curves.

8.7. A curve is to be smoothed with the piecewise cubic polynomials of Eq. (8.3). Show that the resulting curve does not have continuous slope at the knot between segment (i) and segment ($i+1$).

8.8. The function $w = f(z) = az^m$ is to be fit using the linear least squares technique. With the use of logarithms, write Eqs. (8.14) through (8.17) in terms of z, w, a, and m for fitting n data.

8.9. Verify that Eq. (8.29) follows from Eqs. (8.27) and (8.28).

8.10. Set $\tilde{y} = a + bx$. Use the least squares technique to determine expressions for a and b in terms of the given data (x_i, y_i), $i = 1, n$.

8.11. In the implementation of Eq. (8.33), what should be done for the possibility that $x_i = x_k$ or $y_j = y_k$?

8.12. A piecewise, nonparametric cubic polynomial $y = f(x)$ is to be used to smooth a curve through the data points

x	y
−1	0
0	3
1	2
2	0
3	−1

Determine the coefficients $[\mathbf{C}_i]$ for the polynomials. Sketch the resulting curve, and use your equations to determine the maximum y and its location.

8.13. Repeat Problem 8.12 with a piecewise, parametric cubic polynomial, expressed in the notation employed in Eqs. (8.3) and (8.4).

8.14. Repeat Problem 8.12 with a B-spline curve, expressed in 2-D (x-y) form in a manner analogous to Eqs. (8.3) and (8.4).

8.15. A natural cubic spline $y = f(x)$ with relaxed ends is to smooth the four data

x	y
−1	0
0	0
1	2
3	1

(a) Determine the coefficients $c_1, \ldots, c_4$. (b) Determine the coefficients a_1, b_1, and d_1 and write the equation for the first segment of the spline. (c) Determine the magnitude and location of the minimum value of the curve, which occurs in the first segment.

8.16. With the data of Problem 8.15, determine the equation for the first segment of a parametric cubic spline. Then determine the magnitude and location of the minimum which occurs in this segment.

8.17. A tensile test of a steel specimen produced the following data for load versus elongation:

P (kN)	e (mm)
0	0
4	1.2
10	3.0
10	4.0
10	5.0

It is conventional to plot elongation as the independent variable (horizontal axis) and load as the dependent variable. By means of a parametric cubic spline fit to these data with the tension parameter $R = 0$, determine the predicted value of load at $e = 3.1$ mm. Also, check the predicted load for $e = 3.1$ mm with $R = 0.6$ and $R = 1.0$.

8.18. Fit a parametric cubic spline to the following data for $y = f(x)$ in the order given:

x	y
0	0
1	0
1	2
2	2

Determine the predicted values of y at $x = 0.5$ if (a) $R = 0$, (b) $R = 0.33$, and (c) $R = 1.0$.

8.19. For the data of Problem 8.15, determine the straight line fit by least squares. Also determine R^2 and SEE for this fit.

8.20. For the data of Problem 8.15, determine the least squares fit to a parabola. Test the goodness of fit and comment on the meaning of the statistics.

8.21. For the data of Problem 8.17, use the least squares procedure to fit the equation

$$P = a_1 + a_2\sqrt{e}$$

Determine R^2 and check on the predicted value of P at $e = 8$ mm. Comment on this result.

8.22. For y = constant and the following data:

k	x_k	z_k
1	0.85	23.4
2	1.23	18.3
3	1.77	14.0

where $n = 3$, test the gridding algorithm, Eq. (8.33), for the predicted value of z at the point $x = 1.0$ with values of w equal to 0.25, 1.0, 2.0, and 5.0. Tabulate the results and comment on the choice of w.

PROJECTS

NOTE: An available presentation graphics software system may be used as appropriate to execute the projects.

8.1. When a switch is changed on a RC circuit, the voltage and current are given by

$$v(t) = 70 - 95e^{-4t} \text{ (volts)}$$

$$i(t) = 206e^{-4t} \text{ } (\mu\text{amps})$$

where e is the base of natural logarithms. Plot these two functions with linear scales on a common abscissa. Compute approximately 20 evenly spaced points to make a curve which is smooth when the data are connected with straight lines. The abscissa should show sufficient range that the decay of the current is nearly complete.

8.2. The drag force on smooth plates is given by three equations for three different flow regimes:

$$\text{Laminar:} \qquad C_D = \frac{1.33}{\sqrt{\text{Re}}}$$

$$\text{Transition:} \qquad C_D = \frac{0.075}{\text{Re}^{1/5}} - \frac{1700}{\text{Re}}$$

$$\text{Turbulent:} \qquad C_D = \frac{0.075}{\text{Re}^{1/5}}$$

Plot the drag coefficient C_D as a function of the Reynolds number Re on log-log axes for the range $10^4 \leq \text{Re} \leq 10^7$. However, the transition region should be plotted only for the range where C_D is greater than the value of C_D for the laminar region. Use a large enough number of data points that the curves appear to be smooth when the data are connected with straight lines.

8.3. Repeat Project 8.1, except that only five data points each for voltage and current should be computed. Smooth the curve by use of a piecewise parametric cubic polynomial.

8.4. Repeat Project 8.2, except that only five data points should be plotted for each regime. Smooth the data with a cubic spline.

8.5. The following data have been experimentally observed for the contraction coefficient C as a function of area ratio A:

A	0.2	0.4	0.6	0.8	1.0
C	0.632	0.658	0.712	0.813	1.00

Plot the data points. Develop a smooth curve with a cubic polynomial and connect the points.

8.6. Using tables of thermodynamic properties, construct a contour plot of $P = f(V, T)$ for a substance as assigned. Experiment with the value of the weighting factor w.

8.7. Repeat Project 8.6, except that a surface plot is to be constructed.

8.8. Write a program that accepts any arbitrary number of x-y data pairs and fits a best-fit least squares straight line. The data and the line should be plotted in good form.

8.9. Include the capability in the program for Project 8.8 to compute R^2 and SEE for the straight line that is fit.

8.10. Develop a heat conduction solution $T = f(x, y)$ for a square plate with differing temperatures on the edges. Make a contour plot (or a surface plot) of the solution.

8.11. Develop appropriate presentation graphics for laboratory data as assigned.

8.12. Develop appropriate presentation graphics for solutions to analysis problems as assigned.

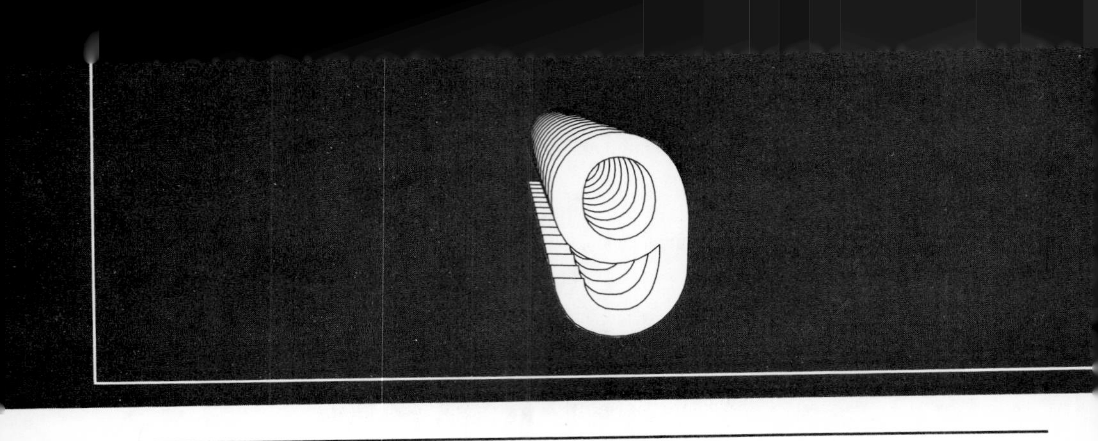

Computer-Aided Drafting

The advent of computer-aided engineering (CAE) systems is revolutionizing work in all fields of engineering practice. CAE encompasses all engineering design activities including drafting, analysis, manufacturing, testing, and modeling. Initially, the acronymn CAD stood for "computer-aided design." However, the accepted use of this acronymn is changing to "computer-aided drafting," which more accurately describes what most CAD systems actually do. Systems which handle design functions in the larger sense may be designated as CADD, for "computer-aided design and drafting" systems, or simply as CAE systems. Another older designation, CAD-CAM, which suggests design and manufacturing, may also be replaced by the term CAE.

This chapter introduces some of the basic functionalities that are found in generic, professional-grade CAD systems. To be considered for professional use, a system should meet the following requirements:

- Include sophisticated enough software and hardware that the system can be cost-justified by time and labor savings.
- Be able to make drawings of the accepted size and complexity.
- Produce results of quality equal to or better than that of the manual drawings previously done.
- Provide data base management facilities for retrieving and merging old drawings and other work.
- Be integrated with other computer-based engineering systems.

It follows that the size of a system needed depends on the application, with attention being given to the abilities for managing complex drawings, executing large enough programs, storing large enough data bases, providing acceptable response speed, and producing high-quality results. Small, personal-computer-based CAD systems are adequate for some applications, while other applications require larger systems.

Traditionally, the distinction has been made between "turnkey" and "do-it-yourself" systems. As with a new automobile, a new turnkey CAD system is set to go by merely turning it on. In turnkey systems, a single vendor furnishes all hardware, software, support, and training necessary. For users who lack computer experience, the turnkey approach minimizes risk.

In contrast, the do-it-yourself approach maximizes flexibility. The hardware and software will come from a number of vendors and may even include the organization's own custom software. With the general growth of computer experience, it is clear that more and more users will be managing their own system integration. In the case of CAD, there are several excellent software systems which run on a variety of workstations.

Important features of CAD systems concern the degree of interactivity and the ease of use. Highly interactive systems quickly and accurately display the result of instructions. Ease of use is related to training time as well as the time and effort required by experienced users to produce a drawing. Important to easy use is the ability to communicate with the program through simple, easily remembered commands, menus, and screen picks (hits).

CAD systems excel at changing, copying, merging, and managing drawings that are already in a data base. As any project moves through the engineering process, many changes will result as the result of the expertise of numerous specialists. As the repository for the most up-to-date information, the CAD data base effectively coordinates the work of the engineering team.

Usually, CAD systems are classified according to the engineering field for which the system has special features such as libraries of symbols, certain techniques for creating drawings, and certain links to analysis and manufacturing. Mechanical CAD, for example, supports dimensioning, has symbols for welding, screw threads, and springs, and as so forth. Such systems may prepare data for numerically controlled machining and stress analysis. Furthermore, CAD systems can be classified as to whether there is a two- or three-dimensional (2-D or 3-D) data base, whether color is supported, whether it is a single-user or multiuser system, and so on.

9.1 USER INTERACTION WITH CAD SYSTEMS

There are several schemes for user control of CAD systems. The logical interactive graphics input devices of Section 1.5—locator, stroke, valuator, choice, pick, and string—classify the methods for user interaction with CAD workstations. The various kinds of interactive hardware found in CAD systems support the logical input devices as follows:

- Keyboards (string, locator, valuator, choice, pick). An alphanumeric keyboard is found on virtually all systems. Many full-time users prefer typing commands on the keyboard, since interaction is faster than for methods which require moving one's hands to some other device. Locator is implemented through cursor control keys, which substitute for the joystick or mouse (see below). For valuator, numbers are typed. Choice is done by typed characters corresponding to certain actions, such as numbered choices on a screen menu. Pick can be made by the typed name of an entity or by use of the cursor.

• Joystick (locator, pick, valuator). The joystick (familiar to video game players), joydisk, or thumb wheels (one for vertical motion and one for horizontal motion) move the graphics cursor around the screen. The cursor may be in the form of a horizontal and vertical line the full width of the screen, or some other marker that moves around on the screen. A "hit" is commonly made with the space bar when the cursor is in the right place.

• Mouse (locator, pick, valuator). Cursor motion can be controlled with a small puck that can be moved on the tabletop or on a special pad. A trackball is essentially an upside-down mouse where the hand replaces the tabletop. A "hit" is made with the buttons provided on the mouse.

• Tablet (stroke, locator, pick, choice). Besides supporting a mouselike function to move the screen cursor, a digitizing tablet is used for copying graphical data into a system. In this application, the old drawing or sketch is secured to the tablet surface, and the puck or stylus is used to select x and y coordinates. The coordinates thus hit are sent by the tablet to the computer system. Not only is the process tedious, but inaccuracy due to paper distortion and human factors can also be a problem.

• Light pen (pick, locator). Technically, the light pen is a pick that can select a segment by correlating it to the time at which it is refreshed. (Recall that with a stroke-refresh graphics device, discussed in Section 1.3, all the vectors on the screen are redrawn several times a second, segment by segment from display memory.) Pick is directly implemented on stroke-refresh devices. The locator function of the light pen determines the x-y coordinates of a pixel or grid point seen by the photomultiplier tube in the light pen. This function can be accomplished on raster as well as stroke-refresh devices. Because users find holding the light pen up to the screen to be fatiguing, and because better, less expensive locating devices are available, light pens are not popular in newer systems.

• Menu (choice). A tablet, an auxiliary keyboard, or a CRT screen can be used to show a menu—a list of CAD commands. Screen menus and the auxiliary keyboard are easily changed under computer control, while printed tablet menus can contain far more details. Menus, of course, help with the problem of remembering commands; the organization and quality of menus are vitally important to the productivity of a CAD system. Some CAD systems permit users to design their own menus in order to accommodate specialized tasks.

• Dial (valuator). Dials are used for input of scalar values to control such operations as the viewing transformations of zooming, panning, and rotation.

Many CAD systems are controlled with a graphics workstation which has a keyboard and a mouse or joystick. Some workstations have even more of the input devices listed above to enhance versatility. The CAD workstation illus-

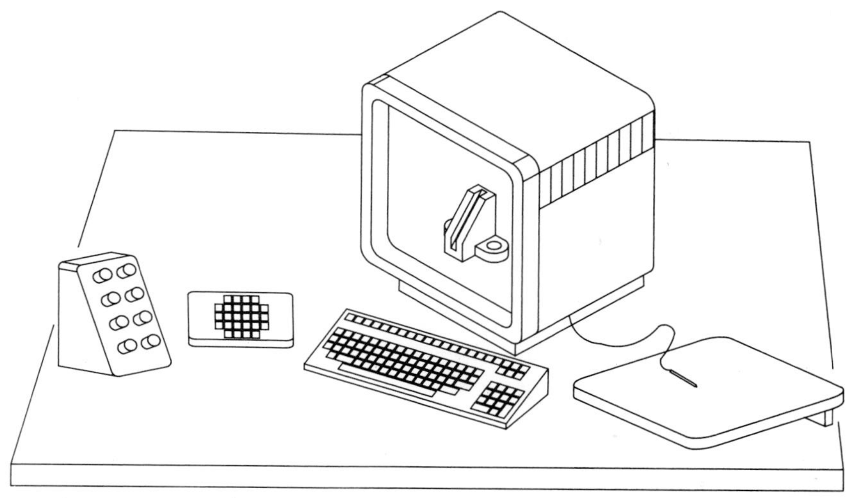

Figure 9.1 Fully equipped CAD workstation with several user interaction devices. Left to right: dials, lighted function keyboard, alphanumeric keyboard, and digitizing tablet with stylus.

trated in Figure 9.1 is a system with many of the aforementioned user input devices. Ongoing research and development in improving interaction with CAD systems deal with projects such as fully automatic digitizers to copy old drawings, voice-actuated systems, user-customized commands, and so forth.

Snapping

An important feature of interactive CAD systems manages the decision of whether or not the user intends to pick some existing point on the graphics screen. In this context, "existing point" includes line segments and their ends, curves, vertices of polygons, and line intersections or, perhaps, a grid point. Because it is almost impossible to hit a point exactly with the locator, snapping is necessary to manage the picking of points, make line intersections meet, and so on. Snapping is the process where the program picks the point nearest to the one located. Although computationally intensive, predictable and logical snapping capabilities are absolutely necessary for a professional-grade CAD system. In snapping, the following occur:

1. The locator function sends the x and y coordinates of the cursor to the processor.
2. The amount of snap tolerance, considered as Δx and Δy (usually $\Delta x = \Delta y$), defines a rectangular region† around x and y. Often snap tolerance can be adjusted under user control.

†If the region for snap tolerancing is circular, the process becomes more computationally intensive.

3. The points in the data base are tested to see if any fall within the predetermined snap tolerance. The first one found or the nearest one is said to be "snapped to." In other words, the point "located" is assigned the coordinates of the existing point. If no point within the snap tolerance is found, the program signals the user to try again or creates a new point, as appropriate.

Efficiency in the snapping algorithm is essential for fast user response. Options under the control of the user that speed up the response include snapping only to predefined evenly spaced grid points, layering (see below), reducing the size of the snap-tolerance zone, and using windowing or zooming (see below) as much as possible. The general strategy here is to reduce the number of points that require testing.

On the other hand, user speed is enhanced by a relatively large snap tolerance, say 0.1 in., since the cursor does not have to be located as carefully. However, a large tolerance can lead to snaps on points other than the one intended. Windowing or zooming is always recommended to prevent such problems; another technique is illustrated in Figure 9.2.

Layers and Views

Organization and layout of a drawing are managed through the selection of separate *layers* or *views*. The two terms have the same meaning. An *active layer* is a part of a drawing which can be drawn on or changed; *inactive layers* appear at the correct place on the workstation screen but are for display only. Management of the drawing into active and inactive parts simplifies user

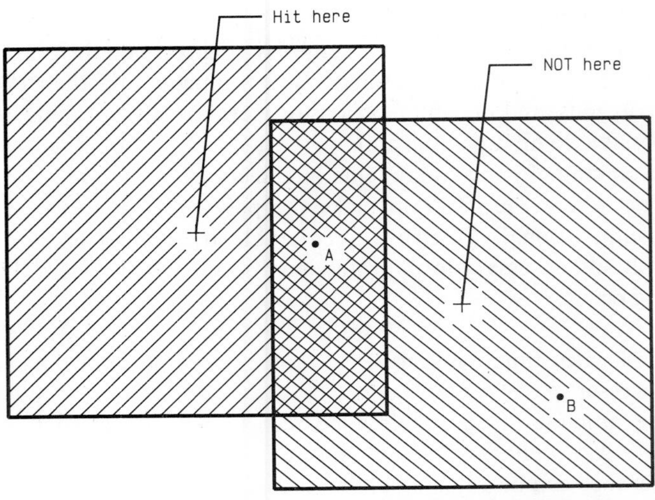

Figure 9.2 User technique to prevent snapping to incorrect point. When the user wishes to pick point *A* and not point *B*, which is nearby, the far side of point *A* is located. The size of the squares represents the snap tolerance.

interaction and provides for faster computer response. One popular microcomputer-based CAD system simplifies plotter management by requiring a separate layer for each different pen color. Some CAD systems permit a definite maximum number of layers that may range from just a few to over a hundred; large systems do not limit the number of layers. With user selection of layers, only the drawing parts of immediate concern are displayed for work.

Printed circuit board (PCB) artwork is a common example which makes good use of layering. In complex applications, both sides (top and bottom surfaces) of the PCBs may contain circuitry, or two or more PCBs may be stacked together and connected by common feedthroughs. Preparation of the artwork is easiest if each surface that contains circuitry is defined as a layer. In the case that the surfaces are separate drawings, it is difficult to check the interconnections. In the case that the surfaces are all on the same drawing but not separated by layers, the complexity is overwhelming, and separation of the individual layers for production masks is difficult if not impossible.

Another useful application of layering is in architectural CAD. The floor plans of a multistory building have utility ducts, stairways, elevators, walls, and columns which match from one floor to another. Obviously, the separate floors should be separate layers. Furthermore, details pertinent to plumbing, heating, electrical, and so forth can be separated into layers to make specialized drawings for the various trades.

Layers do not necessarily have to be stacked one on top of the other. A view and a layer have the same meaning as used here, but the term "view" has specific meaning to certain users in the context of orthographic projection. Here, the top, front, and side of an object are called views—and can be conveniently managed as layers. Another use of layers is for an assembly drawing with many details. Placing each of these details into its own view, perhaps at different scales, facilitates layout of the page by simply relocating the views.

When starting a new drawing, it is good practice to plan for the layers or views that will be needed. Although many simple drawings will be in one layer, complex ones usually benefit by having appropriate multiple layers. One common CAD practice is to put all text in one layer to keep down clutter and speed response.

Window and Zoom

Work on all drawings is expedited by using the viewing transformations of window or zoom as much as possible. Enlargement of the area of interest in the drawing helps the user in many ways. Cursor hits are easier. Clutter is reduced. Text is more readable. Accuracy is improved and time is saved.

Window and zoom have slightly different meanings, although both perform essentially the same functions. To indicate the area of a drawing to be windowed, the user "boxes" the desired portion with the screen cursor, and that portion is subsequently displayed full-screen. Definition of a box ordinarily requires locating two points which describe the opposite corners. In terms of the viewport discussed in Section 2.3, the window function in CAD is the process of selection and enlargement to full screen of a viewport.

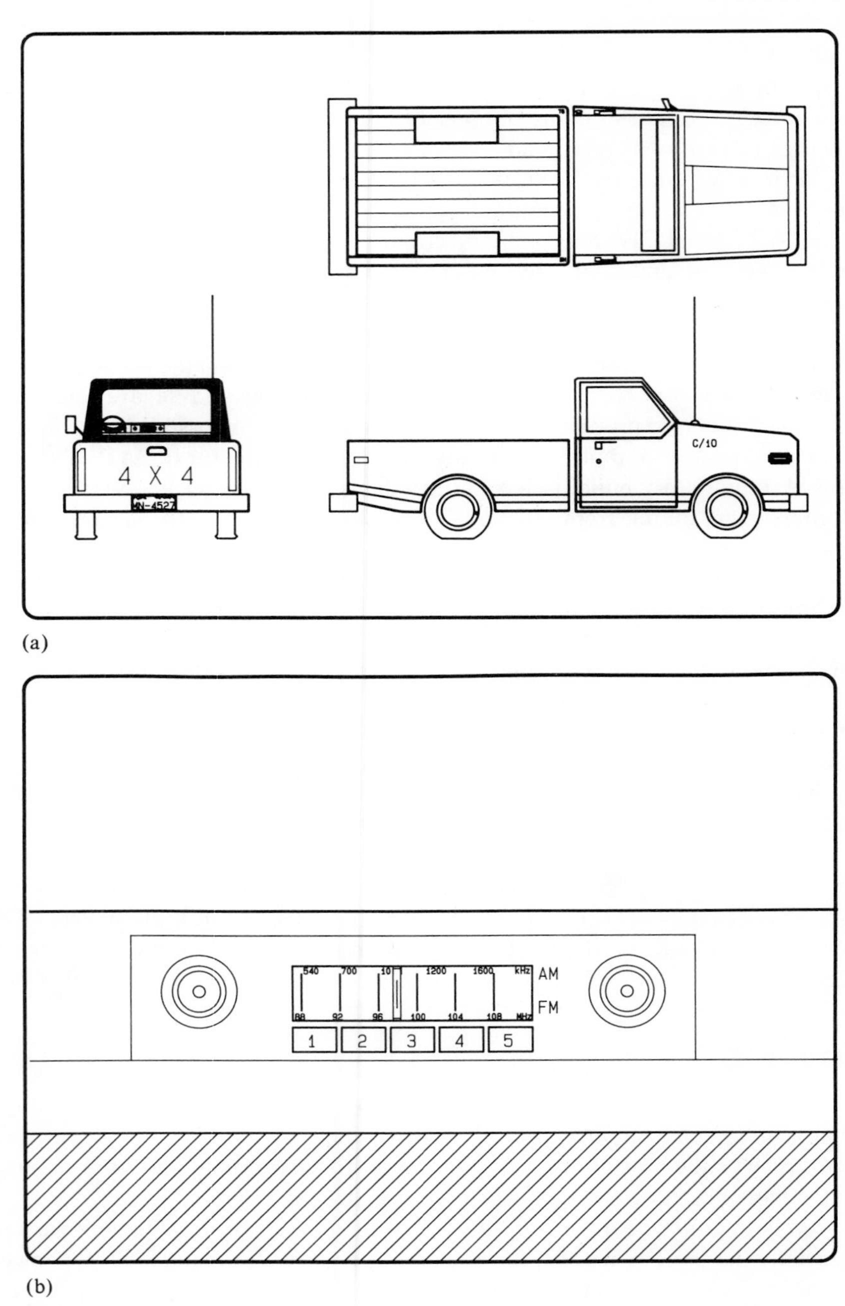

Figure 9.3 Illustration of zoom with the center of scaling located on the radio shown in the rear view. (a) Full screen; (b) zoom at 16×.

Zoom as shown in Figure 9.3 is a direct application of the view transformation for uniform scaling with respect to a center of scaling, Eq. (3.24). The zoom factor is the multiplier by which an area is enlarged. In operation, the user indicates the factor with a valuator and then locates the center of the area to be displayed. A zoom factor which is less than one zooms "out" to show a greater part of the drawing.

Oops

An old saying goes "to err is human." Short of actually redoing the work, CAD systems provide two general strategies for recovery from errors.

One strategy uses system copying facilities, where drawings being modified are a *copy* of the original. If something happens which causes an unrecoverable mess, the work copy can be discarded and the unchanged original retrieved for subsequent work. Users may want to do intermediate "saves" of drawing files if problems are anticipated. Furthermore, the risk of losing drawing files stored on hard or floppy disks is managed by backing up these files on other disks or magnetic tapes. Routine professional practice involves backing up all files daily and removing the tapes or disks to a vault or other safe location.

The other strategy is provided in the menu system of most CAD systems as an "oops" command. Using the oops command immediately after a mistake simply returns the drawing to its former state. For mistakes discovered later, the user has a choice of going to the backup file or changing the drawing.

9.2 TEXT

Text is an important part of drawings. Users can control the size of text, the style of text, and the orientation angle of text. To aid in text placement, options for left-justified, right-justified, or centered text, as well as for matching placement with existing text, should exist. Better CAD systems also have a text editor which permits changing text without simply deleting and retyping.

The typical use, parameters for text size, font, and the like are defaulted to a common standard or may be set by users. To place text on a drawing the user (1) locates the starting point of the text and (2) types the text. The interactivity of the system provides for cases where the text is too long to fit the space, where the text interferes with other items, and so on.

CAD systems differ in the available fonts, as shown in Figures 2.7 and 2.8. In fact, CAD systems are convenient for producing presentation graphics, posters, signs, nameplates, and so forth.

9.3 BASIC DRAFTING OPERATIONS

Commonly, CAD systems are used for drawing lines and curves. Although user interface features vary from one system to another, there is much commonality in the way users access drafting operations such as lines, curves, fillets, chamfers, and crosshatching.

Construction Lines and Points

Construction lines and points assist in the layout and alignment of parts of the drawing but do not appear in the finished copy. Construction entities may either be interspersed within any layer or be contained in their own layer.

Construction points may be arrayed to make a snap grid. Spacing of the points in the grid is set for user convenience. For example, an electrical circuit diagram may be conveniently drawn on a $\frac{1}{2}$-in. grid. This makes it easy to keep all the lines horizontal or vertical and ensures that details and text are uniformly spaced. Another example is a warehouse layout where all aisles and storage areas are assigned to 4-ft centers. Snapping of points located with the screen cursor to grid points facilitates rapid and accurate layout of a drawing.

Construction lines can be used to lay out boundaries, centerlines, and the like. Subsequent points in the drawing snap to construction lines. When the user origin of a drawing is redefined, it may be convenient to place construction lines to mark the new origin. Another use of construction lines is to project drawing features from one view to another.

Lines and Curves

A line is defined by the coordinates of the two ends. The following are two common methods employed in CAD systems for the input of points for definition of lines.

1. Freehand input. In conjunction with a command for "separate line," the starting and ending point are located with the screen cursor. With a "continuous line," the first point and vertices of the series of straight lines are hit in order. These distinctions are shown in Figure 9.4. Freehand input is NOT accurate unless the program is set to snap to existing points or grid points.
2. Typed input. The x and y (and z) coordinates of each point defining the object are entered on the keyboard. The user may have a choice of using

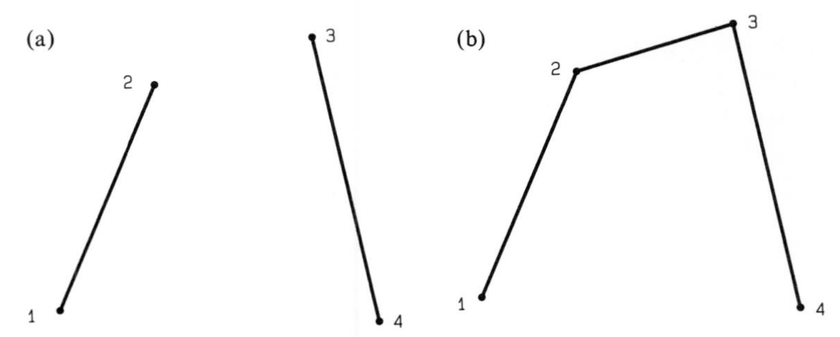

Figure 9.4 Result of freehand drawing resulting from entry of four consecutive points. (a) Separate line; (b) continuous line.

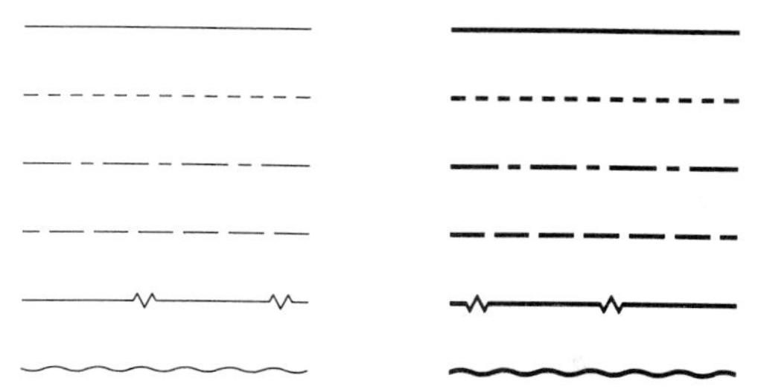

Figure 9.5 Typical line types and weights.

absolute coordinates or relative (incremental) coordinates. There may also be a choice of using rectangular (x and y) or polar (length and angle) coordinates.

Circles may be defined by the center and one point on the radius. Many CAD systems permit specification of a three-point circle, Example 5.2, as an alternative to the center-radius circle. Circular arcs are a special case of the circle, where the points defined are the center, the point (and radius) where the arc starts, and the approximate point where the arc stops.

Other curved lines or splines, discussed in Chapter 5, are defined by locating the control points. Interpolating and approximating functions such as the B-spline and the parametric cubic spline are convenient for drawing smooth free-form curves. Curve smoothing features vary somewhat between the various CAD systems.

Systems often have other features to speed the drafting of lines and curves. Lines can be drawn parallel or perpendicular to existing lines. Multiple parallel lines or concentric circles at a user-defined spacing can be drawn. Construction of a line tangent to a circle or a common tangent between two circles is another convenient feature.

Attributes may be set for lines and curves. Typical attributes include type (solid, dashed, centerline, etc.), color, and weight (where a heavy line is drawn twice or a different pen is used). Some samples of type and weight are shown in Figure 9.5. One can work more rapidly by drawing one kind of line at a time, since extra commands must be executed whenever attributes are changed.

Another technique that speeds line drawing is the use of *trimming* or *relimiting,* shown in Figure 9.6. Here, a line is first made longer than it actually

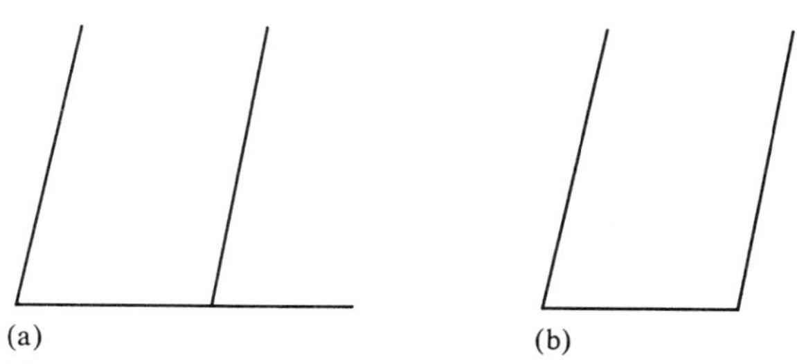

Figure 9.6 Trimming or relimiting a horizontal line: (a) before; (b) after.

needs to be. After intersecting lines are drawn, the extra pieces can be trimmed away or relimited. Typically, the part of a line which extends beyond an intersection with another line can be trimmed back with a single command. This function can also be performed by use of a command to delete a line between two points, but two hits are required. Filleting (see below) with a zero radius is another way to perform the relimit function.

Filleting and Chamfering

Fillets or rounds, Figure 9.7, are easily made by a CAD system. Although the term "fillet" applies to interior angles and "round" to exterior angles, CAD systems use the terms interchangeably. Mathematically, a fillet is a circular arc of a given radius drawn tangent to two lines. Although two such circular arcs exist—one subtending an angle of less than 180° and the other greater than 180°—most CAD systems have no difficulty with the ambiguity. Chamfers are similar to rounds except that the edges are beveled. In practice, a chamfer is defined by the amount the edge is cut back and the angle the chamfered surface makes with one of the surfaces. A chamfer which makes equal angles with the two surfaces is the chord of the arc which defines a round.

In use, filleting or chamfering is done with the following sequence:

1. The radius of the fillet or the size of the chamfer is specified by string input if the default (or previous value) is not satisfactory.
2. The two lines which are to have a fillet or chamfer between them are picked. Some CAD programs may require that these two lines be intersecting lines, others may not.
3. The lines are trimmed (relimited) if the program does not do it automatically.

In filleting and chamfering, as in other CAD operations, it is recommended that the same operations be done at one time. This eliminates issuing the same command over and over. Generally, it is preferable to delay the filleting or chamfering until everything else is drawn. This is particularly true for CAD systems that cannot properly dimension edges that are filleted or chamfered.

Crosshatching

Most CAD systems automatically handle crosshatching, which is the same operation as panel filling described in Section 2.2. Crosshatching requires that a

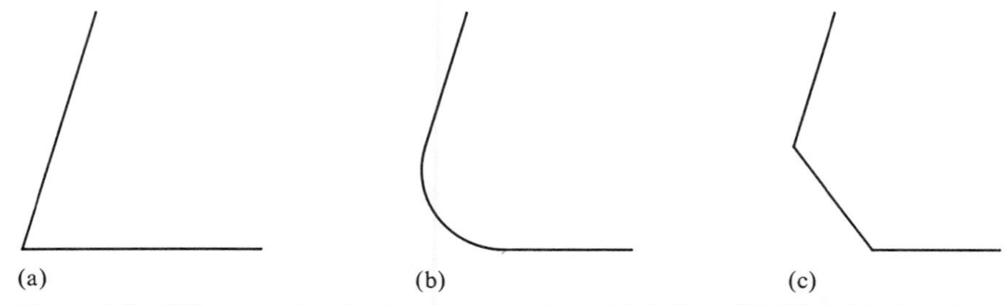

Figure 9.7 Fillet and chamfer between two lines. (a) Before; (b) fillet; (c) chamfer.

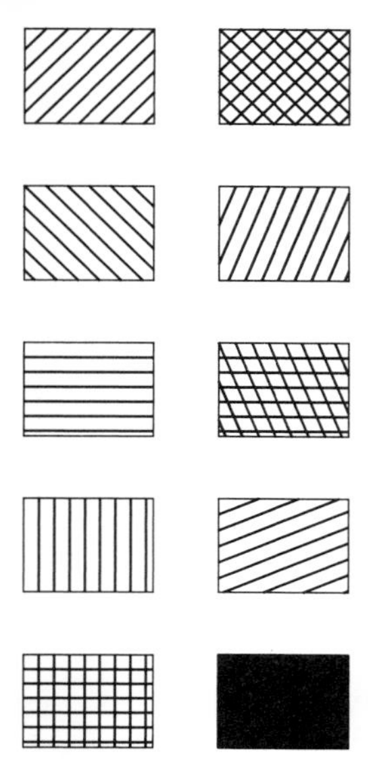

Figure 9.8 Typical crosshatch patterns.

closed contour bound the selected region. Typical choices of crosshatch patterns are shown in Figure 9.8.

The most automatic crosshatching implementations let users pick any point on the closed contour curve. Then the system highlights the contour it found. If the contour is correct, the user initiates the crosshatching. Inner regions within the closed contour can be excluded by picking their contours.

Automatic crosshatching may run into difficulty when there is more than one possible closed contour. Usually, more user input is needed to define the correct boundary. One less "automatic" scheme for defining the region to be crosshatched involves picking all line segments which bound the region. In difficult situations it may be easier to place the crosshatching in a separate layer. Here, the bounding contour might be redrawn, crosshatching performed, and the extra contour deleted to prevent its being drawn twice.

9.4 DIMENSIONING

The term *dimensionally accurate data base* means that the numerical values of the world coordinates of points making up the drawing are stored. Typically, CAD systems store numbers in 32-bit real format, providing about six places of precision. From the data base, a CAD program can compute the distance between two points, the diameter of a circle, the radius of an arc, an angle between two lines, and other data needed for dimensions.

To produce a linear dimension, the two points being dimensioned are picked and any point on the dimension line is located. If the two points do not lie on a horizontal or vertical line, there is a choice of whether the actual distance, the horizontal component, or the vertical component is to be dimensioned.

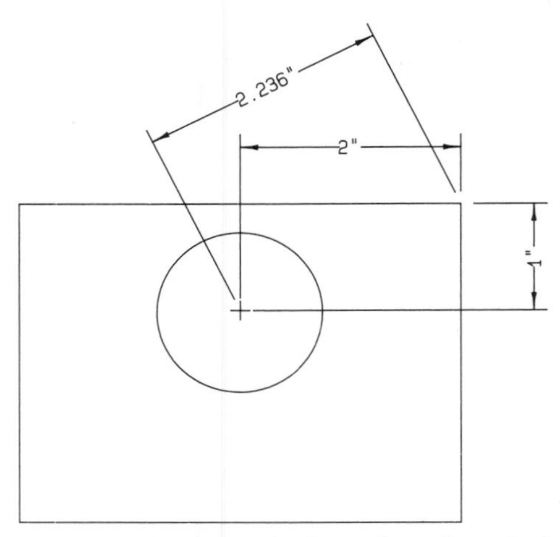

Figure 9.9 Actual, horizontal, and vertical dimension lines. The two points picked to be dimensioned are the center of the circle and the upper right-hand corner of the rectangle.

Usually, the default is the actual distance. In Figure 9.9, the three choices are shown.

To dimension the diameter of a circle or the radius of an arc, only two points are needed: One is any point on the curve to be dimensioned and the other is the location of the dimension text.

To dimension the angle between two lines, three points are needed: The first two indicate the lines and the third locates the dimension text. Some CAD systems may require that the two lines be picked in counterclockwise order to give the angle less than 180°; others may simply default to the smaller angle.

The dimensioning system described above is properly designated as *semi-automatic dimensioning,* although some CAD vendors call this functionality *fully automatic dimensioning.* The latter term should be reserved for systems that do full and proper dimensioning without any additional user action—sophistication in a computer program that would be very difficult to attain. *Manual dimensioning* means that the user must enter the line work and the text for the dimension. Manual dimensioning would be used in cases where the data base is not dimensionally accurate.

A dimensionally accurate data base facilitates the conversion between U.S. customary units and SI units. In doing such a conversion, the user should be cognizant of the precision implied by the dimensions. For example, if dimensions in a mechanical drawing are given to the nearest 32nd of an inch, the new drawing should show precision to the nearest millimeter.

In the following example, some basic functions involved in line drawing and dimensioning are illustrated. In view of the widely different CAD systems available and the diversity of techniques employed by skilled users on any one system, many variations in the procedure to produce even a simple drawing are possible.

EXAMPLE 9.1

A control box is to be made from a purchased aluminum sheet metal box, with three 2-in.-diameter holes punched for meters, and four $\frac{3}{8}$-in.-diameter holes punched for switches. Outline the steps in making the drawing of the control box.

Solution. (NOTE: The following procedure outlines the order of steps with a generic CAD system. Study of the user's manual for a specific system or, preferably, demonstration by an experienced user doing this drawing is an effective means for system familiarization and for comparison between different systems doing the same drawing.)

As a beginning step, the size of paper and the resultant scale of the drawing are arbitrarily set up as B-size and full scale. These choices are indicated to the CAD system at the time of start-up. Also, the drawing must be given an appropriate name. After start-up is finished, the drawing menu can be accessed. Where reference is made to "issuing commands," the user types commands or makes menu selections as appropriate to the system being used.

1. Because the front and side view of the control box will be shown in the finished drawing, a mental idea is formed as to where the two views will be drawn on the paper. Because this is a relatively simple object, the decision is made *not* to set up two separate layers (views) in the data base.

2. The command for a construction snap grid with $\frac{1}{2}$-in. centers is given. This is a convenient grid spacing for fast entry of the drawing, since all the dimensions are to the nearest $\frac{1}{2}$ in.

3. The line drawing command is used to make the 4 in. by 8 in. outline of the front view of the box. The default line type, a single solid line, is used. The points defining the corners of the box are located with the snap grid.

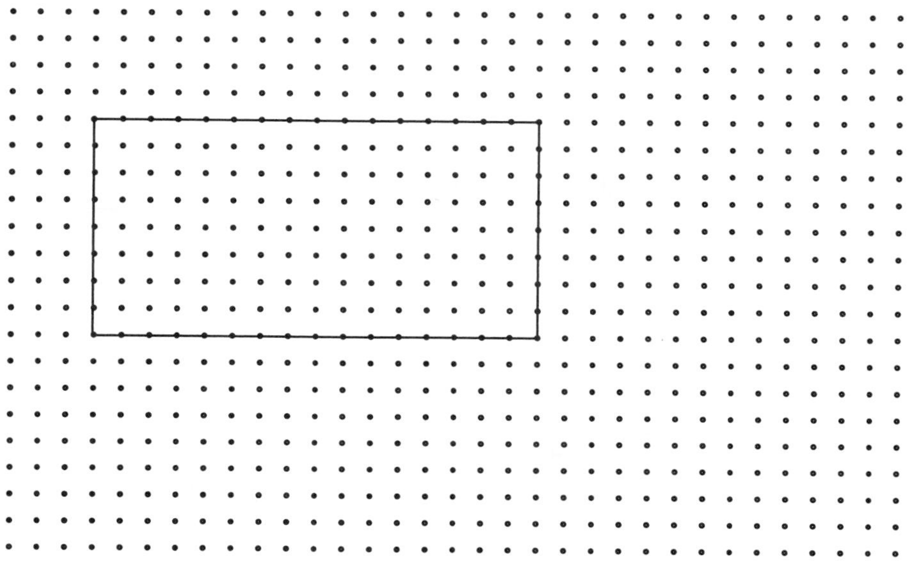

4. Circles for all the holes, centered on the snap grid, are drawn next.

5. The snap grid is used to locate the end view, where the depth of the purchased box is 2 in.

6. The line type is changed to long dash–short dash, and the centerlines are drawn.

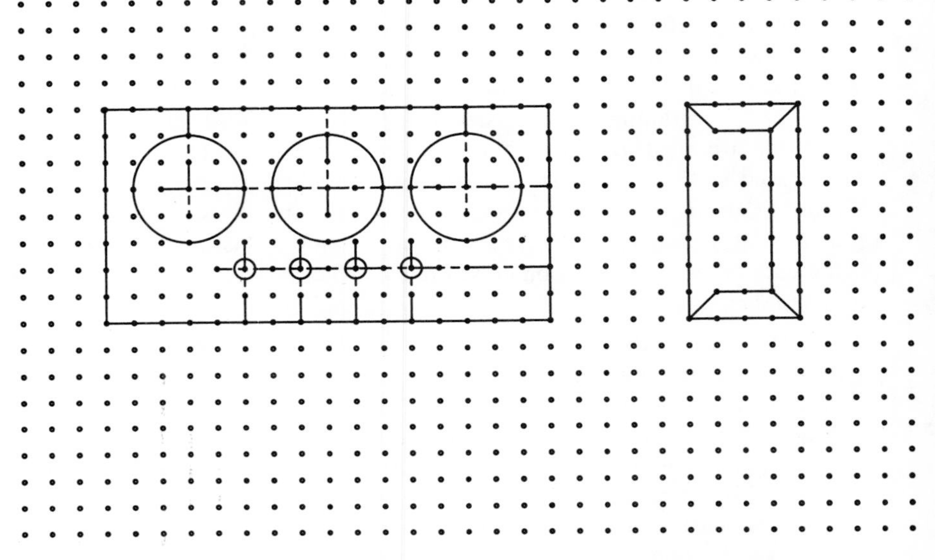

7. The text size, 0.2 in., is selected. All linear dimensioning is done at the same time. For each dimension, the two points being dimensioned are hit and a point on the dimension line is located. It is convenient to use the $\frac{1}{2}$-in. snap grids for alignment of the dimension lines.

8. Circle dimensions are done by picking the edge of the circles to be dimensioned and locating the lower left corner of the dimension text.

9. Text is added to the drawing by locating and typing the text when prompted.

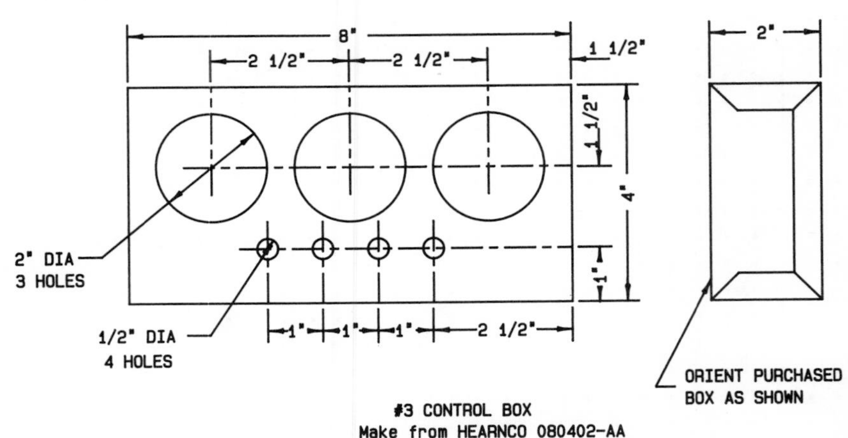

10. The drawing file is closed. A plot file is created and sent to the plotter. Retention of the drawing file is very important, since this allows rapid alteration of the original drawing.

9.5 COPY

The ability to reproduce selected drawing entities in the data base is an essential feature of all CAD systems. This useful function has two forms: (1) *copy,* where one or more duplicates are made and the original is preserved, and (2) *drag* or move, where a duplicate is made and the original is deleted. Both forms use essentially the same software.

With the copy function, the user selects the part of the drawing to be copied or dragged. The selection might be accomplished by picking the segments to be copied, or it might be done by drawing a polygon around the area to be copied. After making the selection, the user indicates where to put the copy.

A versatile implementation of the copy function uses the six-point transformation of Section 3.6 for one, two, or three pairs of points. For a copy defined by one pair of points, a selected point on the object is picked, and then a "copy-to" point is located or picked. The program performs translation of the copy to the new point without any rotation or scaling.

A copy defined by two pairs of points uses the second point on the original object to define a corresponding point on the copy. Enough information is available from two points to make a copy which is translated, rotated, and scaled.

A copy with three pairs of points uses the extra information to permit shearing in addition to translation, rotation, and scaling. One use of the three-point-pair copy is construction of isometric drawings, as in Example 9.4. A system that supports construction of oblique or axonometric pictorial views (Sections 4.5 and 4.6) without the benefit of a 3-D data base is called a "$2\frac{1}{2}$-D" CAD system. Such transformations to make a pictorial view may be built in or may be a user-defined macro.

Another mode of copy is mirror reflection, which aids in constructing drawings of objects with mirror symmetry. The transformation matrices for mirror reflection are given in Section 3.4. With mirror reflection, it is necessary to pick the portion of the drawing to be copied and to indicate the mirror axis. Experienced CAD users look for the opportunity to use mirror reflection whenever possible, since this is a great time-saving option.

Multiple copies at some fixed parallel distance or at some fixed angle of rotation are other time-saving features for replication of repeated details. Here, the user must indicate the number of copies to be made and how the copies are placed relative to each other. Example 9.2 illustrates the use of mirror reflection and rotated multiple copies.

Lines that are not entirely in the region defined for copying are called truncated lines. In the copy facility of some CAD systems, the user has a choice of whether truncated lines are "included" or "ignored" when being copied. This option permits the user to leave out partial lines that are unwanted in the copy.

Copying parts of archival drawings requires additional management facilities in the CAD system. Typically, the old drawing will be brought up on the screen and the part of the drawing to be transferred will be picked just as in a regular copy command. Here, the picked area will be stored as a temporary file (buffer) by the system. When the new drawing is restored to the screen, the contents of the buffer may be placed (drawn) at the desired location. This ability to copy parts of old drawings is very useful, for example, in making detailed parts drawings from an existing assembly drawing.

EXAMPLE 9.2

Use the copy commands of a CAD system to draw a flat part with repeated symmetry. In this example, the part is one of the flat metal sides of a 7-in. audio tape reel, pictured below.

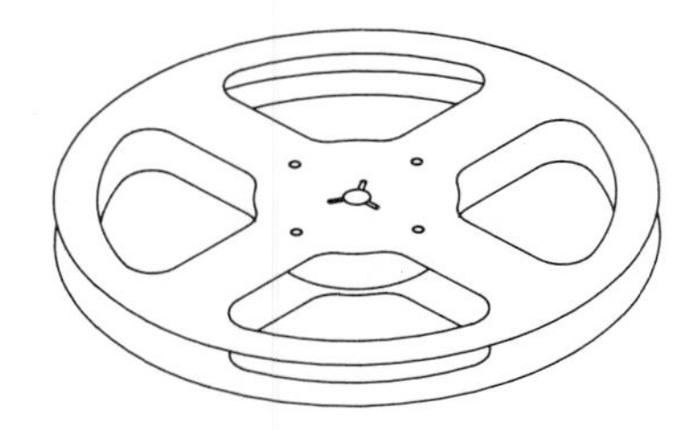

Solution

1. A name is decided for this drawing, and the system is initialized to make use of half scale on A-size (8.5 × 11 in.) paper.

2. The origin is relocated to the approximate center of the paper to serve as the center of the part. This location is shown by the cross (+) in the screen views of the drawings.

3. Because there are four symmetrical spokes, one-eighth of the part is all that is drawn initially. The center hole and keyway, which have different symmetry, will be done later. For convenience, the user zooms into an area just large enough for one of the spokes with the origin centered near the left edge. The window facilitates work on small details.

4. Two construction lines which define a 45° sector with a 3.5-in. radius are drawn. Details of this step include (a) a command to activate construction lines, (b) drawing a line from the origin 3.5 in. horizontally, and (c) drawing another line from the origin with a length of 3.5 in. and an angle of 45°.

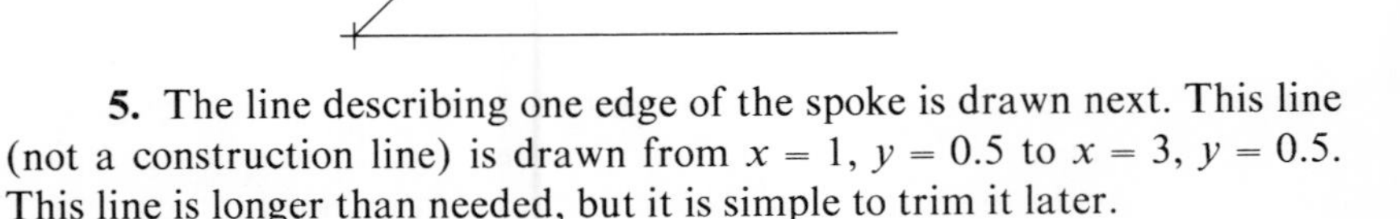

5. The line describing one edge of the spoke is drawn next. This line (not a construction line) is drawn from $x = 1$, $y = 0.5$ to $x = 3$, $y = 0.5$. This line is longer than needed, but it is simple to trim it later.

6. Three circular arcs with centers at the origin are drawn between the two construction lines. The radius for the bottom of the hole is 1.1875 in., the radius for the top of the hole is 3.0 in., and the radius for the outside is 3.5 in. The arcs should end near the straight line for the edge of the hole.

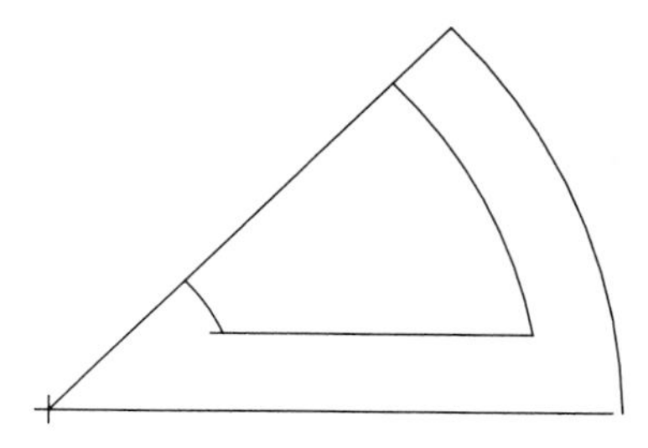

7. The fillet radius for the edges of the holes, 0.375 in., is set and fillets are drawn between the arcs and the lines defining the edges of the hole. The system automatically trims the remaining segments of line and arc.

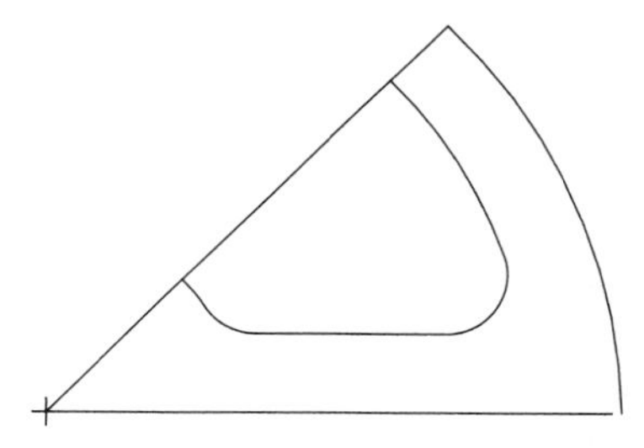

8. A mirror copy around the horizontal construction line is made to complete one spoke. All of the lines describing the eighth of the part are picked.

9. A 0.125-in.-diameter hole for the hub rivet is drawn with its center at $x = 1, y = 0$.

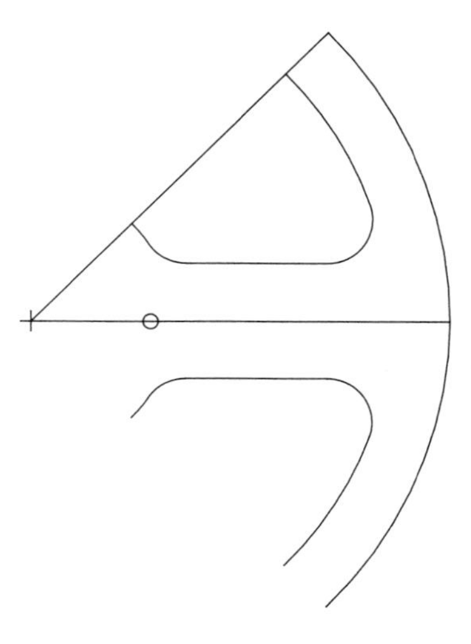

10. Three more copies of everything drawn so far are needed to complete the drawing. Zooming out to full screen is helpful. A copy rotate command for three copies is issued, with the center of rotation at the origin and a 90° angle of rotation for each copy.

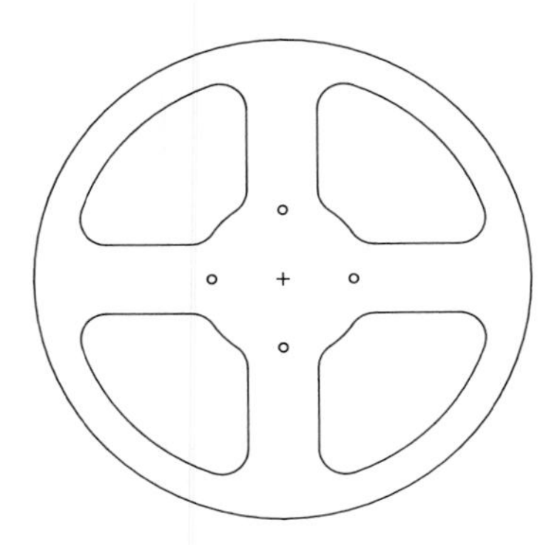

11. Next, the center hole with three symmetrical keyways is drawn. It is convenient for the user to zoom up so that the area covered by the screen is about 0.4 in. square, with the origin centered on the left.

12. A construction line is drawn from the user origin at an angle of 60° with a length of $\frac{5}{32}$ (0.15625) in. This defines half of one-third of a circle and a hole diameter of $\frac{5}{16}$ in.

13. An arc (not construction) with its center at the origin is drawn from the tip of the construction line of step 12 clockwise down to the existing horizontal construction line of step 4.

14. A line to define the straight edge of the keyway is drawn from $x = 0$, $y = 0.03125$ to $x = 0.3125$, $y = 0.03125$. (The 0.03125-in. dimension makes the finished keyway $\frac{1}{16}$ in. wide.)

15. The end of the keyway is drawn. This is a quarter circle arc with its center at $x = 0.3125$, $y = 0$ drawn clockwise from the right end of the keyway edge to the centerline.

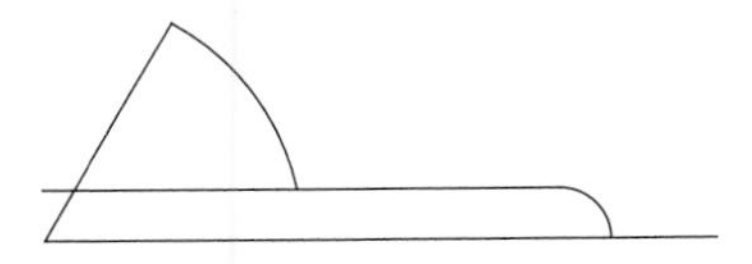

16. The unwanted parts of the straight line and arc are removed. This may be done by a relimit command or a command to delete a line between two points.

17. The half keyway is now ready to mirror around the horizontal axis of symmetry. Only the keyway, and not the rest of the drawing, is mirrored.

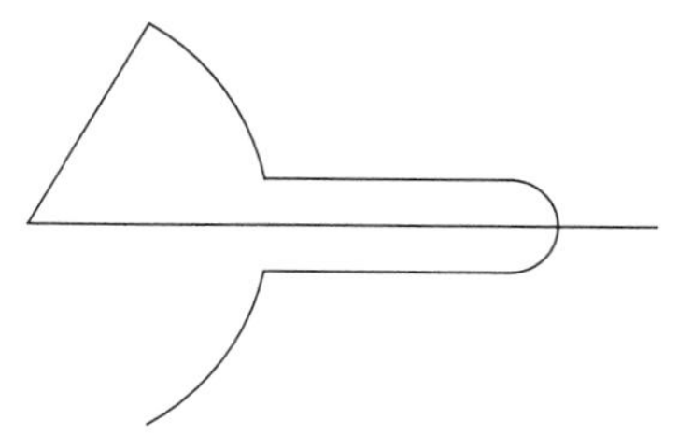

18. Two copies of the detail of step 17 are made with a rotation of 120° around the origin. The entire result can be seen by zooming out.

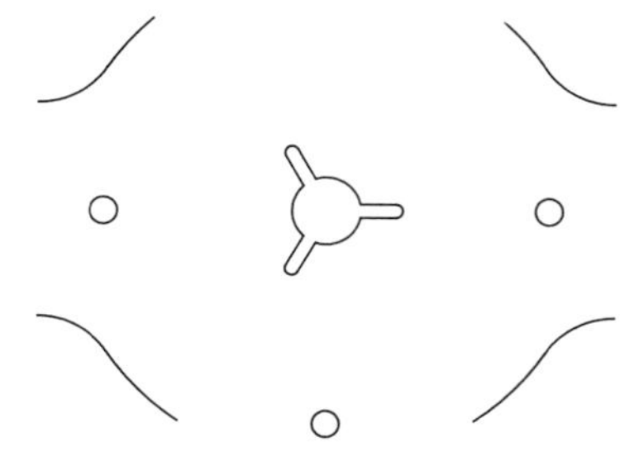

19. The entire part is next viewed full-screen.

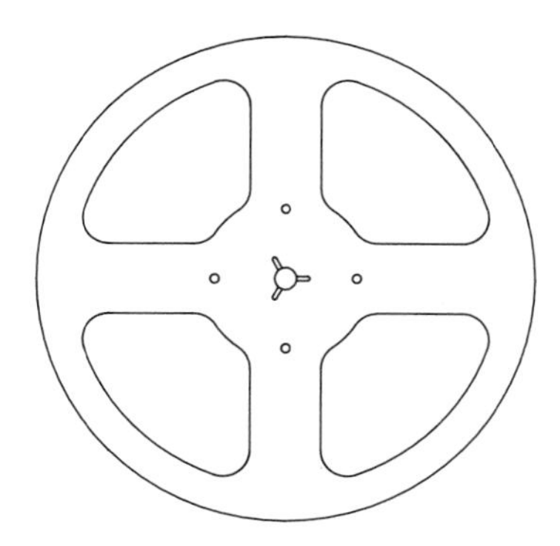

20. Dimensions and notes are added as appropriate. The details of these operations are similar to those in Example 9.1.

This completes the detail part drawing for the flat side piece of the tape reel. This drawing would be kept in a drawing data base from which it could be recalled for modification or for incorporation in other drawings.

9.6 SYMBOL STORAGE AND RETRIEVAL

Segments, as defined in Section 2.8, can be extracted and placed in a buffer for copying into a new drawing. The contents of the buffer could just as well be copied into a file which can be retrieved as needed. A *symbol* is the term used in

CAD systems for a segment, which is any group of lines, curves, text, and so forth. Retrieval is done by identifying the symbol and locating its placement on the drawing. Actually, the same software used for copying may be used to manage symbols.

Symbols enhance versatility and speed of CAD systems. Some systems support a symbol protocol which prompts the user for text to be inserted. Such a symbol could be used for a title block with a company logo and variable information such as drawing name and date. This application eliminates the need for preprinted paper. Symbol libraries customize systems for specific applications. For example, architectural CAD systems contain symbol libraries with such details as doors, windows, and foundation sections. Another example, the use of electrical CAD symbols, follows.

EXAMPLE 9.3

Use symbols to draw a circuit for an electronic filter. Label values of the components.

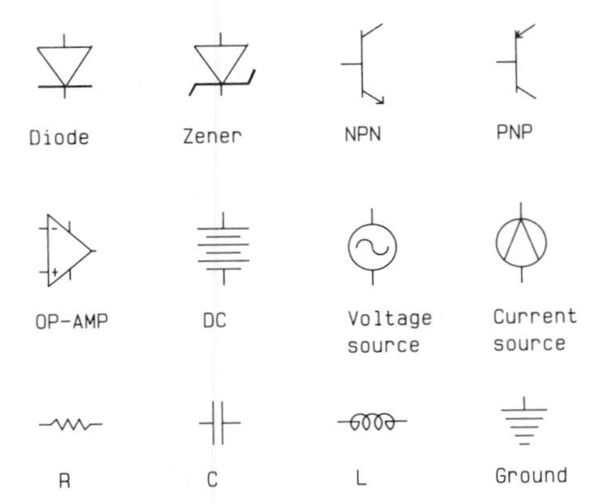

Solution. Previously created and stored in the CAD system data base are symbols for the components and sources shown.
The system is started with A-size paper selected.

1. A construction snap grid with 10-mm spacing is set up, selected as an appropriate spacing for the planned drawing.

2. Symbols for the various components are snapped on the grid. Ends and connection points on the symbols match grid points.

3. Line drawing commands are actuated to complete the connections between the symbols.

4. Necessary text is added to the drawing.

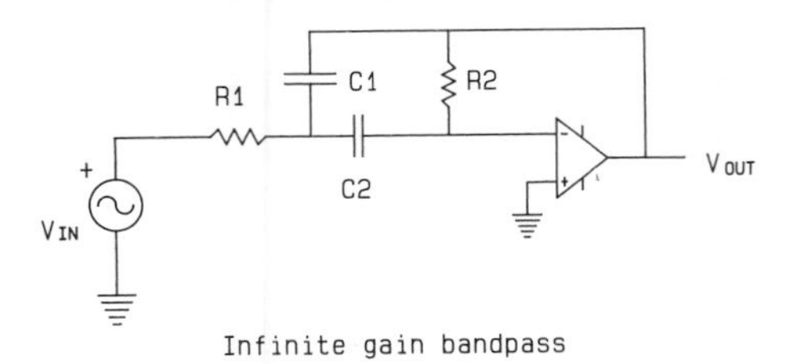

Infinite gain bandpass

9.7 OTHER CAD FEATURES

Special features and ease of use distinguish one CAD system from another. Evaluation of CAD systems for possible purchase should balance cost with overall performance. The following examples of special features found in CAD systems have the potential for increasing performance in selected applications.

Macros

A *macro,* or a *parametric program,* brings together any number of separate functions as a group. In CAD, user-defined macros permit consolidating a sequence of commands into a single command. A symbol is a macro in the sense that an arbitrary sequence of automatically executed drawing commands can be executed with minimum user input. While a symbol represents the consolidation of several drawing commands into a single command, macros found on some CAD systems have even greater capabilities.

Such macros permit the entry of program language statements in addition to executing drawing commands. In this way, macros can make design decisions such as the size or numbers of components needed and then draw the result. Use of macros offers significant time savings in design work with routine content. An example is drawing plans for warehouses constructed from modular components. In this case, a user-defined macro could be invoked which starts by prompting for input of the dimensions of the building. The macro determines the needed quantity and size of roof trusses as well as the other structural components and shows the results.

Three Dimensions

An important difference between CAD and solid modeling programs is the method used to store data. In CAD, the mode of data storage emphasizes the operations to produce the drawing, whereas solid modeling uses a geometric description of the part. To keep the information compactly and give fast response, CAD data bases contain the name of an operation (e.g., "draw line") and the necessary Cartesian coordinates to locate the result of the operation. In contrast, a solid modeling data base will necessarily be much more verbose.

A few CAD systems are classified as 3-D, because x-y-z Cartesian components are stored, and the points can be manipulated with the transformations of Chapter 4. Generally CAD systems have 2-D data bases, in part because 2-D (x-y) Cartesian components require one-third less storage than 3-D components. Furthermore, 2-D systems are inherently less complex.

The pseudo-3-D function known as $2\frac{1}{2}$-D performs viewing transformations that result in a pictorial representation within the constraints of a 2-D data base. Some user intervention is necessary to produce the correct result, since the $2\frac{1}{2}$-D representation does not contain enough information to remove hidden surfaces or to rotate the model for viewing from different directions in space.

Specific methods for producing pictorials vary greatly among various systems. Example 9.4 shows the use of the six-point transformation (Section 3.6) for manual construction of an isometric representation. Some systems contain macros that automate this process to some extent; the task is routinely done with a 3-D CAD system.

Measurement

Frequently, it is necessary to determine a geometric quantity in the course of executing a design. Possible items to be measured or analyzed include perimeter, area, length of an arc or a line, radius of a circular arc, angle between two lines, or distance between two points. Algorithms for the computation of many common geometric properties are given in Chapter 7.

With measurement of surface area of a part made from sheet metal, for example, it is an easy matter for the user to determine the weight of the part. Length measurement features can be used to estimate the cost of utility lines in a subdivision layout. The distinction between a length measurement and a dimension is that a measurement is communicated to the user's workstation dialog area, while a dimension is made part of the drawing.

Property Take-Off

Some CAD programs assign attributes to symbols, lines, and so forth and tabulate the amount of usage of each such entity in the drawing. The total length drawn of lines with a certain identifier, the number of times a certain symbol is used, and the like can be recorded in a data base.

This data base is convenient for the preparation of bills of material, cost estimates, and other tabular data. A computer-produced bill of materials is shown in Figure 9.10. When changes are made in the drawing, the bill of material is automatically updated.

Clearly, property take-off can be integrated with measurement facilities to develop many useful automatic features in design work.

Analysis and Checking

Other features may be available that help the design engineer work faster and better. (a) Mechanical assembly drawings can be checked automatically for fitting and for interference from the buildup of tolerance. (b) Structural drawings can be converted to geometric information for a finite element analysis

COMPONENT BILL OF MATERIALS

QUANT	DWGREF	DESCRIPTION	VENDOR	VENDOR#	UNIT COST	TOTAL COST
1	B1	BOARD ASSEMBLY	CUSTOM ELEC	PO-4532	12.00	12.00
2	L1	LAMP, 1W, 9V	LAMPCO	SS1033	0.45	0.90
2	R1	RESISTOR 200 OHM, 1 WATT	OMC	0-200-1-C	0.12	0.24
2	R2	RESISTOR 5 KOHM, .5 WATT	OMC	0-5K-0.5-C	0.18	0.36
1	S1	SWITCH, SPDT	SWITCHCRAFT	A345-1-1	1.36	1.36
1	S2	RELAY	RLC	99R-475698	3.88	3.88
1	T1	TRANSFORMER 12:1, 10 W	GAUSS	GG10-9V	7.40	7.40

TOTALS
PARTS COUNT 10
COST $26.14

Figure 9.10 Example bill of materials produced by a CAD system.

program. (c) Electrical circuit drawings can be interfaced to a circuit analysis program to predict performance.

Integration of CAD with CAM

After the detailed design is completed, engineering specialists in manufacturing develop the tooling and production techniques. Because production considerations usually require design changes, some give and take is inevitable before all parties arrive at a final design.

The communication of information between design and manufacturing functions is greatly expedited by sharing access to the same data base. The geometric information in CAD drawings may be all that is needed to generate production tooling. For example, the data prepared for printed circuit boards can be used for plotting the artwork used for the photoetching process. The CAD data base can also be used to generate a set of numerical control instructions for drilling the holes in the board.

EXAMPLE 9.4

Starting with the dimensioned drawing of the spacer plate shown, use the six-point transformation capability of a CAD system to produce a wire-frame isometric drawing. Remove the hidden lines. Determine the volume.

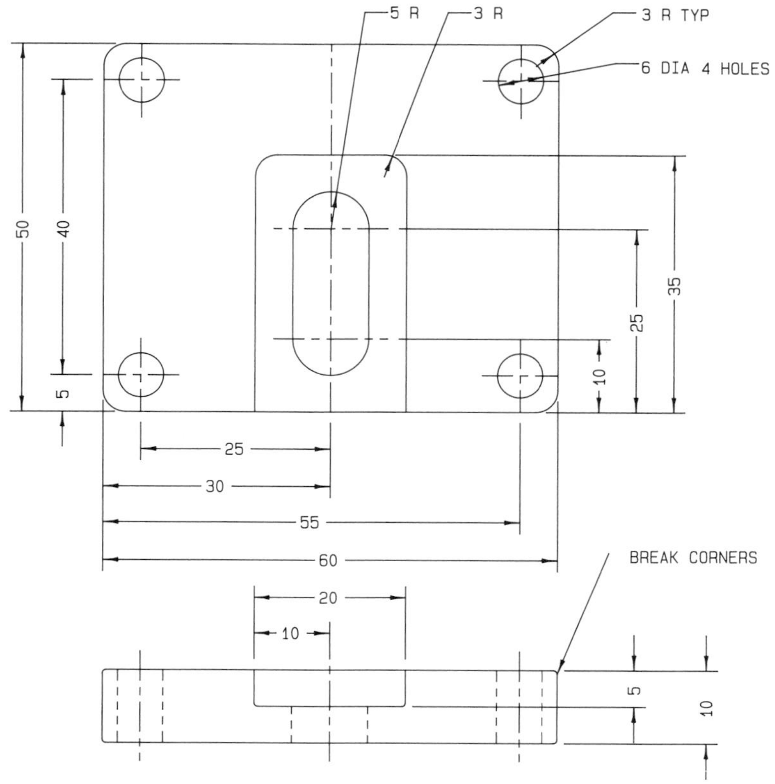

Solution. (NOTE: The method given is somewhat system-dependent.)

1. The dimensions, notes, and centerlines are removed from the existing drawing.

2. Construction axes, formed from a pair of construction lines with some convenient length (arbitrarily 10 mm in this example) at right angles, are drawn on the top view in an uncluttered location. The location selected is convenient for projection of the inset area.

3. Three sets of construction axes are needed in the isometric view for the three levels of the part, which are 5 mm apart. The length of the axes in the isometric view is 82% of those in the original for the reasons given in Section 4.5. Hence, the three sets of axes are drawn with construction lines 8.2 mm long at $\pm 30°$ to the horizontal spaced 4.1 mm apart vertically.

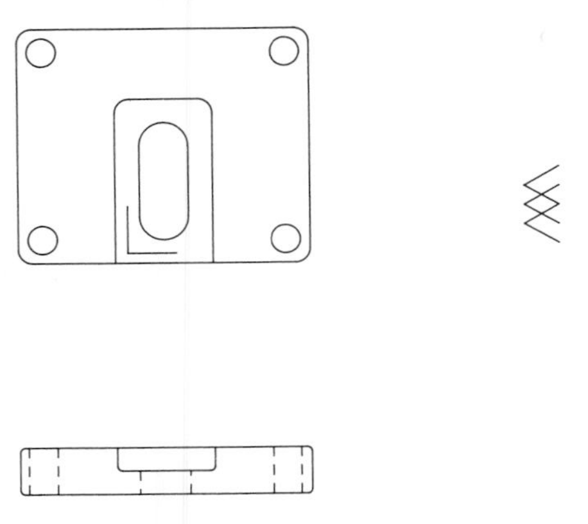

4. The appropriate lines in the top view are picked to be copied. The six-point transformation of Section 3.6, implemented in some CAD programs, performs translation, scaling, rotation, and shearing. Here, the intersection and the ends of the construction axes define the three key points that define the copy.

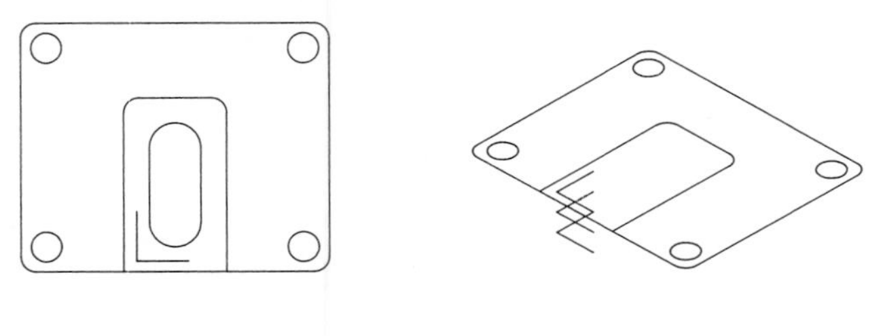

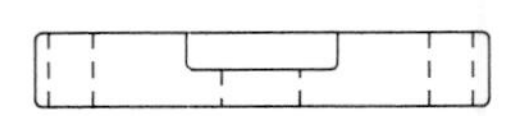

5. The appropriate lines for the middle level of the isometric view are next picked for copying. The three-point copy operation is repeated, except that the middle set of copy-to axes is used.

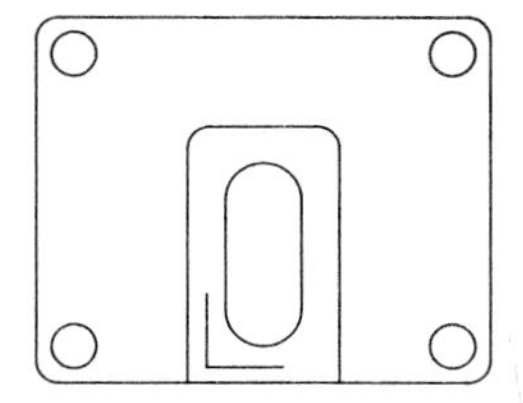 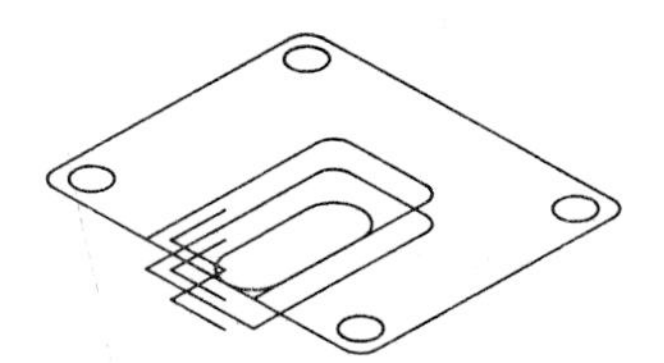

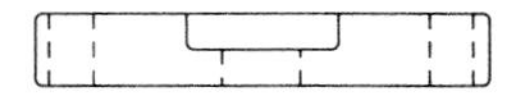

6. The appropriate lines for the bottom level of the isometric view are next picked and copied.

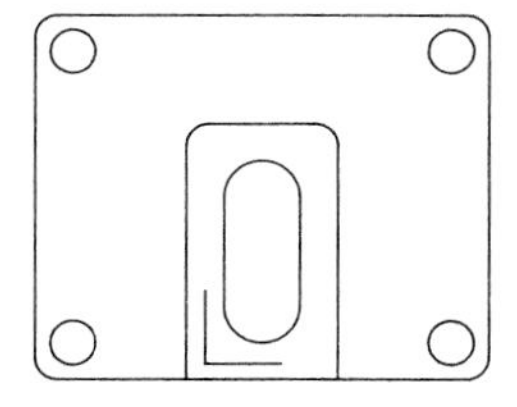 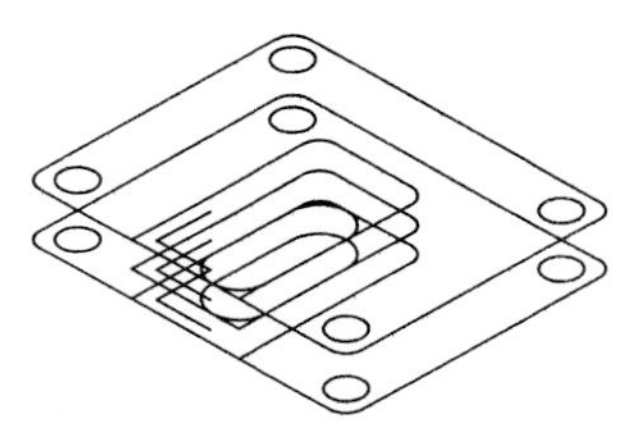

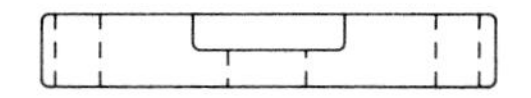

7. The wireframe drawing is completed by deleting the extraneous lines which resulted from the copying process and by drawing the necessary vertical lines. As a convenience, most of these can be drawn by a command for a common tangent to two curves. Otherwise, the user should zoom up on the various parts to aid in carefully locating the end points. The construction lines are no longer needed and can be deleted.

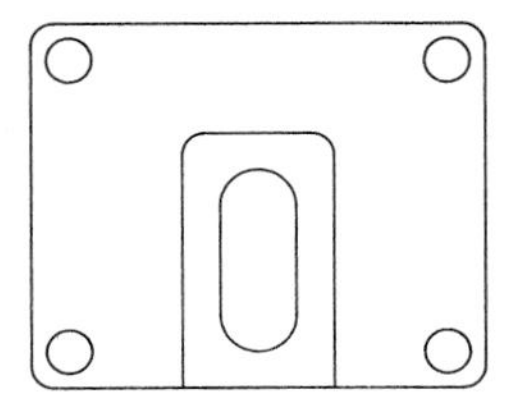 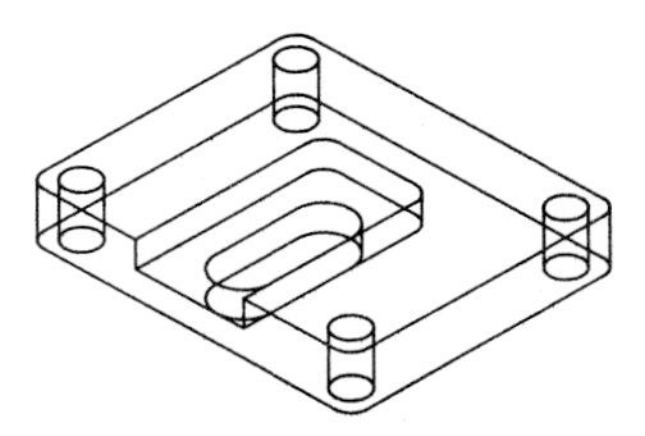

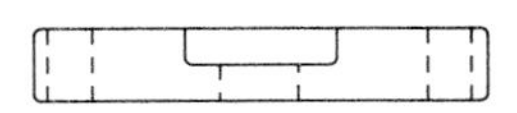

8. Commands to delete lines and curves can now be used to remove the hidden lines.

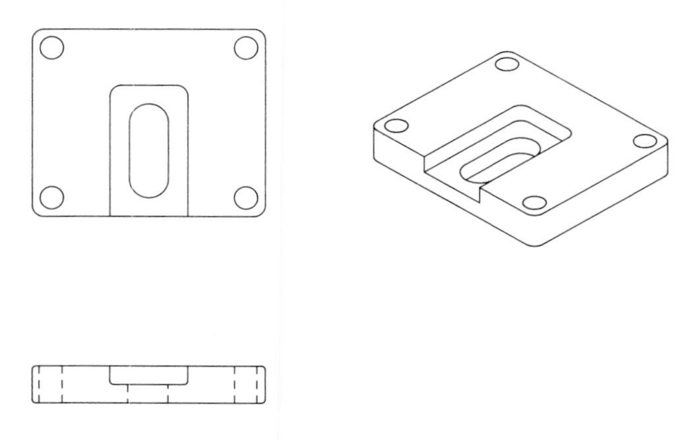

9. The 2-D CAD system used here computes areas bounded by contours picked by the user. The area of the top surface with the four holes and slot deleted is 2650.64 mm^2. The area of the reduced section with the slot deleted is 467.598 mm^2. Hence, the volume of the part is $10 \times 2650.64 - 5 \times 467.598 = 24{,}168.4$ mm^3, which, in view of the precision of the given information, should be reported as either 24,000 mm^3 or $2.4\ (10)^{-5}\ m^3$.

9.8 CLOSURE

The design function is an interactive process which involves conceptualization, communication, analysis, and testing. The rudiments of an integrated CAE system in Figure 9.11 illustrate such an approach to product development.

The control role of CAD is evident in the design process. For mechanical parts other than simple flat designs, 3-D solid models are useful. One example in which solid modeling is used is the design of molds for plastic parts.

The CAE functions represented by the blocks in Figure 9.11 often are performed by separate software systems. Although it may be difficult, integration of these functions is clearly desirable. Many CAE systems partially bridge the gaps; an example is a finite element analysis system with facilities for model preparation and interpretation. Integration of the design and manufacturing functions is more complex because of the diverse nature of CAM software, which may include numerical control (NC), manufacturing resources planning (MRP), process planning and control, group technology, and robotics.

Total incorporation and compatibility among many CAE functions, sometimes known as *computer-integrated manufacturing,* or CIM, is rapidly progressing to improve industrial productivity. This type of vertical integration requires even more highly specialized CAE systems for specific engineering disciplines.

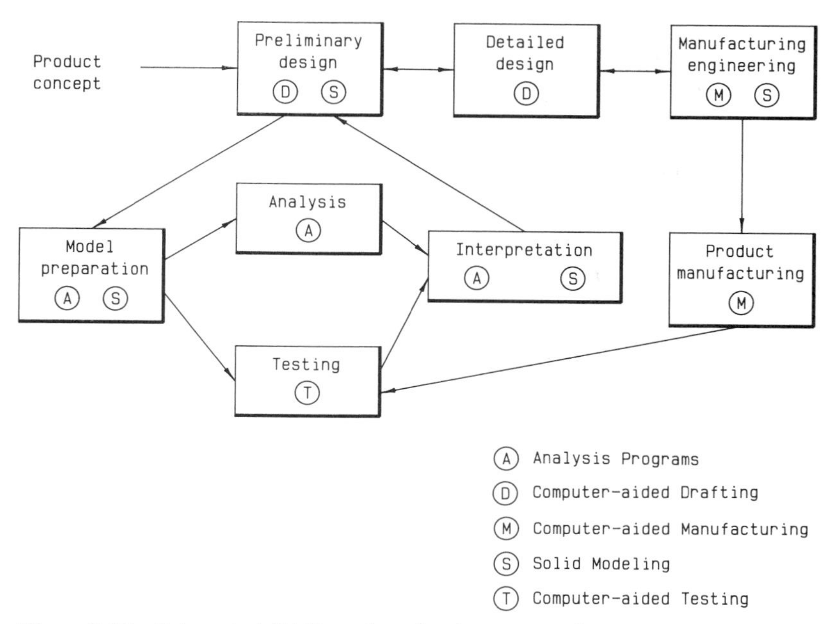

Figure 9.11 Integrated CAE product development cycle.

PROJECTS

NOTE: The following projects are intended to facilitate learning about the basic operations available on the user's CAD system. The documentation with the CAD system should be consulted for guidance.

9.1. Demonstrate different ways to construct a line segment at right angles to an existing line segment. Consider the cases where the existing line is parallel to the x and y axes and where the line is oblique.

9.2. Demonstrate different ways to construct from a given point a line tangent to a circle. Repeat, but construct the common tangent to two circles. Be aware that there is more than one tangent line in each case; tell how your CAD system controls the choice of one.

9.3. Construct two concentric circles of known radius and measure the area between the two with the area-measuring feature of a CAD program. Check the result algebraically. If the result is not particularly accurate, try the same process with two rectangles. Comment on the reasons for the lack of accuracy.

9.4. Draw an object as assigned with rotational symmetry. Use the ability of the CAD program to copy and rotate.

9.5. Draw a map showing directions to some point within a city. Label names on streets and highways. Indicate the best route to use.

9.6. Draw a flowchart of a computer program as assigned. Use standard flowchart symbols from a symbol file in your system.

9.7. Produce an electrical schematic drawing as assigned. Use a symbol library if available.

9.8. Draw a floor plan for a residence. Use symbols to show details such as electrical features and plumbing fixtures. Separate the drawing into layers for the various trades.

9.9. Draw the elevations for a residence. Use crosshatching or shading as appropriate to show texture and so forth.

9.10. Draw a landscaping plan for a residence. Use symbols for the various plants selected. Include a legend which describes the symbols.

9.11. Lay out a research laboratory floor plan as assigned. Show connection of essential utilities.

9.12. Draw the orthographic projections of an airplane as selected. Draw only half the plane, and use mirror-image features to produce the other side.

9.13. Draw a vehicle as assigned. Use the symbols and mirroring facilities for production of repeated details.

9.14. Draw, with a 120° grid system, a pictorial view of your CAD workstation. Omit small details. Produce text pointing out essential components of the system.

9.15. With a 120° grid system, draw a pictorial view of the exterior of a building. If feasible, use symbols or multiple copy facilities to produce repeated parts such as windows. Omit small details.

9.16–9.20 For the mechanical parts shown in isometric views below, select appropriate dimensions and produce dimensioned orthographic views. Use available facilities to reproduce an isometric, oblique, perspective drawing, or other pictorial drawing.

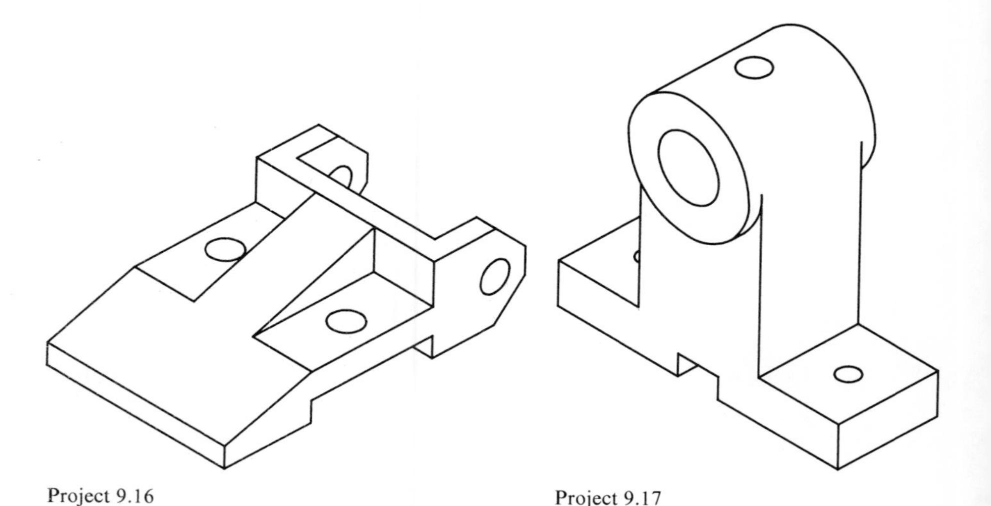

Project 9.16 Project 9.17

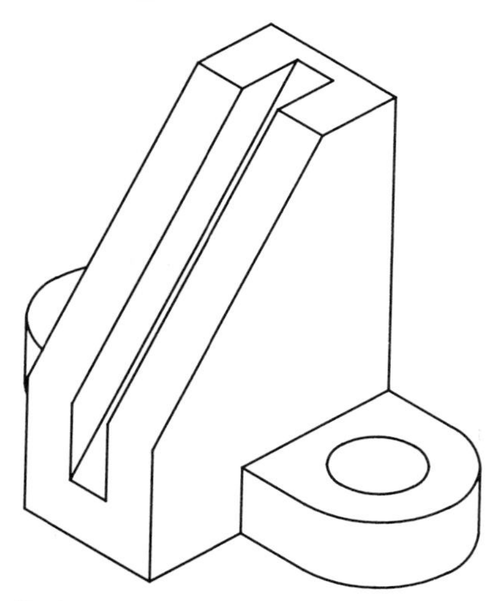

Project 9.18

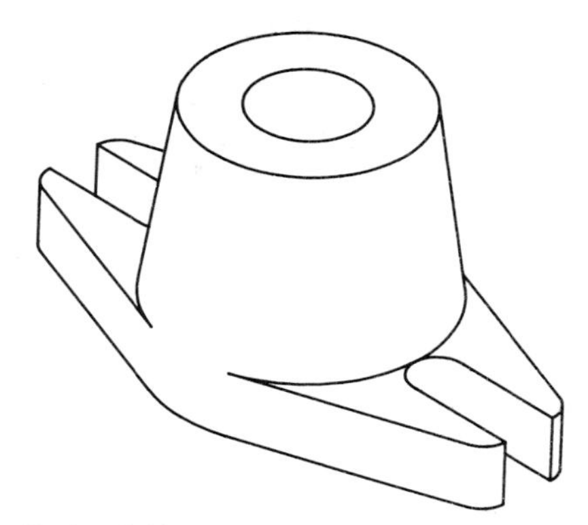

Project 9.19

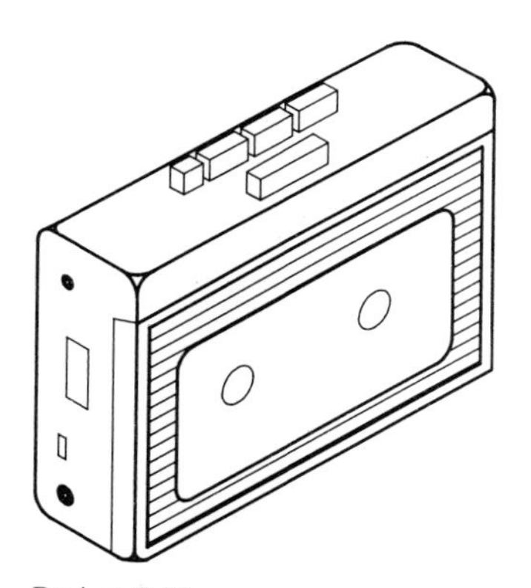

Project 9.20

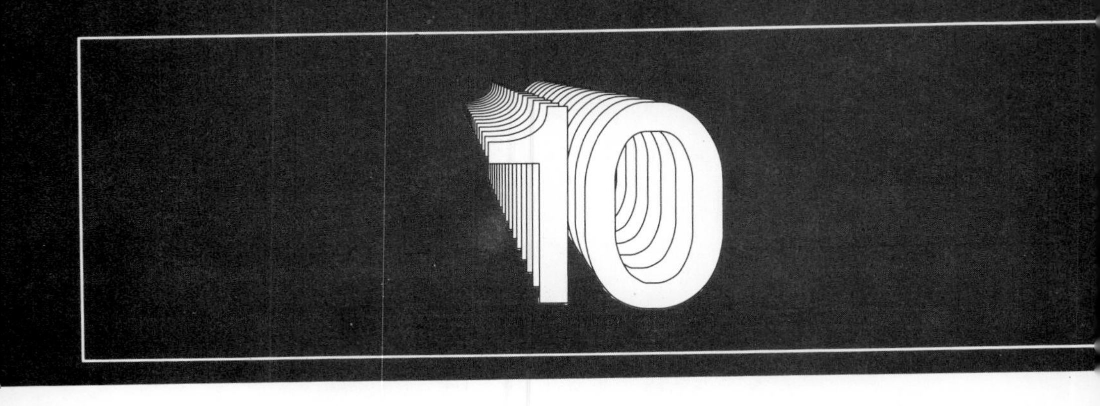

Solid Modeling

Solid modeling deals with the design and representation of real-world objects such as structures, machines, components, and assemblies of parts. From solid models, realistic three-dimensional (3-D) renderings of an object can be produced. Realism can be aided by light-source shading, representation of surface texture, perspective projection, and color. Sophisticated techniques for rendering are used for applications such as aircraft flight simulation and image production for entertainment.

Besides furnishing realistic images, solid models are used for analysis. Surface area, volume, moments of inertia, and other geometric properties can be computed (Chapter 7). Finite element models can be constructed to analyze stress, deflection, vibration, field intensity, or temperature distribution. Kinematic models of mechanisms and robots can be checked with animated solid models. Mechanical parts assemblies can be checked for fit and interference. Layout of piping or other components can be done to ensure that the design is physically realizable.

In manufacturing, solid models form the basis for generation of programs to control lathes, milling machines, and other machine tools. Furthermore, the development of programming for robots can be verified with solid models. Solid modeling plays a central role in computer vision and artificial intelligence.

Characteristics of solid models include the following:

Three-dimensional. The model may be viewed from any direction for visualization and study.

Complete. The model contains sufficient information for all intended applications, including information about the interior of the model, such as whether it is solid or hollow.

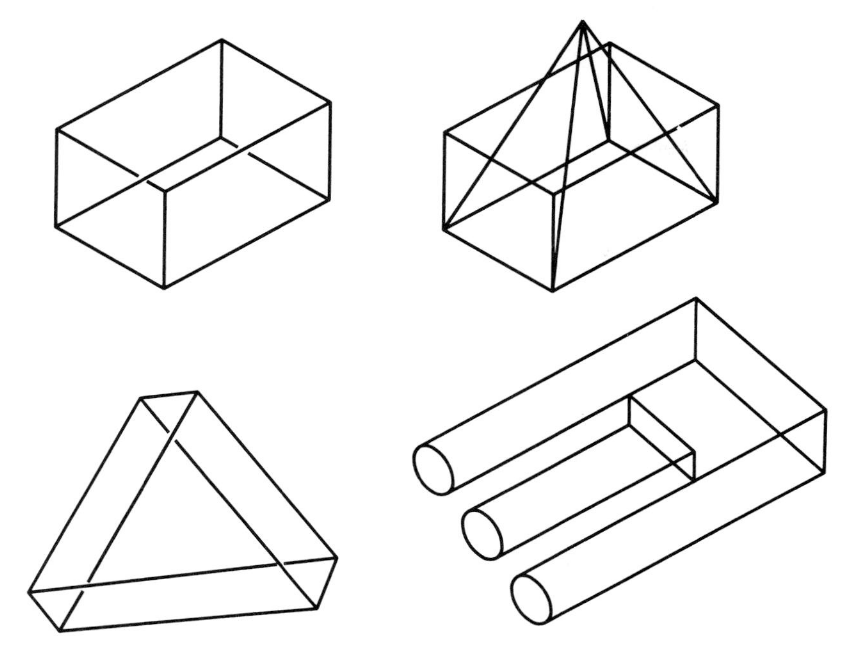

Figure 10.1 Ambiguity is impossible in a true solid model. Shown here are examples of nonsense models.

Unambiguous. The model ensures that the object can be physically realized. Thus, missing surfaces, dangling edges, or the examples in Figure 10.1 are not possible.

Systematic. The 3-D shape definitions are done with some type of rational operation such as sweeping out, combining primitive shapes, or evaluating a mathematical expression.

Solid models can be constructed in a variety of ways, as explained in the following section. Some of the techniques which follow are deficient in one or more of the characteristics listed above. Still, such a model may be adequate for the intended tasks.

10.1 CLASSIFICATION OF SOLID MODELING SYSTEMS

Solid modelers differ in how the user builds models and how the models are represented in the data base. Although these characteristics are somewhat interrelated, the scheme of internal storage of the model serves best for classification purposes.

Usually, a model is defined by two principal types of data, geometric and topological, a simple example of which is shown in Figure 2.5. Geometric data are conveyed in a points matrix or other structure containing world coordinates. Topological data are lists of connectivities among the geometric elements.

Wireframes

The oldest, simplest, and most widely used modeling scheme is the *wireframe,* which is based on showing the edges of objects as lines. Typically, the storage is arranged as *vertices,* which are the geometric data, and *edges,* which are the topological data. To visualize curved surfaces, any number of extra lines may be added. The name "wireframe" is appropriate because the objects appear to be constructed of wires.

Since wireframe models are simply lines connecting vertices, minimal sophistication in code and display hardware is required. Transformations by scaling, translation, rotation, and perspective are easily performed. Interpretation of wireframe models requires more skill on the part of the user, although visualization is aided by use of the perspective tranformation, as is shown in Figure 10.2.

Besides the difficulty in visualization of wireframe models, the models are not necessarily complete and unambiguous. Nonsense models, such as those Figure 10.1, are possible with wireframes. Because the model consists only of points and lines, removal of hidden lines is complicated by the ambiguity posed in Figure 10.3. Furthermore, wireframes contain too little information for evaluation of area and volume properties. The lack of information also precludes applications such as interference checking and tool path generation.

Another disadvantage of wireframe models is the amount of data required to describe the model. For example, a simple rectangular block can be described by its height, width, length, the *x-y-z* world coordinates of one point, and the three direction cosines of one line—a total of nine quantities. The wireframe model of this object requires the *x-y-z* world coordinates of the eight vertices and a topology list for the 12 edges each with two quantities—a total of 48 quantities. As shapes become more complex, the amount of information needed to store a wireframe increases rapidly.

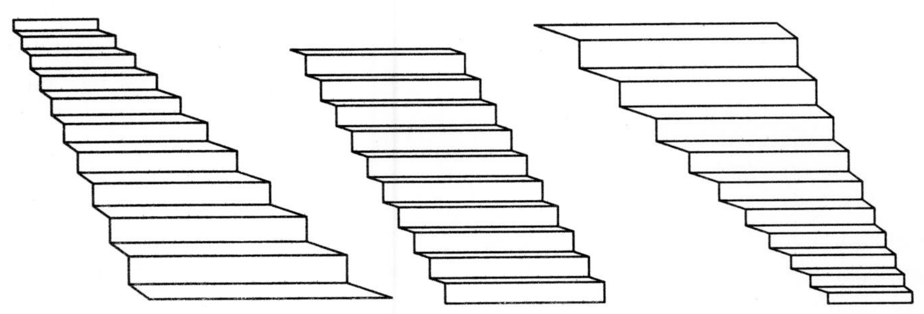

Figure 10.2 Wireframe model of a stairway. The perspective tells whether the stairway is seen from above or below. The model is defined by points and topology matrices.

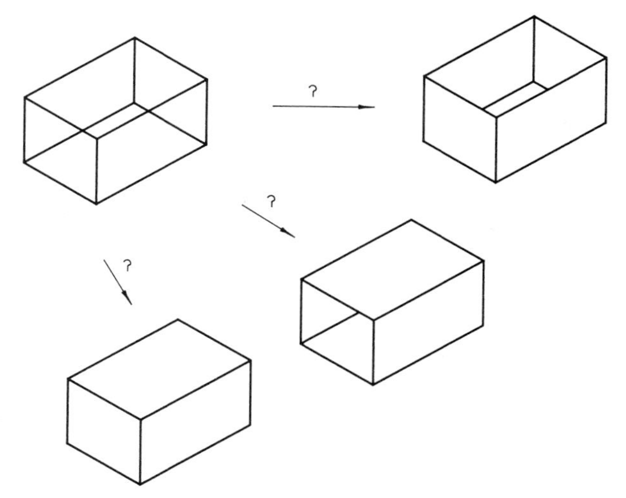

Figure 10.3 Ambiguity posed by wireframe models in hidden-line elimination.

Boundary Representation

A *boundary representation* or *B-rep,* which represents the enclosing surface of the object, can be visualized as the result of stretching a thin skin over a wireframe model. Rather complex surfaces such as an aircraft fuselage can be built up as a set of flat polygonal faces, where a *face* is defined as a bounded surface. A B-rep of higher quality results from using faces of curved surface patches, such as the parametric cubic patches in Section 6.2.

The boundaries describing an object may be segmented into faces, edges, and vertices in many different ways, as shown in the example in Figure 10.4. B-reps recognize that part *topology* and part *geometry* can be separated.

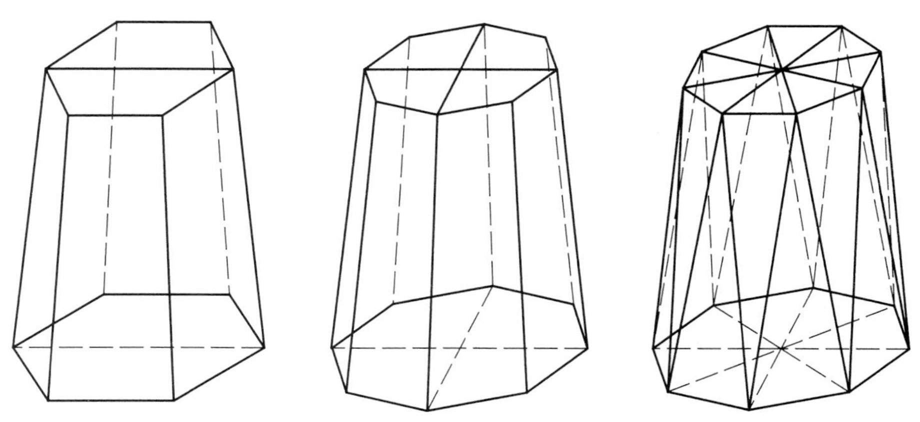

Figure 10.4 The B-rep of the frustum of a cone can be accomplished with a tessellated model having an arbitrary number of vertices, edges, and faces.

Topology describes how faces are bounded by edges, how edges are shared by faces, and how vertices are shared by edges. Topology tells how the object is connected, but not its size. Geometry describes the location and the size of the topological elements. More often than not, design changes of a B-rep model involve only geometry.

If the faces are flat and the edges straight, the boundary representation is the special case known as a *faceted* or *tessellated* model, such as the model shown in Figure 10.4, where the faces are either quadrilaterals or traingles. Some interactive solid modeling systems first create an approximation with a faceted model, which requires relatively little computation for display. For the final rendition, the model is smoothed or "skinned," a process that takes somewhat more computation.

A *polyhedron* is an array of polygons which completely enclose a region in space, with only two polygons sharing any edge. A *simple* polyhedron, examples of which are shown in Figure 10.4, does not have any holes or passages. For such polyhedra, the Euler relationship[36] states

$$V + F - E = 2 \tag{10.1}$$

where V, F, and E are the numbers of vertices, faces, and edges, respectively. For *nonsimple* polyhedra, the parameter G is used to describe the *genus* or the number of *handles*. The genus of a torus is 1 and that of a plate with two holes is 2, for example. (For simple polyhedra, the genus is 0.) The Euler relationship modified for nonsimple polyhedra is

$$V + F - E + 2G = 2 \tag{10.2}$$

By maintaining the Euler relationships, a program can ensure that the resulting model remains unambiguous.

Models are built in B-rep by piecing together faces to enclose the space occupied by the object. In solid modeling systems, the definition of surfaces is aided by *sweep* operations. A linear sweep translates a surface to form an extruded shape with constant thickness. A rotational sweep forms a model with axial symmetry. A sweep along an arbitrary curve produces a more complex solid.

The use of curved surface geometry such as parametric cubic or B-spline surface patches is advantageous in constructing models of sculptured surfaces. Curved surfaces thus defined required somewhat less storage than a tessellated model. Furthermore, the rendering of surfaces bounded by curved faces is somewhat more accurate than Gouraud shading, Eqs. (2.22) through (2.24), which is commonly used for tessellated models. An example of the more accurate representation afforded by parametric cubic geometry, Plate 10.5, uses the computed Gaussian curvature (the rate of change of curvature) to check the surface of a solid model of a turbine blade. In this application, the lack of continuity in Gaussian curvature just barely discernible in the color-shaded surface plot would adversely affect performance.

There are several advantages of boundary representation as compared to wireframe models. Surface definition permits removal of hidden surfaces and addition of enhancements such as the light-source shading and texture. Machine tool paths can be described since the boundary is the surface being worked. Interference can be detected by intersection of surfaces comprising the parts of an assembly, such as the valve of Plate 10.6. Data storage requirements are potentially less than for wireframe models, particularly if the efficiencies afforded by using curved surface patches are used.

There is, however, the disadvantage that B-rep models do not provide information about the interior of the model. This means that evaluation of mass and other volumetric properties must be based on assumptions such as homogeneity. Furthermore, the generation of cutaway views (such as Plates 10.6 and 10.8) are solid finite element models (Section 10.2) is much more difficult with B-rep models.

Constructive Solid Geometry

A building-block approach, where primitive shapes are combined, is known as *constructive solid geometry* or *CSG*. Primitives such as those shown in Figure 10.5 are solid models in themselves, having mass and occupying a region in space. The simple primitives are related to machine tools—for example, the cylinder results from turning. For this reason, CSG has a natural affinity for describing machined parts.

The primitives are combined by the *Boolean operators,* where

Union ($\cup$) combines two primitives.

Difference ($-$) subtracts one primitive from another.

Intersection ($\cap$) defines a region common to both primitives.

The union and intersection operators are commutative; that is $A \cup B = B \cup A$, and $A \cap B = B \cap A$. The difference operator is not commutative, $A - B \neq B - A$.

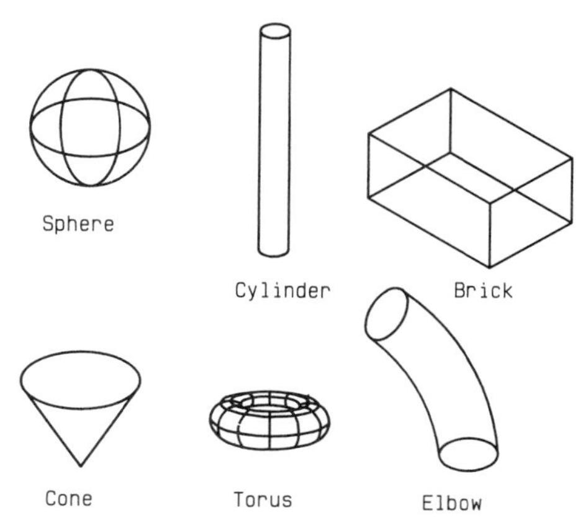

Figure 10.5 Examples of primitives for constructive solid geometry modeling.

Model construction starts with selecting and transforming each primitive. Modeling transformations such as scaling, translation, and rotation fix size and spatial position of the primitives before application of the Boolean operators. Thus, for example, a plate with a hole is modeled by (1) transforming a brick primitive to give the correct geometry, (2) transforming the geometry of a cylinder primitive, and (3) subtracting the transformed cylinder from the transformed brick. The CSG construction of a more complicated part is shown in Figure 10.6.

A model which is described only by a sequence of Boolean operations is said to be an *unevaluated* model. The evaluation process involves computation of the intersections to determine the new vertices, edges, and faces. To permit rendition of the model, algorithms known as the *boundary evaluator* determine the B-rep of the surface. Evaluation of CSG models is computationally intensive; and an unevaluated model requires a great deal less storage than an evaluated one. Some commercial modeling systems generate and store the B-rep information as the modeling progresses to permit maximum interactivity.

Since CSG models are true solid models, requirements for completeness and ambiguity pose no problem. However, the finite nature of available primitives makes modeling of arbitrarily sculptured objects such as engine intake manifolds difficult with CSG. For this and other reasons, commercial solid modeling systems combine features from both CSG and B-rep.

Analytic Solid Modeling

A newer development in solid modeling systems is called *analytic solid modeling* or *ASM*.[11] Such systems are an extension to B-rep with the addition of mathematically described solids, such as the hyperpatch in Section 6.5. Compared to CSG, solid models of arbitrarily sculptured shapes are made more easily with ASM. The evaluation of the boundaries of hyperpatches makes use of the fact that these boundaries are parametric surfaces which degenerate to a B-rep when either u, v, or w is constant at 0 or 1. In a manner analogous to surface patches, hyperpatches can be combined to give smooth representations of arbitrarily shaped solids.

The primitives used for CSG can, or course, be represented as hyperpatches. Thus, as ASM system can have a CSG-like user interface which recognizes the union, intersection, and difference of hyperpatches. The simple example of a plate with a hole shown in Figure 10.7 uses the difference between a brick and a cylinder. The starting primitives consist of one hyperpatch for the brick, which represents the plate, and four hyperpatches combined together† for the cylinder, which represents the hole. The difference operation follows: (1) the curves which represent the intersection between the surfaces of the hyperpatches defining the brick and the cylinder are determined, (2) new faces are constructed with the intersection curves and the appropriate original edges, (3) the appropriate old and new faces are used to define new hyperpatches, and (4) the

†In Section 5.3 the need for two or more parametric cubic curves to form a circle is discussed. The same reasons apply to forming circular cylinders from hyperpatches.

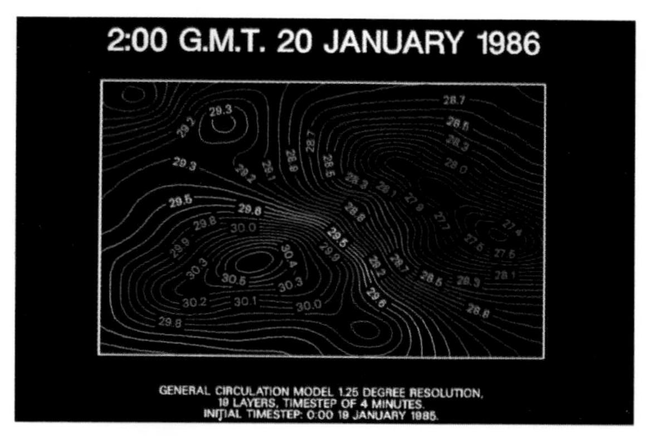

Plate 10.1 Contour plot enhanced with color. (Courtesy of ISSCO.)

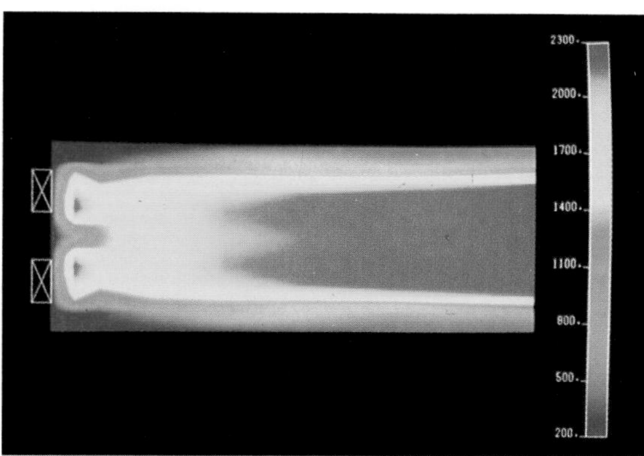

Plate 10.2 Color-shaded contour plot of computed steady-state temperature distribution in a flame propagating through a combustion chamber. (Courtesty of Patran Division, PDA Engineering.)

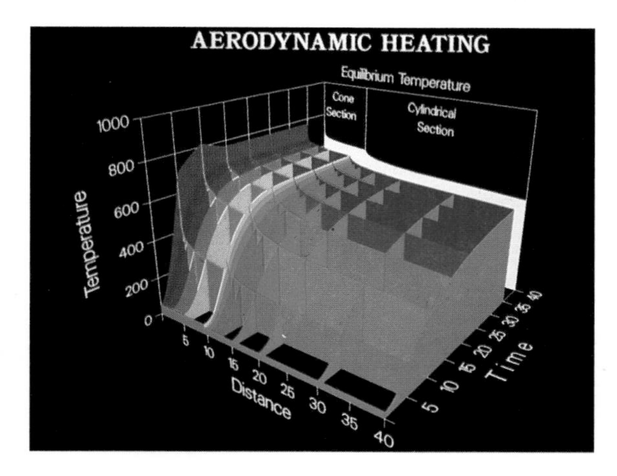

Plate 10.3 Surface plot enhanced with color. Axes convey numerical values. (Courtesy of ISSCO.)

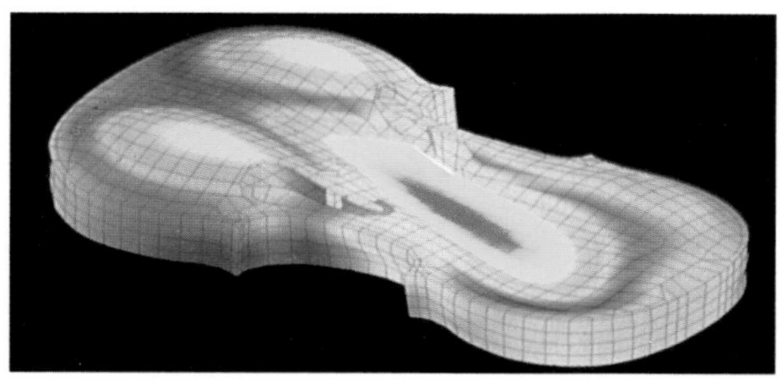

Plate 10.4 Color-shaded surface plot showing the ninth mode of vibration in a violin as predicted by a finite element analysis. (Courtesy of Patran Division, PDA Engineering.)

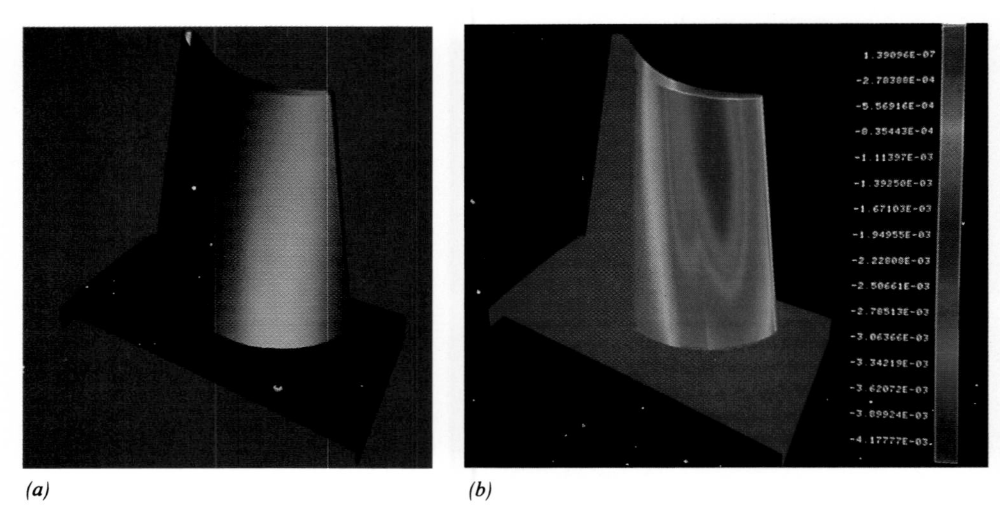

Plate 10.5 Smoothness of turbine buckets is essential for optimum performance. A model described with parametric cubic geometry permits calculation of Gaussian curvature, which is based on the second derivative. *(a)* Surface appears to be perfectly smooth in image produced with Phong and Gouraud shading which is based on first derivatives. *(b)* Contour plot of Gaussian curvature projected onto curved surface shows an area of possible flow perturbation where green fringe penetrates cyan. (Courtesy of Patran Division, PDA Engineering.)

Plate 10.6 A true solid model such as this hydraulic actuator valve allows cutaway views, checking of interferences, and simulation of mechanical action. (Courtesy of PAFEC, Inc.)

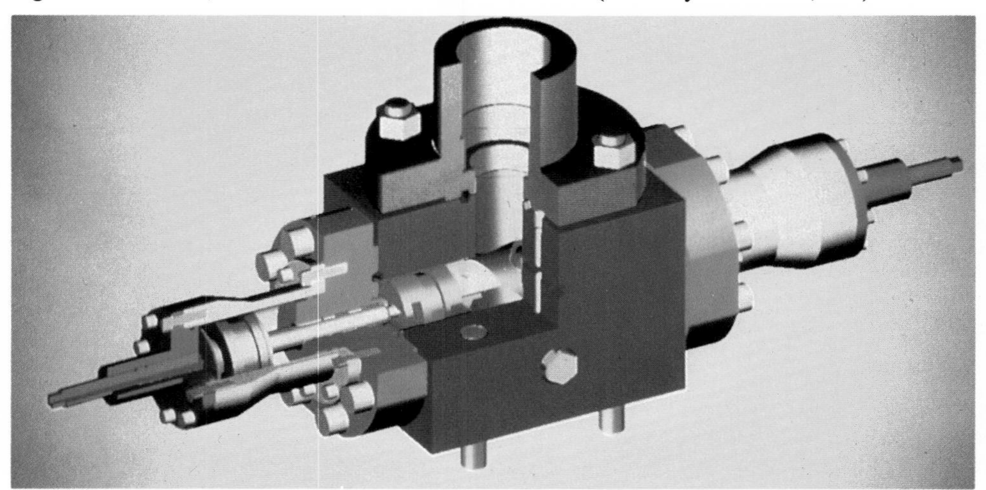

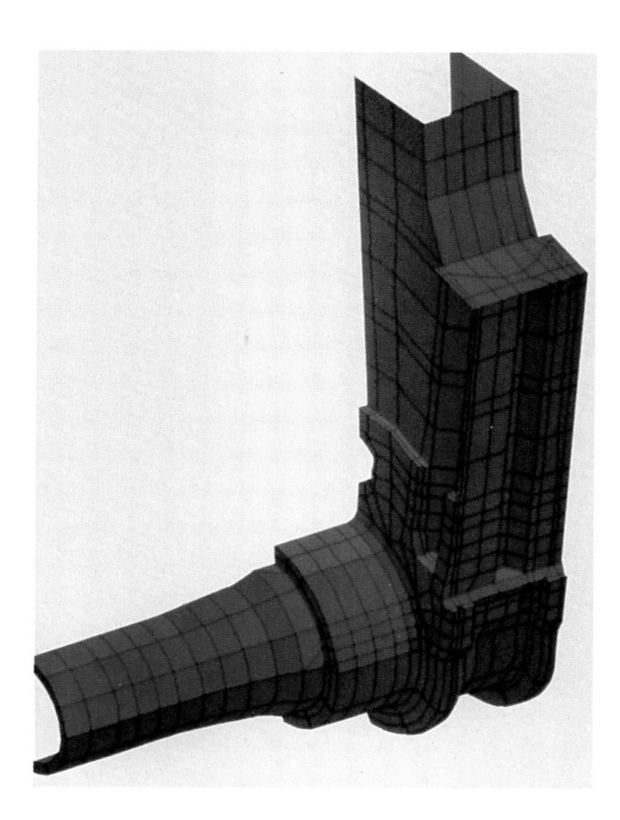

Plate 10.7 Model of a component approximated with flat-plate finite elements. Symmetry permits including only half the part in the model. (Courtesy of PAFEC, Inc.)

Plate 10.8 *(below)* Combination of a shaded image and a color-shaded contour plot to show cooling stresses in a casting. (Courtesy of Patran Division, PDA Engineering.)

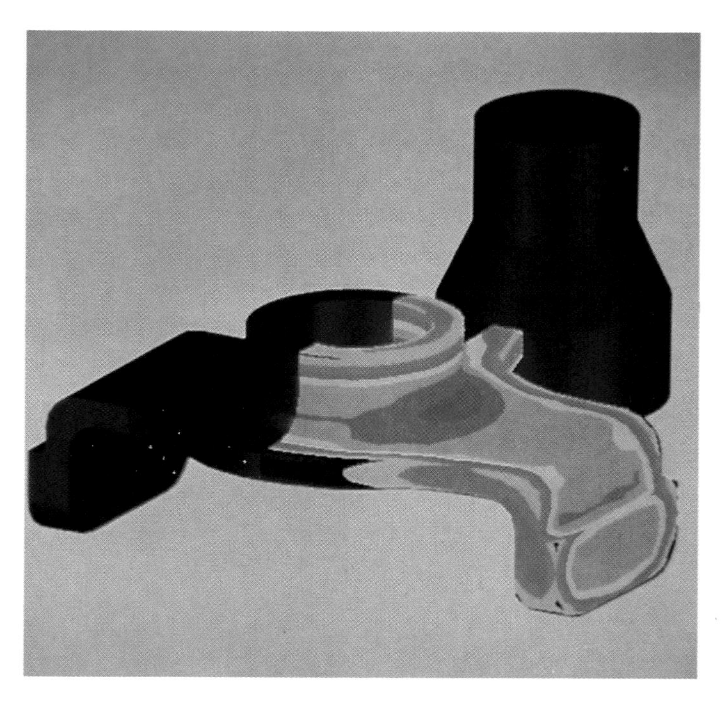

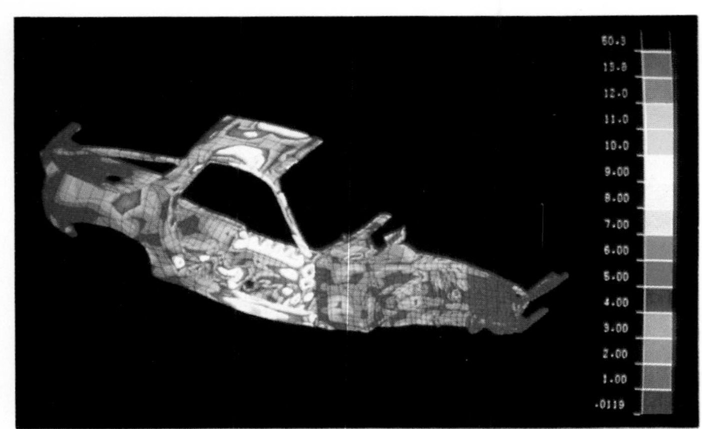

Plate 10.9 Deformation analysis of a sports car body resulting from a finite-element model with 4,360 thin-plate elements. Color-coded areas of high deformation indicate problem spots to the designer. (Courtesy of Patran Division, PDA Engineering.)

Plate 10.10 Results of a transient finite-element thermal analysis of a a section of a rocket canister at two different times during burn. *(a)* Temperature distribution in early part of burn. *(b)* Temperature distribution a minute later shows not only that the casing is near melt temperature, but that the thermal protection for the propellant is adequate. (Courtesy of Patran Division, PDA Engineering)

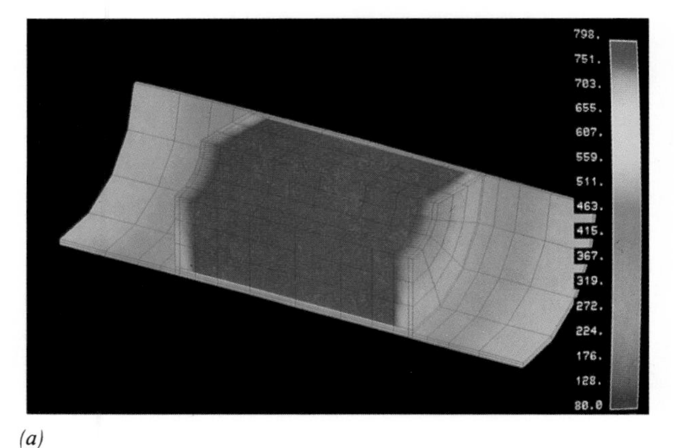

(a)

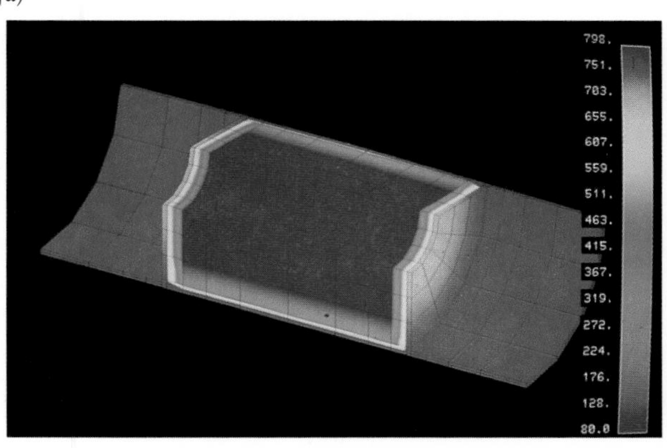

(b)

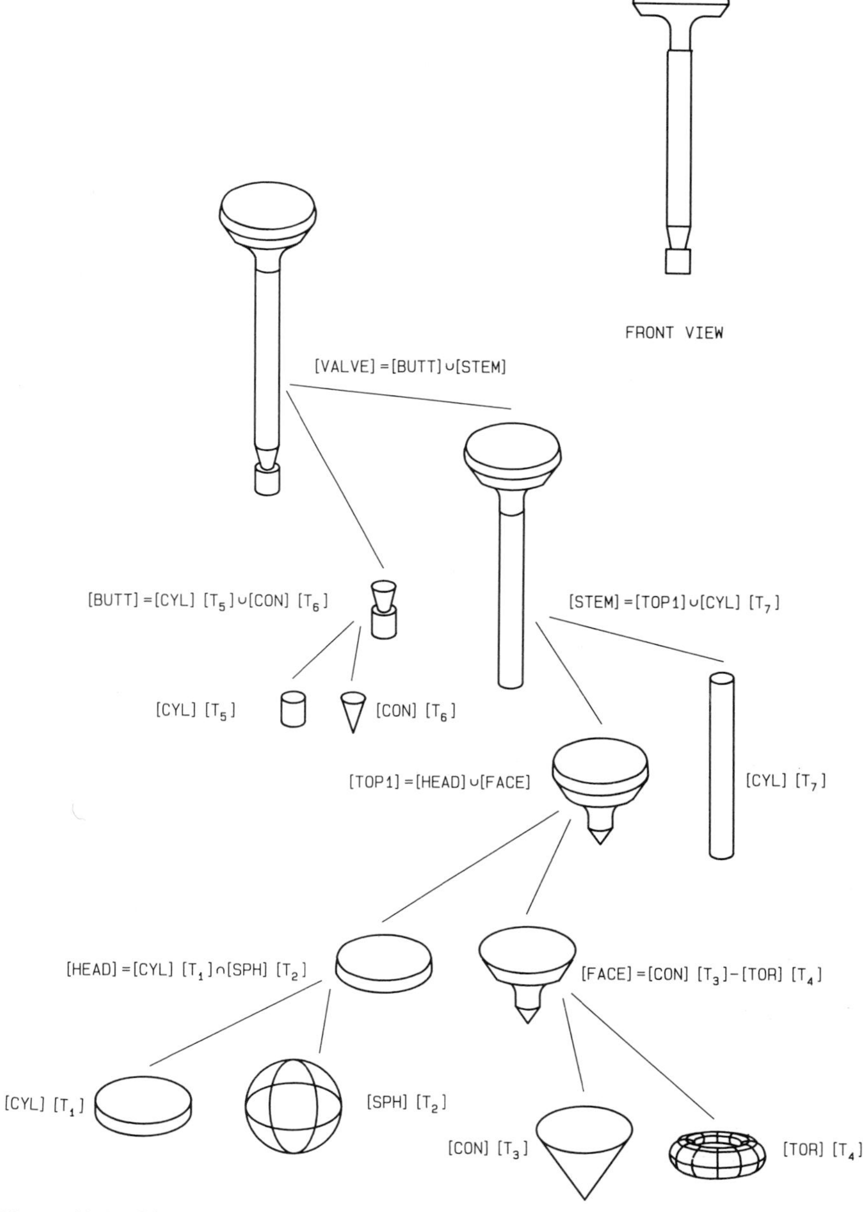

Figure 10.6 Binary tree representation of the CSG of a small engine valve.

hyperpatches defining the original brick and the cylinder are discarded. Incorporation of CSG operations into ASM is difficult, and users may find that other methods are better for generating models in ASM systems.

Construction of solids by the sweep method is particularly attractive in ASM. A curved line can be generated by sweeping a point along a parametric

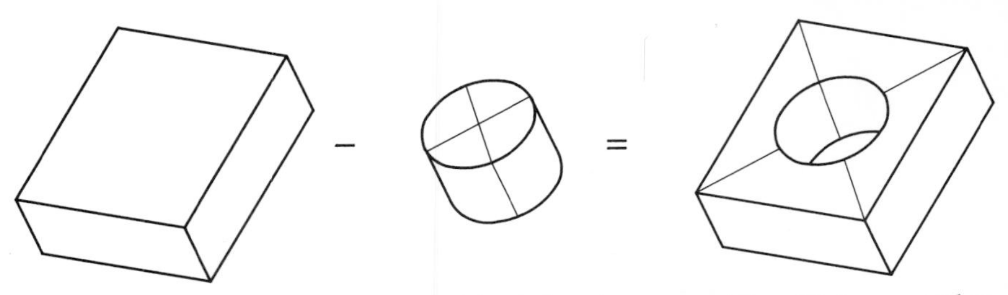

Figure 10.7 Construction of a plate with a hole using parametric cubic hyperpatches. The four hyperpatches in the final result are defined partly by faces from the original elements and partly by faces determined from the intersections.

path, a curved surface by sweeping a curved line, and a curved solid by sweeping a curved surface. Sculptured solids may be described by a set of points, curves, or surfaces, as illustrated in Figure 6.15.

The storage requirements and the computational intensity of ASM are comparable to those of B-rep systems. Besides incorporating the advantages of B-rep systems, ASM has the additional advantage of representing solids in a way which is consistent with that of curves and surfaces. Because the solid is described by mathematical functions, the geometric properties can be evaluated routinely with the methods of Chapter 7. However, the use of Boolean operations can be much more difficult in ASM than in CSG. Furthermore, operations on surfaces are more cumbersome with ASM than with B-rep systems.

The foregoing overview of solid modeling should suggest many engineering and architechural applications in which solid modeling is useful. Commercial solid modeling systems often are hybrid systems containing features from the various types of systems listed here; selection of a system should be based on the support of the intended applications. Described next are two typical engineering uses of solid models, finite element analysis and mechanical dynamics.

10.2 FINITE ELEMENT ANALYSIS

An important computer-aided engineering (CAE) tool, finite element analysis provides approximate solutions to continuum mechanics problems with arbitrary geometries. To introduce the concept, a very simple example is considered. Estimating the displacements and stresses in the rocker arm loaded as shown in Figure 10.8(a) is important in achieving a good design. For analysis, the

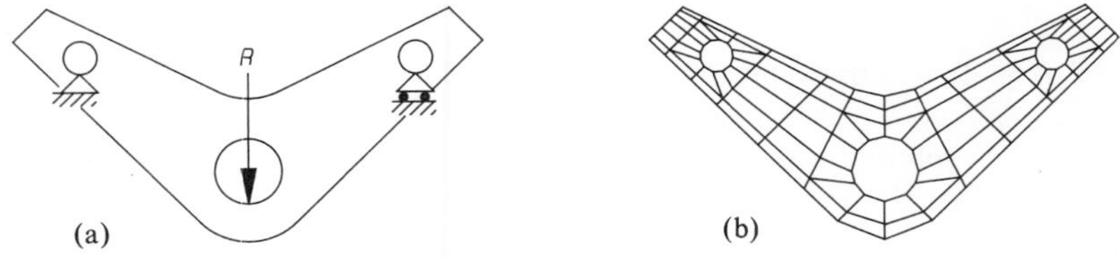

Figure 10.8 Finite element approximation to solve for displacements and stresses in a rocker arm. (a) Component and loading; (b) finite element mesh.

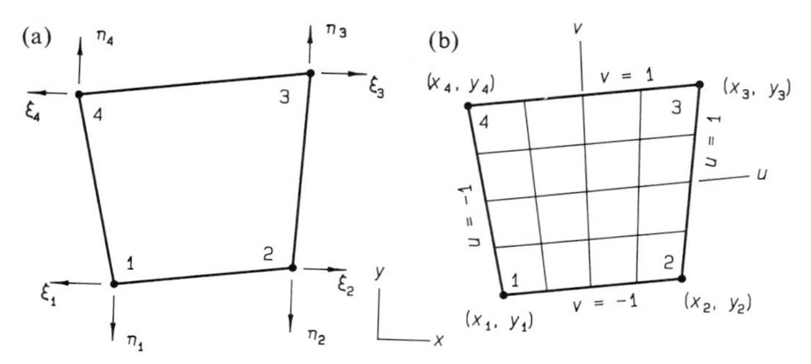

Figure 10.9 A four-node isoparametric plane element. (a) Nodal displacements; (b) geometry.

component can be considered as a 2-D part and divided into quatrilateral finite elements, connected to each other at the *nodes* as shown in the *mesh* of Figure 10.8(b). Finite element meshes can be developed from a B-rep or solid model of the component.

Because the loading and the geometry of the component in Figure 10.8 is planar, all deformations will lie in the x-y plane. Accordingly, each node has a displacement vector which is described by its components in the x and y directions, ξ and η, as shown in Figure 10.9(a) for a single element. The *shape function* matrix [**N**] describes the interpolation for any quantity within the element in terms of the values of the same quantities at the nodes, which are numbered from 1 to 4. The coordinates of a point within an element are

$$[x \quad y] = [x_1 \quad y_1 \quad x_2 \quad y_2 \quad x_3 \quad y_3 \quad x_4 \quad y_4]\,[\mathbf{N}] \tag{10.3}$$

where [**N**] is determined below. Furthermore, the same shape function may be used to relate the displacement of a point within the element to the x and y components of the displacement ξ_i and η_i at the nodes, or

$$[\xi \quad \eta] = [\xi_1 \quad \eta_1 \quad \xi_2 \quad \eta_2 \quad \xi_3 \quad \eta_3 \quad \xi_4 \quad \eta_4]\,[\mathbf{N}] \tag{10.4}$$

In this formulation there are two degrees of freedom per node, since two quantities at the node are necessary and sufficient to describe the magnitude and direction of the displacement vector in two dimensions. The element used here is called *isoparametric,* which means the same [**N**] is used for geometry and displacement formulations.

The parametric variables (u, v) define a local coordinate system as shown in Figure 10.9(b) to determine the form of [**N**]. It can be demonstrated that

$$[\mathbf{N}] = \frac{1}{4}\begin{bmatrix} N_1 & 0 & N_2 & 0 & N_3 & 0 & N_4 & 0 \\ 0 & N_1 & 0 & N_2 & 0 & N_3 & 0 & N_4 \end{bmatrix}^T \tag{10.5}$$

where $N_1 = (1 - u)(1 - v)$

$N_2 = (1 + u)(1 - v)$

$N_3 = (1 + u)(1 + v)$

$N_4 = (1 - u)(1 + v)$

Based on the interpolation provided by the shape function, the strains on the interior of the element may be calculated from strain-displacement relationships. Use of the elasticity equations that relate stress and strain provides an expression for *element stiffness,* a matrix which, when multiplied by the matrix of displacements of the nodes, produces the nodal forces corresponding to the displacements. When all the finite elements are merged together, or assembled, the interior nodes are shared by four elements, and the nodes along the boundaries are usually shared by two elements. The assembly yields a system of linear equations of the form

$$[\mathbf{K}][\Delta] = [\mathbf{P}] \tag{10.6}$$

where [**K**] is the overall stiffness matrix, which incorporates geometry and material properties, [Δ] is the matrix of unknown nodal displacements, and [**P**] is the matrix of loads acting on the component. In a plane problem such as the one under consideration here, for n nodes there are $2n$ simultaneous linear equations in $2n$ unknowns. After solution for [Δ], stresses can be determined by differentiation of the displacements. Using the results of the analysis, the design of the part can be improved by changing the geometry of the high-stress regions.

In the foregoing example, there are two degrees of freedom (dof) per node. Had this been a 3-D stress analysis problem, the addition of the z component of displacement would make three dof per node. Because most of the detail has been omitted here, use of one or more of the many references[15,47,60] on finite element theory is highly recommended.

Solid modeling is an extremely useful tool for generating and verifying finite element meshes, such as the example shown in Plate 10.7. At the outset, consideration should be given to whether a 2-D or 3-D model is needed for the analysis, since 2-D analysis is much easier and more economical to perform. Many engineered structures have axisymmetric geometry and loading, which means that a 2-D mesh of the cross section is all that is needed. In structures such as pressure vessels, where the wall thickness is small compared to other dimensions, *shell* elements are a good choice, since there are far fewer total degrees of freedom than with full 3-D *brick* elements. For members which carry axial or axial/bending loads, there are *truss* and *beam* elements. Many other types of specialized elements are available.

Instead of developing their own finite element programs, most engineers use commercial systems such as ANSYS, NASTRAN, and ABACUS for analysis of stress and vibration, heat transfer, electric fields and networks, fluid flow, and so on. Commercial systems have extensive element libraries which aid in modeling the system accurately and economically. Among the available elements are ones for 2-D and 3-D problems, elements for specific structural shapes, elements with straight and curved edges and surfaces, elements for different kinds of analyses, and many others.

The finite element method is well developed for both steady-state and transient problems. Components made of nonlinear materials, such as materials with temperature-dependent properties in heat conduction, can be routinely analyzed. The region being studied can have arbitrary shape, boundary conditions, materials, loading, and so on.

As is the case in any design activity, engineers involved in finite element modeling require specialized training and experience. Typical of many pitfalls are so-called distorted elements, where the relative sizes of linear dimensions and angles produce numerical difficulties. As an example of avoiding distortion, the plane linear isoparametric element in Figure 10.9 should have angles in the range of roughly 30° to 150° and should have the length of the longest side no more than five times the length of the shortest side. An experienced analyst will place nodes closer together in areas where the variable of interest (displacement temperature, etc.) is rapidly changing. Thus, because of the expected stress concentration, nodes are closer together around the holes in the part of Figure 10.8.

Finite element analysis finds wide use in engineering design. The development of solid modeling has been driven, in part, by applications in the generation and checking of finite element meshes. Data presentation graphics, including x-y plots as well as contour and surface plots, are extremely useful for displaying results from finite element analyses. The color-shaded surface plots in Plates 10.8 through 10.10 are examples of using finite element analysis to guide design studies involving heat transfer, stress, and vibration.

10.3 MECHANICAL DYNAMICS

Solid models are ideal for studying the dynamics of mechanisms, machine tools, and robots. Solid models furnish the basis not only for kinematic analysis of motion, but also for the dynamic analysis of stress and deflection.

In manufacturing, numerical control[27,45] (NC) and robotics[1,54] are two well-developed applications of solid modeling for kinematic simulation. Other applications of computing in manufacturing, such as scheduling, monitoring equipment, inventory control, management information systems, and quality control, drive the trend toward organizing all information in an integrated data base.

Computer Numerical Control

Automation of machine tools began at Massachusetts Institute of Technology in the early 1950s with the development of a servo-controlled milling machine. An important result of the MIT work is the widely used parts programming language APT, for "Automatically Programmed Tools." APT language statements are computer-processed to produce instructions usable by NC machine tools. Another widely used NC language is COMPACT II. These and other NC languages have drivers for many different makes of machine tools. Thus, the same part program can be made to work on any of several NC machines.

In the earliest NC systems, the instructions to drive the machine tools were punched on paper tapes. The NC machine played back the instructions from its paper tape reader and interpreted them through its hard-wired controller. Problems with such older systems included difficulty in correcting errors, lack of flexibility, and unreliability. The newer technologies of computer numerical control (CNC) and direct numerical control (DNC) permit rapid reprogramming of machine controls, adjustment of machine settings based on in-process inspection, and diagnostics. The distinction between CNC and DNC is that DNC has the additional functionality of communication with a central computer system. The advent of low-cost reliable computing has made the older, inflexible NC machines all but obsolete.

Setting up for manufacturing with numerically controlled machines in a CAE environment involves the following steps:

1. Design. Solid modeling, CAD, and analysis systems are used to produce geometric and topological information in the data base.
2. Process planning. A manufacturing specialist determines the sequence of production operations.
3. Part programming. Starting with the geometric definition of the part which is already in the data base, the part programmer develops tool selection, speed, and feed rates, as well as workpiece fixturing and cutter-path definitions.
4. Verification. Interactive computer graphics verification permits the part programmer to correct any mistakes and to improve the process. Animated graphics permits observation of the entire machining sequence. Verification software includes viewing transformations that permit zooming up and changing viewpoint. Making a part from plastic foam in the actual NC machine is another common verification technique.
5. Production. Shop personnel set up the machine tools and management movement of workpieces to and from machines.

Good candidates for NC machining include parts which are (1) complex, (2) produced in small quantities, (3) made by removing large amounts of material, (4) held to close tolerances, (5) subject to frequent engineering design changes, and (6) costly to remake if ruined by a mistake.

Numerical control is routinely used for machine tools that do milling, boring, turning, drilling, and grinding. Besides supporting machining applications, NC has been applied to welding and flame cutting, laser bean cutting, stamping, wire wrapping, assembly, and other related processes.

Robotics

Robots are mechanical manipulators which move materials, workpieces, and other devices through variable programmed motions. Industrial uses of robots include material handling, assembly, spray painting, and welding. There are similarities between numerically controlled machines and robots, because both types of systems work with mechanical positioning of tools in 3-D space.

As a mechanical manipulator, an industrial robot is similar to the human arm. Many arrangements of the moving parts of robots have been developed for specialized uses. One of the more common robot configurations is the jointed arm, Figure 10.10. The arm is mounted on a rotating base which provides a full

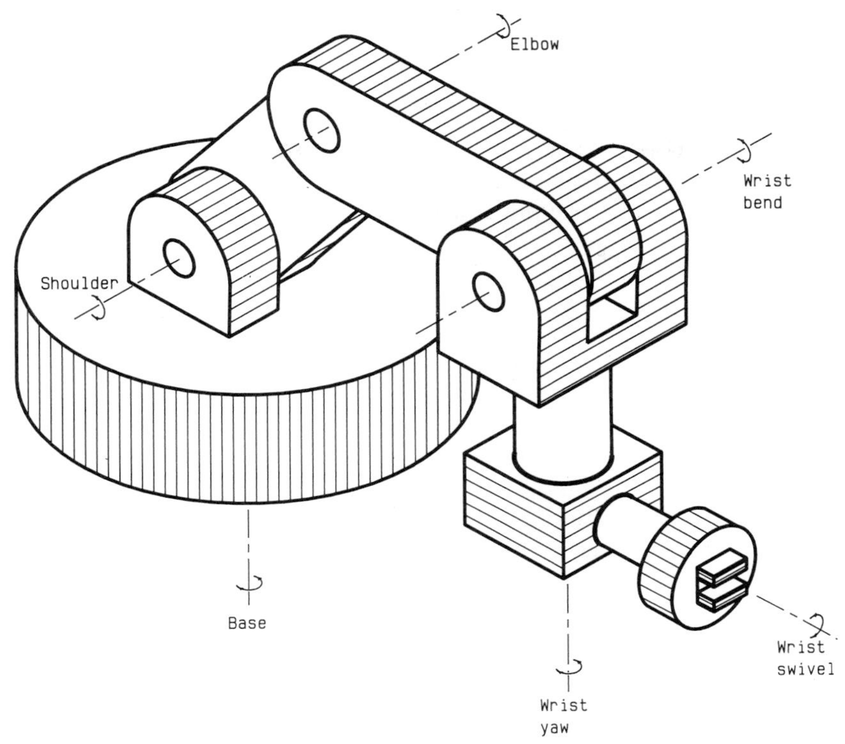

Figure 10.10 Motions in a typical robot.

circle of working area. Rotating joints correspond to the shoulder, elbow, and wrist. Analogous to the hand, the end effector can be selected to adapt the robot to a variety of tasks. The motion of the end effector can be described with all or some of the following six degrees of freedom:

Vertical traverse: motion of the arm up and down, done in the shoulder and elbow joints.

Radial traverse: motion of the arm in and out, done in the shoulder and elbow joints.

Rotational traverse: circular motion of the arm done with rotation in the base.

Wrist bend: rotary motion of the wrist up and down.

Wrist yaw: rotary motion of the wrist to left and right.

Wrist swivel: roation of the end effector relative to the wrist assembly.

Because determining the motion of each separate joint is not easy, most robotic systems have automated ways to create motion instructions. One of the most popular ways to program robots, *teaching,* requires that the operator take the robot through its sequence under manual control. Each motion is recorded into memory for later playback in normal operation. Although simple and direct, teaching has the disadvantages of being time-consuming and tying up production facilities.

In contrast, there are several advantages to programming robots with *off-line* programming. Besides not disrupting production, off-line programming allows robotic systems to be integrated into the factory computer communications network in the same way as DNC systems. Robot programming languages permit systematic description of the motions, distances, and end effector actions. With interactive graphics and solid modeling, computer simulation of the robot program can be viewed as it is being developed. Furthermore, since robot programming is under computer control, the motion sequence of the robot can be optimized for more efficient operation.

Robots have clear advantages in replacing human workers where there are repetitious, heavy, and hazardous tasks to be performed. Robots can be effectively utilized in applications such as spray painting, welding, handling radioactive materials, lifting heavy objects, and picking and placing parts. More advanced robotic systems with decision-making capability and vision systems will be able to deal with nonroutine situations such as inspection and alignment of parts.

10.4 CLOSURE

Computers excel at tasks that humans do poorly and vice versa. Computers can perform arithmetic at a rate that staggers the imagination; humans can arrive at creative and intuitive decisions that may never be possible with computers. Clearly, the most effective uses of computing are those which augment and expand human capabilities.

The incorporation of computer graphics in computer-aided engineering strongly supports the synergistic relationship between the engineer and the computer. When routine tasks such as data plotting, drafting, analysis, preparation of pictorials, retrieval of information, and control of machines are performed by computer, the engineer can work at higher levels of cognition. As a result, the engineer gains the ability and resources to better analyze, design, develop, create, operate, and manage. Well documented in modern history is the fact that expanded productivity of engineering professionals has a multiplying effect on the betterment of living for all people.

PROJECTS

Use of a solid modeling system is suggested.

10.1. Make a model of the exterior of a residence. A minimum model would have a surface representation of the exterior walls and roof. If possible, use your model to determine the area of the roof, exterior walls, the floors, and the enclosed volume.

A more elaborate model would have the interior walls and floors also represented and would allow a "walk-through" of the inside.

10.2. Design a cast wheel for an automobile. Make use of symmetry features in the construction of the model. Use the solid modeling system to determine the mass of the wheel and the mass moment of interia with respect to the axis of rotation.

10.3. Design a decorative bottle to contain 0.75 liter of beverage. The bottle should be wider than it is deep so as to look larger on the shelf. The solid modeling system should check that the bottle has the required fluid capacity. Also, the volume of glass should be computed; it may be assumed that the wall is of constant thickness so that surface area may be used.

10.4. Design a spiral staircase for a residence. The distance between floors should be 9 feet. The size and height of the treads should conform to customary architectural practice.

10.5. Design an axisymmetric cylindrical pressure vessel with 2:1 ellipsoidal heads and constant wall thickness of 0.25 in. to contain 1500 $in.^3$ Use the solid modeling system to ascertain the capacity and the amount of material used for the pressure vessel.

10.6. Design a ceramic insulator to hold an electric transmission cable. Ascertain that the electrical and mechanical performance of the part is adequate.

10.7. Design a cam to perform a specific function. Select an appropriate cutter size and generate the tool path. Verify the tool path with computer graphics.

10.8. Develop a simulation for a robot arm to pick and place a workpiece. Show the simulation with computer graphics.

10.9. Build a solid model of an airplane or rocket. Compute selected properties such as surface area of the model and volume occupied. Optionally, assume a homogeneous solid and compute the center of mass and the principal moments of inertia.

10.10. Perform a finite element stress analysis of the pressure vessel in Project 10.5, assuming the internal pressure is 500 psig. Prepare data presentation graphics of the section of wall which is highly stressed, and improve the design if necessary.

Appendixes

Review of Vector and Matrix Algebra

A *vector* is a quantity having magnitude and direction which obeys the laws of vector algebra. It is common to express a vector in terms of its Cartesian components, where, for example, the vector **F** is given by $\mathbf{F} = F_x\mathbf{i} + F_y\mathbf{j} + F_z\mathbf{k}$. Vector quantities are in **bold** type while scalar quantities are in *italic* type. The vectors **i**, **j**, and **k** are unit vectors which indicate the x, y, and z directions. The *magnitude* of **F** is given by

$$|\mathbf{F}| = F = \sqrt{F_x^2 + F_y^2 + F_z^2} \tag{A.1}$$

The vector $\mathbf{n} = \mathbf{F}/|\mathbf{F}|$ is a *unit vector,* so named becuase it has unit magnitude. Its use is to indicate direction.

A *matrix* is a rectangular array of elements which obeys the laws of matrix algebra. A matrix with m rows and n columns is called an m by n matrix. If $m = n$, the matrix is said to be *square.* A matrix with just one row is called a *row matrix,* while a matrix with just one column is called a *column matrix.* Row and column matrices are also known as "vectors" since the components of a vector can be expressed as a row (or column) matrix. The convention will be followed that a vector is written as a row matrix, for example,

$$\mathbf{F} = F_x\mathbf{i} + F_y\mathbf{j} + F_z\mathbf{k} \quad \text{or} \quad [\mathbf{F}] = [F_x \quad F_y \quad F_z] \tag{A.2}$$

A matrix is denoted with square brackets and bold type, such as [**A**]. The elements of a matrix may be scalars, vectors, or other matrices. The element of [**A**] in row i and column j is denoted by A_{ij} if it is a scalar quantity, $\mathbf{A}_{ij}$ if it is a vector quantity, and $[\mathbf{A}_{ij}]$ if it is a matrix. Sometimes, for brevity, the square brackets may be dropped from the notations of matrices. A single subscript often is chosen for row and column matrices.

The *determinant* is a scalar quantity defined for square matrices with real, scalar elements. The determinant of a 2×2 matrix is written out as

$$\det \begin{bmatrix} A_{11} & A_{12} \\ A_{21} & A_{22} \end{bmatrix} = \begin{vmatrix} A_{11} & A_{12} \\ A_{21} & A_{22} \end{vmatrix} = A_{11}A_{22} - A_{12}A_{21} \tag{A.3}$$

For a 3×3 matrix the determinant is computed as

$$\begin{vmatrix} A_{11} & A_{12} & A_{13} \\ A_{21} & A_{22} & A_{23} \\ A_{31} & A_{32} & A_{33} \end{vmatrix} = A_{11} \begin{vmatrix} A_{22} & A_{23} \\ A_{32} & A_{33} \end{vmatrix} - A_{12} \begin{vmatrix} A_{21} & A_{23} \\ A_{31} & A_{33} \end{vmatrix} + A_{13} \begin{vmatrix} A_{21} & A_{22} \\ A_{31} & A_{32} \end{vmatrix} \tag{A.4}$$

which illustrates expansion by *minors*. With an n by n array, the minor of the element A_{ij} is the determinant of the $(n - 1)$ by $(n - 1)$ submatrix with the ith row and the jth column removed. The sign of the term is *positive* if the sum $i + j$ is *even* and *negative* if the sum $i + j$ is *odd*.† Expansion of higher-order determinants is done with a successive expansion by minors.

Two matrices with the same number of rows and columns can be added or subtracted by adding or subtracting the corresponding elements. For example, the two row matrices $[\mathbf{A}] = [A_1 \quad A_2 \quad A_3]$ and $[\mathbf{B}] = [B_1 \quad B_2 \quad B_3]$ combine into new matrices $[\mathbf{C}]$ and $[\mathbf{D}]$ as

$$[\mathbf{C}] = [\mathbf{A}] + [\mathbf{B}] = [A_1 + B_1 \quad A_2 + B_2 \quad A_3 + B_3] \tag{A.5a}$$

and

$$[\mathbf{D}] = [\mathbf{A}] - [\mathbf{B}] = [A_1 - B_1 \quad A_2 - B_2 \quad A_3 - B_3] \tag{A.5b}$$

Addition and subtraction of vectors follow the same rule. If a matrix is subtracted from itself, the *null* or *zero* matrix results, so named because all its elements are zero.

Multiplication of a matrix by a scalar λ is accomplished by multiplying each element of the matrix by λ. Thus for the $m \times n$ matrix $[\mathbf{B}]$, if

$$B_{ij} = \lambda A_{ij} \qquad i = 1, m; \quad j = 1, n \tag{A.6a}$$

then

$$[\mathbf{B}] = \lambda[\mathbf{A}] \tag{A.6b}$$

This operation also applies to components of vectors.

The operation defining the *product* between two matrices arises from the way in which simultaneous linear equations are expressed in matrix notation. For example, consider the system

$$\begin{aligned} a_{11}x_1 + a_{12}x_2 + a_{13}x_3 &= y_1 \\ a_{21}x_1 + a_{22}x_2 + a_{23}x_3 &= y_2 \\ a_{31}x_1 + a_{32}x_2 + a_{33}x_3 &= y_3 \end{aligned} \tag{A.7a}$$

which in matrix notation is written as

$$\begin{bmatrix} a_{11} & a_{12} & a_{13} \\ a_{21} & a_{22} & a_{23} \\ a_{31} & a_{32} & a_{33} \end{bmatrix} \begin{bmatrix} x_1 \\ x_2 \\ x_3 \end{bmatrix} = \begin{bmatrix} y_1 \\ y_2 \\ y_3 \end{bmatrix} \tag{A.7b}$$

†The sign can be determined from the formula $(-1)^{(i+j)}$.

or

$$[\mathbf{A}][\mathbf{X}] = [\mathbf{Y}] \tag{A.7c}$$

The scheme for matrix multiplication is consistent with replicating the nine elements on the left-hand side of the system Eq. (A.7a) from the product $[\mathbf{A}][\mathbf{X}]$ in Eq. (A.7c).

In general, the product $[\mathbf{A}][\mathbf{B}]$ of an m by n matrix $[\mathbf{A}]$ and an n by p matrix $[\mathbf{B}]$ is an m by p matrix $[\mathbf{C}]$ with the elements defined by

$$C_{ij} = \sum_{k=1}^{n} A_{ik} B_{kj} \qquad i = 1, m; \quad j = 1, p \tag{A.8}$$

As an example, the matrix product

$$\begin{bmatrix} 1 & 2 & 0 \\ 0 & -1 & 3 \end{bmatrix} \begin{bmatrix} 2 & 1 \\ 1 & 0 \\ 3 & 1 \end{bmatrix} = \begin{bmatrix} (1 \times 2 + 2 \times 1 + 0) & (1 \times 1 + 0 + 0) \\ (0 - 1 \times 1 + 3 \times 3) & (0 - 0 + 3 \times 1) \end{bmatrix} = \begin{bmatrix} 4 & 1 \\ 8 & 3 \end{bmatrix} \tag{A.9}$$

Note that *conformability* requires that the number of columns in the first matrix be equal to the number of rows in the second. In general,

$$[\mathbf{A}][\mathbf{B}] \neq [\mathbf{B}][\mathbf{A}] \tag{A.10}$$

even if conformability is not violated.

The *transpose* of an m by n matrix $[\mathbf{A}]$, denoted by $[\mathbf{A}]^T$, is an n by m matrix formed by interchanging the rows and columns of $[\mathbf{A}]$. The elements of $[\mathbf{C}] = [\mathbf{A}]^T$ are

$$C_{ij} = A_{ji} \tag{A.11}$$

In other words, if A_{ij} is the element in the ith row and jth column of $[\mathbf{A}]$, this same element occupies the jth row and the ith column of $[\mathbf{A}]^T$. If the transpose of a square matrix equals the original matrix, the matrix is said to be *symmetric*. Furthermore, the transpose of the sum of two matrices is

$$([\mathbf{A}] + [\mathbf{B}])^T = [\mathbf{A}]^T + [\mathbf{B}]^T \tag{A.12}$$

and the transpose of the product of two matrices is

$$([\mathbf{A}][\mathbf{B}])^T = [\mathbf{B}]^T[\mathbf{A}]^T \tag{A.13}$$

The *identity* matrix $[\mathbf{I}]$ is an m by m (square) matrix where the diagonal elements $(i = j)$ are all unity and the off-diagonal elements $(i \neq j)$ are all zero. This matrix has the property that

$$[\mathbf{A}][\mathbf{I}] = [\mathbf{A}] \tag{A.14}$$

A matrix $[\mathbf{A}]$ is said to be orthogonal if

$$[\mathbf{A}][\mathbf{A}]^T = [\mathbf{I}] \tag{A.15}$$

where $[\mathbf{I}]$ is the *identity* matrix.

The operations for computing vector multiplication can be expressed by matrices and determinants. The *dot product* or *scalar product* between two vectors $\mathbf{A}$ and $\mathbf{B}$ is a scalar such that

$$\mathbf{A} \cdot \mathbf{B} = |\mathbf{A}||\mathbf{B}| \cos\theta = AB\cos\theta \tag{A.16}$$

where θ is the angle between **A** and **B**. In terms of Cartesian components,

$$\mathbf{A} \cdot \mathbf{B} = A_x B_x + A_y B_y + A_z B_z \tag{A.17a}$$

which in matrix notation is written as the product

$$[\mathbf{A}][\mathbf{B}]^T = [A_x \quad A_y \quad A_z] \begin{bmatrix} B_x \\ B_y \\ B_z \end{bmatrix} \tag{A.17b}$$

The dot product expresses the component of **A** in the direction of **B** or vice versa. When **A** and **B** are nonzero and $\mathbf{A} \cdot \mathbf{B} = 0$, **A** is perpendicular to **B**.

The *cross product* or *vector product* between vectors **A** and **B** is defined by

$$\mathbf{A} \times \mathbf{B} = AB \sin \theta \, \mathbf{n} \tag{A.18}$$

where θ is the angle between **A** and **B** and **n** is a unit vector normal to the plane of **A** and **B**. Note that θ and **n** obey the right-hand rule, where the direction of **n** corresponds to the thumb and the sense of θ corresponds to the fingers. The determinant representation of the cross product is convenient for Cartesian components,

$$\mathbf{A} \times \mathbf{B} = \begin{vmatrix} \mathbf{i} & \mathbf{j} & \mathbf{k} \\ A_x & A_y & A_z \\ B_x & B_y & B_z \end{vmatrix} = (A_y B_z - A_z B_y)\mathbf{i} - (A_x B_z - A_z B_x)\mathbf{j} + (A_x B_y - A_y B_x)\mathbf{k} \tag{A.19}$$

From Eq. (A.19) it is apparent that $(\mathbf{B} \times \mathbf{A}) = -(\mathbf{A} \times \mathbf{B})$. In matrix notation, the cross product between vectors **A** and **B** is

$$[A_x \quad A_y \quad A_z] \begin{bmatrix} 0 & -B_z & B_y \\ B_z & 0 & -B_x \\ -B_y & B_x & 0 \end{bmatrix} = [A_y B_z - A_z B_y \quad -A_x B_z + A_z B_x \quad A_x B_y - A_y B_x] \tag{A.20}$$

The area of a triangle with **A** and **B** forming two sides is $|\mathbf{A} \times \mathbf{B}|/2$.

The dot and cross products can be combined in the scalar triple product,

$$(\mathbf{A} \times \mathbf{B}) \cdot \mathbf{C} = (A_y B_z - A_z B_y)C_x - (A_x B_z - A_z B_x)C_y + (A_x B_y - A_y B_x)C_z \tag{A.21}$$

It can be shown from Eq. (A.21) that

$$(\mathbf{A} \times \mathbf{B}) \cdot \mathbf{C} = \mathbf{A} \cdot (\mathbf{B} \times \mathbf{C}) \tag{A.22}$$

The scalar triple product may be expressed as the determinant

$$(\mathbf{A} \times \mathbf{B}) \cdot \mathbf{C} = \begin{vmatrix} A_x & A_y & A_z \\ B_x & B_y & B_z \\ C_x & C_y & C_z \end{vmatrix} \tag{A.23}$$

Another way to compute the scalar triple product is with matrix multiplication

$$(\mathbf{A} \times \mathbf{B}) \cdot \mathbf{C} = [A_x \quad A_y \quad A_z] \begin{bmatrix} 0 & -B_z & B_y \\ B_z & 0 & -B_x \\ -B_y & B_x & 0 \end{bmatrix} \begin{bmatrix} C_x \\ C_y \\ C_z \end{bmatrix} \tag{A.24}$$

A matrix divided into smaller blocks or *submatrices* is known as a *partitioned matrix*. The dimensions of the submatrices must be such that the rows and elements match. The 4 by 3 matrix [**P**] can be partitioned, for example, where each row is itself a matrix,

$$\begin{bmatrix} P_{11} & P_{12} & P_{13} \\ P_{21} & P_{22} & P_{23} \\ P_{31} & P_{32} & P_{33} \\ P_{41} & P_{42} & P_{43} \end{bmatrix} = \begin{bmatrix} [\mathbf{P}_1] \\ [\mathbf{P}_2] \\ [\mathbf{P}_3] \\ [\mathbf{P}_4] \end{bmatrix} = \begin{bmatrix} \mathbf{P}_1 \\ \mathbf{P}_2 \\ \mathbf{P}_3 \\ \mathbf{P}_4 \end{bmatrix} \tag{A.25}$$

where the notation $[\mathbf{P}_1] = \mathbf{P}_1 = [P_{11} \quad P_{12} \quad P_{13}]$ (a vector), etc.

For a *nonsingular* square matrix [**A**], there exists an *inverse* written as $[\mathbf{A}]^{-1}$ such that the product

$$[\mathbf{A}]^{-1}[\mathbf{A}] = [\mathbf{C}][\mathbf{A}] = [\mathbf{I}] \tag{A.26}$$

The elements of the inverse matrix $[\mathbf{A}]^{-1} = [\mathbf{C}]$ can be found by means of *cofactors*. The cofactor A_{ij}^c is the minor of A_{ij} times its sign, which can be computed as $(-1)^{(i+j)}$. Then, the elements

$$C_{ij} = A_{ji}^c / |\mathbf{A}| \tag{A.27}$$

where $|\mathbf{A}|$ is the determinant of [**A**]. If $|\mathbf{A}| = 0$, the matrix is singular and the inverse does not exist. In computer programs, numerical methods such as Gaussian elimination[32,59] are used in preference to the method of cofactors.

EXAMPLE A.1
Using the method of cofactors, invert the matrix

$$[\mathbf{A}] = \begin{bmatrix} 1 & 3 & 0 \\ 1 & 0 & 2 \\ -1 & 0 & 1 \end{bmatrix}$$

Solution. The determinant $|\mathbf{A}|$ is

$$|\mathbf{A}| = 1(0) - 3(1 + 2) + 0 = -9$$

The cofactors are

$$A_{11}^c = (-1)^{(1+1)} \begin{vmatrix} 0 & 2 \\ 0 & 1 \end{vmatrix} = 0$$

$$A_{12}^c = (-1)^{(1+2)} \begin{vmatrix} 1 & 2 \\ -1 & 1 \end{vmatrix} = -3$$

$$A_{13}^c = (-1)^{(1+3)} \begin{vmatrix} 1 & 0 \\ -1 & 0 \end{vmatrix} = 0$$

$$A_{21}^c = (-1)^{(2+1)} \begin{vmatrix} 3 & 0 \\ 0 & 1 \end{vmatrix} = -3$$

$$A_{22}^c = (-1)^{(2+2)} \begin{vmatrix} 1 & 0 \\ -1 & 1 \end{vmatrix} = 1$$

$$A_{23}^c = (-1)^{(2+3)} \begin{vmatrix} 1 & 3 \\ -1 & 0 \end{vmatrix} = -3$$

$$A_{31}^c = (-1)^{(3+1)} \begin{vmatrix} 3 & 0 \\ 0 & 2 \end{vmatrix} = 6$$

$$A_{32}^c = (-1)^{(3+2)} \begin{vmatrix} 1 & 0 \\ 1 & 2 \end{vmatrix} = -2$$

$$A_{33}^c = (-1)^{(3+3)} \begin{vmatrix} 1 & 3 \\ 1 & 0 \end{vmatrix} = -3$$

The required inverse is thus, by Eq. (A.27),

$$\frac{1}{|\mathbf{A}|} \begin{bmatrix} A_{11}^c & A_{21}^c & A_{31}^c \\ A_{12}^c & A_{22}^c & A_{32}^c \\ A_{13}^c & A_{23}^c & A_{33}^c \end{bmatrix} = \begin{bmatrix} 0 & 1/3 & -2/3 \\ 1/3 & -1/9 & 2/9 \\ 0 & 1/3 & 1/3 \end{bmatrix}$$

A check of this result can be made by multiplying $[\mathbf{A}]^{-1}[\mathbf{A}]$,

$$\begin{bmatrix} 0 & 1/3 & -2/3 \\ 1/3 & -1/9 & 2/9 \\ 0 & 1/3 & 1/3 \end{bmatrix} \begin{bmatrix} 1 & 3 & 0 \\ 1 & 0 & 2 \\ -1 & 0 & 1 \end{bmatrix} = \begin{bmatrix} 1 & 0 & 0 \\ 0 & 1 & 0 \\ 0 & 0 & 1 \end{bmatrix}$$

which correctly returns the identity matrix.

PROBLEMS

A.1. Determine the unit vector in the direction of $\mathbf{P} = 5\mathbf{i} + 6\mathbf{j} - \mathbf{k}$.

A.2. Determine the unit vector in the direction given by the row vector $[10 \quad -2 \quad 4]$.

A.3. Determine the unit vector indicating the direction connecting the points $(-1, 2, 4)$ and $(3, 2, 0)$.

A.4. Determine the distance between the points $(2, 0, 6)$ and $(3, -4, 2)$.

A.5. Given the two vectors $\mathbf{A} = 3\mathbf{i} - 4\mathbf{j}$ and $\mathbf{B} = 6\mathbf{j} + 8\mathbf{k}$, determine the angle formed from **A** to **B**.

A.6. Determine the unit vector normal to the plane formed by vectors **A** and **B** given in the problem above.

A.7. Given the vectors $\mathbf{R} = 10\mathbf{i} - 4\mathbf{j}$ and $\mathbf{S} = 2\mathbf{i} + 3\mathbf{j} - 2\mathbf{k}$, determine the component of **R** in the direction of **S**.

A.8. Determine a unit vector normal to the plane defined by the points $(0, 3, 1)$, $(2, 1, 0)$, and $(1, 0, 2)$.

A.9. Given the vectors $[\mathbf{A}] = [4 \quad 2 \quad 1]$, $[\mathbf{B}] = [4 \quad 0 \quad 6]$, and $[\mathbf{C}] = [6 \quad 0 \quad 7]$, determine (a) the scalar triple product $\mathbf{A} \times \mathbf{B} \cdot \mathbf{C}$ and (b) the vector triple product $(\mathbf{A} \times \mathbf{B}) \times \mathbf{C}$, and the vector triple product $\mathbf{A} \times (\mathbf{B} \cdot \mathbf{C})$.

A.10. Show that $(\mathbf{A} + \mathbf{B}) \cdot (\mathbf{A} - \mathbf{B}) = |\mathbf{A}|^2 - |\mathbf{B}|^2$. Illustrate this result with an appropriate sketch.

A.11. Show that $(\mathbf{A} + \mathbf{B}) \times (\mathbf{A} - \mathbf{B}) = 2\mathbf{B} \times \mathbf{A}$.

A.12. Show that the volume of a parallelepiped with three edges defined by the triad **A**, **B**, and **C** can be computed from the scalar triple product.

A.13. Given the two matrices **A** and **B**

$$[\mathbf{A}] = \begin{bmatrix} 1 & 3 & 3 \\ 2 & 0 & 1 \\ 0 & -2 & 1 \end{bmatrix} \qquad [\mathbf{B}] = \begin{bmatrix} 0 & 1 & 2 \\ 3 & 1 & 3 \\ 0 & 3 & -1 \end{bmatrix}$$

Evaluate the expression $([\mathbf{A}][\mathbf{B}] - [\mathbf{B}][\mathbf{A}])$.

A.14. Is the matrix

$$\begin{bmatrix} \cos\theta & \sin\theta & \sin\theta \\ \sin\theta & \cos\theta & \cos\theta \\ \sin\theta & \cos\theta & 1 \end{bmatrix}$$

(a) symmetric, (b) orthogonal, (c) able to be inverted?

A.15. For the matrix

$$\begin{bmatrix} \cos\theta & -\sin\theta & 0 \\ \sin\theta & \cos\theta & 0 \\ 0 & 0 & 1 \end{bmatrix}$$

determine (a) the transpose, (b) the inverse, (c) whether the matrix is orthogonal.

A.16. Invert the matrix

$$\begin{bmatrix} 1 & 3 \\ 2 & -1 \end{bmatrix}$$

and check your result.

A.17. Invert the matrix

$$\begin{bmatrix} a & 0 & 1 \\ 0 & -1 & a \\ 0 & 1 & a \end{bmatrix}$$

using the method of cofactors. Check your results by multiplying the inverse by the original matrix.

A.18. Given the two matrices

$$[\mathbf{C}] = \begin{bmatrix} 1 & 0 & 1 \\ 0 & -1 & 2 \\ 3 & 1 & 0 \end{bmatrix} \qquad [\mathbf{F}] = \begin{bmatrix} 2 & 1 \\ 0 & 1 \\ 0 & 1 \end{bmatrix}$$

Demonstrate that $([\mathbf{F}]^T[\mathbf{C}]^T) = ([\mathbf{C}][\mathbf{F}])^T$.

A.19. Use matrix multiplication to expand the scalar triple product $\mathbf{A} \cdot (\mathbf{B} \times \mathbf{C})$ where the three vectors are described by their Cartesian components.

A.20. Determine the inverse of the matrix

$$\begin{bmatrix} \cos\theta & \sin\theta & 0 & 0 \\ -\sin\theta & \cos\theta & 0 & 0 \\ 0 & 0 & 1 & 0 \\ 1 & -2 & 0 & 1 \end{bmatrix}$$

A.21. Determine the inverse of the matrix

$$\begin{bmatrix} \cos\theta & \sin\theta & 0 & -r_x \\ -\sin\theta & \cos\theta & 0 & -r_y \\ 0 & 0 & 1 & -r_z \\ 1 & 0 & 0 & 1 \end{bmatrix}$$

Typical Routines for a Graphics Tools Library

The following subroutine library describes the one developed in the Engineering Science Interactive Graphics Laboratory at the University of Wyoming. The source code is written in machine-independent FORTRAN. The routines have been used to build several packages for presentation graphics and solid modeling applications.

Color routines

BKGCLR	GINCLR
CLRMAP	LINCLR
CLRRST	TEXCLR

Curve plotting

AXES	POLLIN
MARKER	POLMRK

Drawing routines

ARC	MOVABS
CIRCLE	MOVE
CLIP	MOVREL
DRAW	NOCLIP
DRWABS	POLGON
DRWREL	POLLIN
DSHLIN	POLMRK
MARKER	

Graphics device routines

BELL	NOSTRK
GRINIT	SETGIN
GRSTOP	STROKE
LOCATE	VIEWPT
NEWPAG	WINDOW

Hardware panel filling

BEGIN
CIRCLE
ENDPAN
FILPAN
POLGON

Math routines

MATCPY
MATINV
MATMLT
MATSOL
MATTRN

Modeling/viewing transformations

PERSP/VPERSP
ROTX/VROTX
ROTY/VROTY
ROTZ/VROTZ
ROT3D/VROT3D
SCALE/VSCALE
TRAN/VTRAN

Segment routines

SEGCL
SEGCR
SEGDEL
SEGINC
SEGMOD
SEGMOV
SEGVIS

Text routines

SYMBOL
TEXANG
TEXCLR
TEXSIZ
TEXT

Routines are listed alphabetically below.

ARC(RAD,X0,Y0,THETA1,THETA2,IFINE)

Input: RAD—radius of arc, real
X0,Y0—coordinates of center of arc, real
THETA1, THETA2—start angle, end angle (degrees), real (x-axis is 0)
IFINE—number of line segments making up a complete circle, integer
Output: A circular arc is drawn.

AXES (X0,Y0,STRING,NCHAR,AXSLEN,IANGLE,FIRST,DELTA,FLAST, NUMDIG,LABTIC,ISKIP)

Input: X0,Y0—coordinates of axes intersection, (world coordinates), real
STRING—descriptive title on axis, character
NCHAR—number of characters in title, integer
+ = title on counterclockwise side of axis,
− = title on clockwise side
AXSLEN—length of axis, (world coordinates), real
IANGLE—angle in degrees to rotate axis, integer
(0, 90, 180, and 270 are valid)
FIRST—the initial data value to put on the axis
DELTA—data step size, real
FLAST—the last value to put on the axis, real
NOTE: FIRST, FLAST, and DELTA are in data units.
NUMDIG—number of digits after decimal point, integer

LABTIC—number of tick marks per step, integer
ISKIP—(1 or 0) if ISKIP = 1, then omit the first number on the axis
If ISKIP = 0, then print the number, integer

BEGIN(X,Y,IB)

Input: X,Y—reals
IB—integer
Output: Graphics move to user position X,Y and start of a panel definition.

After calling BEGIN, use DRAWs to define the panel boundary. End sequence with ENDPAN. If IB is 1, a boundary will be drawn, otherwise a boundary will not be drawn.

BELL

Causes terminal bell to ring.

BKGCLR(ICOL)

Input: ICOL—integer
Output: Sets background screen color for terminals only. ICOL is defined as in LINCLR. The new color will not take effect until the screen is cleared.

CIRCLE(RAD,X0,Y0,IFINE,IB,IPAN)

Input: RAD—radius of circle, real
X0,Y0—coordinates of center of circle, real
IFINE—number of line segments making up circle, integer
IB—border drawing descriptor. If IB = 0, no boundary is drawn.
If IB = 1, boundary is drawn in current LINCLR, integer.
IPAN—pattern type (see FILPAN)
Output: Circle is drawn with or without panel fill. NOTE: If both IB and IPAN are 0, nothing will appear.

CLIP

Causes clipping of subsequent MOVEs and DRAWs. This will slow down the plot, but will ensure that no lines are drawn outside terminal or plotter boundaries. The default is to clip.

CLRMAP(ICOL,IHUE,ILIT,ISAT)

Input: ICOL,IHUE,ILIT,ISAT—integers
Output: ICOL hue, lightness, and saturation changed to IHUE, ILIT, and ISAT. No action is taken for plots.

Ranges: 0<ICOL<7 (System dependent)
0<IHUE<360
0<ILIT<100
0<ISAT<100

CLRRST

Restores color map to preset defaults.

DRAW(X,Y)

Input: X,Y—reals
Output: Graphic

Draws a line from current position to position X,Y (user units).

DRWABS(IX,IY)

Input: IX,IY—integers
Output: Graphic

Draws a line from current position to screen coordinates IX,IY. System dependent.

DRWREL(IX,IY)

Input: IX,IY—integers
Output: Graphic

Relative draw of IX,IY from current position to new position in screen coordinates. System dependent.

DSHLIN(ILINE)

Input: ILINE—integer
Output: Sets line type for subsequent line drawing. See the chart below for line patterns.

	. . . .	—.—.—	- - - -	— — —	—..—..—	—.—.—	— — —
Line style:	1	2	3	4	5	6	7

ENDPAN

Output: Panel closed and filled. Last in sequence of BEGIN, DRAW, . . . , DRAW, ENDPAN.

FILPAN(IPAN)

Input: IPAN—integer
Output: Sets a pattern with which a polygon will be filled in conjunction with the BEGIN and ENDPAN routines.

Effect of IPAN:

−1 to −7	Solid fill in the color index of |IPAN|
0	Background color
1 to 16	Hatch fill
17 to 50	Undefined (error)
50 to 174	Pattern fill

NOTE: These capabilities are hardware-dependent.

GINCLR(IHUE,ILIT,ISAT)

Input: IHUE,ILIT,ISAT—integers
Output: Sets the color of the graphics cross hairs according to IHUE, ILIT, and ISAT, as in CLRMAP.

GRINIT(ITERM,IPLOT,IPAPER)

Input: ITERM,IPLOT,IPAPER—integers (see below)

Required as the first graphics routine called in the program. The workstations options including the following:

ITERM

= 0 No screen graphics. Use for alphanumeric terminal
= 4105, 4107, etc. for graphics terminal

IPLOT

= 0 No plotter file
= 7550, 7580, 7470, 7475, etc. for plotter

IPAPER

= 1 A size paper (8.5 × 11 in.)
= 2 B size paper (11 × 17 in.)
= 3 C size paper (17 × 22 in.)
= 4 D size paper (22 × 34 in.)

When IPLOT is not equal to 0, a file HPP.PLT appears in user's directory. Spool this file to the plotter by using the appropriate command.

GRSTOP

Required as last routine called in a program. Terminates all graphics operations.

LINCLR(ICOL)

Input: ICOL—integer
Output: Color set for lines and markers drawn subsequently.

The parameter ICOL is defined as follows:

0 = black (on screen, nothing on paper),
1 = white (black on paper), 2 = red, 3 = green, 4 = blue, 5 = cyan,
6 = magenta, 7 = yellow. Can be changed for terminals by CLRMAP.

LOCATE(PX,PY,IDAT)

Output: PX,PY—location of selected point in world coordinates, real
IDAT—ASCII value of key hit for selection, integer

This routine actuates the cross hair cursor and reports the x-y coordinates when a key is struck on the keyboard. No action is taken for plotters. The cross hairs are moved with the joydisk. For the tablet this routine works in conjunction with SETGIN.

MARKER(X,Y,IMARK)

Input: X,Y—reals
Output: Graphic

MARKER places a marker at the point X,Y. The marker type (below) set by IMARK, integer

1	2	3	4	5	6	7	8	9	10
+	+	⁎	0	X	□	◇	⊡	◈	⊠

MATCPY (A,B,N1,N2)

Input: A—N1 × N2 real matrix
N1,N2—dimensions of A, integers
Output: B—N1 × N2 real matrix

MATCPY sets [**B**] = [**A**]

MATINV(A,N,DET)

Input: A—N × N real matrix. Destroyed and replaced with the inverse of A.
N—order of A-integer

Output: DET—determinant of A. If DET = 0, A has no inverse.

MATINV calculates the inverse of matrix A.

MATMLT(A,B,C,N1,N2,N3)

Input: B—real matrix dimensioned N1 × N2
C—real matrix dimensioned N2 × N3
N1,N2,N3—integers

Output: A—real matrix dimensioned N1 × N3.

MATMLT multiplies B by C and stores the result in A. B and C may be the same matrix but A must be distinct.

MATSOL(A,B,J,NB,N1)

Input: A—real matrix (N1 × N1) of coefficients
B—real matrix (N1 × NB) of right-hand sides. Must be doubly dimensioned.
J—order of matrix A, integer
NB—column dimension of B, integer
N1—row dimension of A, integer (J ≤ N1 ≤ 100)

Output: B—solutions of X in AX = B, real.

MATSOL uses Gaussian elimination with full pivoting to solve simultaneous linear equations with one or more right-hand sides.

MATTRN(A,B,N1,N2)

Input: A—N1 × N2 real matrix
N1,N2—dimensions of A and B

Output: B—N2 × N1 real matrix containing the transpose of A

MATTRN computes the transpose of the matrix A and then places the result in the matrix B.

MOVABS(IX,IY)

Input: IX,IY—integers

Output: Graphic move to absolute screen or plotter coordinates IX,IY. System-dependent.

MOVE(X,Y)

Input: X,Y—reals

Output: Graphic move to coordinate position X,Y(user units).

MOVREL(IX,IY)

Input: IX,IY—integers

Output: Graphic move relative to current position. The current location will be the old location plus IX and IY [e.g., when the current position is 100,100 in screen units, a call to MOVREL (50,100) will result in a new position of 150,200]. System dependent.

NEWPAG

NEWPAG clears dialog and graphics screens.

NOCLIP

Turns off clipping. Use for faster drawing.

NOSTRK

Sets tablet so that points will be sent to the calling program only when a button is pushed. (Holding a button down continuously has no effect.) (Default)

PERSP(P,NP,PX,PY,PZ)

Input: P—NP × 3 array of X,Y,Z triplets, real (destroyed)
NP—row dimension of P, integer
PX,PY,PZ—negative reciprocal of X,Y,Z viewing distances, real.
Ordinarily, PX = PY = 0, PZ = small negative number
Output: P with perspective modeling transformation performed. Subsequent rotation of P produces meaningless results. Use VPERSP if such a viewing transformation is needed.

POLGON(P,NP,NLOW,NHI,IB,IPAN)

Input: P—matrix of X,Y,Z points, real
NP—row dimension of P, integer
NLOW—number of the point to start the polygon definition at, integer
NHI—number of the point to stop the polygon definition at, integer
IB = 0 no boundary drawn, IB = 1 boundary drawn, integer
IPAN—type of panel fill. See FILPAN for complete description, integer
Output: The points in P between NLOW and NHI are drawn as a polygon.

POLLIN(P,NP)

Input: P—matrix of X,Y,Z triplets, real
NP—row dimension of P, integer
Output: A line is drawn between points in P (Polyline)

POLMRK(P,NP,IMARK)

Input: P—matrix of X,Y,Z triplets, real
NP—row dimension of P, integer
IMARK—marker type, integer (see table under MARKER)
Output: Markers are drawn at points in P (Polymarker).

ROTX(P,NP,THETA,Y0,Z0)

Same as ROTZ except that rotation is around an axis parallel to the X-axis through Y0,Z0.

ROTY(P,NP,THETA,X0,Z0)

Same as ROTZ except that rotation is around an axis parallel to the Y-axis through X0,Z0.

ROTZ(P,NP,THETA,X0,Y0)

Input: P—NP × 3 array of X,Y,Z triplets, real (destroyed)
NP—row dimension of P, integer
THETA—angle of rotation in degrees, real
X0,Y0—coordinates of axis of rotation, real
Output: Modeling transformation with P rotated THETA degrees around an axis parallel to the Z-axis through X0,Y0.

Use for 2-D rotation around an axis perpendicular to the screen. Use VROTZ if a viewing transformation is required.

ROT3D(P,NP,C1,C2,C3,THETA)

Input: P—NP × 3 array of X,Y,Z triplets, real (destroyed)
NP—row dimension of P, integer
C1,C2,C3—direction cosines of axis of rotation (through origin) with respect to X,Y, and Z, real
Note that C1**2 + C2**2 + C3**2 = 1
THETA—rotation angle in degrees, real

Output: P rotated in 3-dimensional space as a modeling transformation.

For rotation around a coordinate axis as a modeling transformation, use ROTX, ROTY, or ROTZ. For a viewing transformation, use VROT3D.

SCALE(P,NP,SX,SY,SZ,X0,Y0,Z0)

Input: P—NP × 3 array of X,Y,Z triplets, real (destroyed)
NP—row dimension of P, integer
SX,SY,SZ—X,Y,Z scaling factors (<1 makes smaller or −1 makes mirror image), real
X0,Y0,Z0—the X,Y,Z coordinates the center of scaling (i.e. the FIXED point), real

Output: P—the scaled points matrix

This is a modeling transformation. For a viewing transformation use VSCALE.

SEGCL

Ends (closes) the definition of a segment.

SEGCR(ID,X0,Y0)

Input: ID—number of new segment to be created. This number must not be currently associated with a segment and must be in the range 0 to 32767. Integer
X0,Y0—location of the pivot point of the segment (the point about which all rotations and translations are done) in world coordinates, real.

Output: Segment ID is opened for definition.

NOTE: To create a segment, call SEGCR to initialize the definition of the segment, then use graphics routines such as MOVE and DRAW to define the segment. When the segment is finished, call SEGCL to close the definition of the segment. Enclose any code between SEGRCR and SEGCL to place an image into local segment memory and allow local pan and zoom.

SEGDEL(ID)

Input: ID—number of segment to be deleted, integer

Output: Segment ID is deleted. Set ID equal to −1 to delete all segments.

SEGINC(ID)

Input: ID—number of an existing segment to include in a segment definition, integer

Output: A copy of segment ID is included in the current segment definition. This subroutine must be called during a segment definition.

SEGMOD(ID,SX,SY,THETA,X0,Y0)

Input: ID—number of segment to be modified, integer
SX,SY—scaling factors in the X and Y directions, real
THETA—rotation angle in degrees from initial definition, real
X0,Y0—new location of pivot point in world coordinates, real
Output: Segment ID is redrawn after being moved, rotated, and scaled. Calls to SEGMOD are not cumulative; each translation, rotation, and scale is done from the original segment definition. SEGMOD may be used for animation of screen images.

SEGMOV(ID,X0,Y0)

Input: ID—number of segment to be moved, integer
X0,Y0—new location of pivot point in world coordinates, integer
Output: Segment ID is moved to location X0,Y0. Calls to SEGMOV are cumulative; each translation, rotation, and scale is done from the current segment definition. SEGMOV may be used for animation of screen images.

SEGVIS(ID,IV)

Input: ID—number of existing segment, integer
IV = 0 invisible, IV = 1 visible, integer
Output: Segment ID is either redrawn or erased, based on IV.

SETGIN(IGIN)

Input: IGIN—integer

This routine defines the method of cursor control used with LOCATE. The valid options are:

IGIN = 1—control is from the keyboard. (default)
= 2—control is from the tablet.

STROKE

Sets tablet so that a stream of points is continuously sent back as long as a button is depressed.

SYMBOL(X0,Y0,ISIZE,STRING,IANGLE,NCHAR)

Input: X0,Y0—point to place text at, in world coordinates, real
ISIZE—size of text (1, 2, or 3) as explained in TEXSIZ, integer
STRING—alphanumeric text to write to screen, character (or integer)
IANGLE—angle from the horizontal to write text at. Valid angles are 0, 90, 180, and 270. Integer
NCHAR—number of characters in STRING, integer
Output: The string STRING is drawn on the screen.

TEXANG(IANGLE)

Input: IANGLE—integer with value 0, 90, 180, or 270
Output: Sets angle of graphics text to be written.

TEXCLR(ICOL)

Input: ICOL—integer
Output: Sets graphics text color according to ICOL as in LINCLR above.

TEXSIZ(IHT)

Input: IHT—integer in range 1, 2, or 3
Output: Sets graphics text size, with 1 being smallest text size.

NOTE: Approximate sizes on 13-inch screen and A-size paper: 1: 3 mm (1/8 in) 2: 7 mm (1/4 in) 3: 10 mm (3/8 in)

TEXT(NCHAR,STRING)

Input: NCHAR—number of characters in STRING, integer
STRING—array of ASCII characters
Output: Writes graphics text starting at current position.

An example of a call is CALL TEXT(16,'THIS IS A STRING').

TRAN(P,NP,X0,Y0,Z0)

Input: P—NP × 3 array of X,Y,Z triplets, real (destroyed)
NP—row dimension of P, integer
X0,Y0,Z0—the amount of translation in the X,Y, or Z direction, real
Output: P—translated points matrix

VIEWPT(XL,XH,YL,YH)

Input: XL,XH.YL,YH

This routine defines the active viewport on the screen, with size determined by the input arguments. The arguments range between 0.0 and 1.0 with (0.0,0.0) being the lower left corner and (1.0,1.0) being the upper right corner.

VPERSP(P,NP,PX,PY,PZ,VP)

Input: P—same as PERSP except P not destroyed
Output: VP—NP × 3 points matrix when the perspective viewing transformation is performed, real. If VP is rotated, the results are garbage.

VROTX(P,NP,THETA,Y0,Z0,VP)

Input: Same as ROTX except P not destroyed
Output: VP—NP × 3 rotated points matrix, real

VROTY(P,NP,THETA,X0,Z0,VP)

Input: Same as ROTY except P not destroyed
Output: VP—NP × 3 rotated points matrix, real

VROTZ(P,NP,THETA,X0,Y0,VP)

Input: Same as ROTZ except P not destroyed
Output: VP—NP × 3 rotated points matrix, real

VROT3D(P,NP,C1,C2,C3,THETA,VP)

Input: Same as ROT3D except P not destroyed
Output: VP—NP × 3 rotated points matrix, real

VSCALE(P,NP,SX,SY,SZ,X0,Y0,Z0,VP)

Input: P—same as SCALE except P not destroyed
Output: VP—NP × 3 scaled points matrix, real

VTRAN(P,NP,X0,Y0,Z0,VP)

Input: Same as TRAN except P not destroyed
Output: VP—NP × 3 translated points matrix, real

WINDOW(XL,XH,YL,YH)

Input: XL,XH,YL,YH—reals

This routine scales user data to the graphics device. The user coordinates XL correspond to the left side, XH to the right side, YL to the bottom, and YH to the top. The default window called by GRINIT is (0.0, 133.0, 0.0, 100.0) Note: for undistorted drawings (i.e., circles not ellipses) set (XH − XL)/(YH − YL) = 1.33.

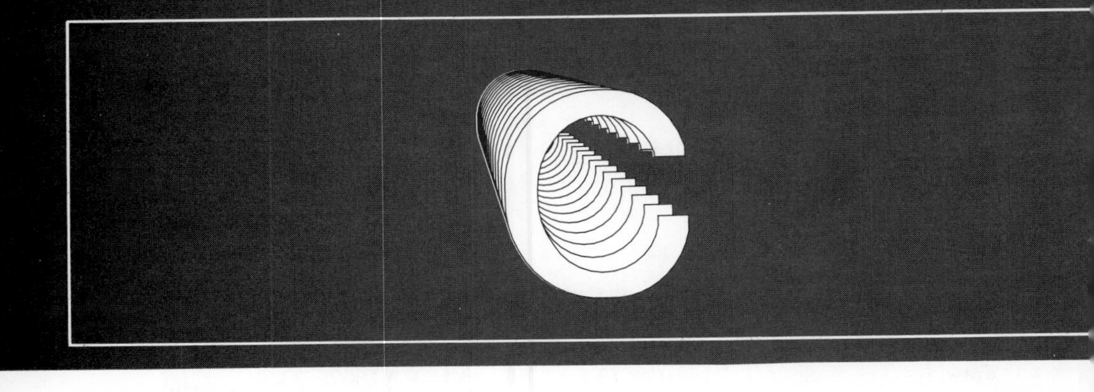

Glossary of Computer-Aided Engineering Terms

A-size paper 8½ × 11 inch paper

ACM Association for Computing Machinery. Professional society.

Ada A high-level programming language used by the Department of Defense.

AI Artificial intelligence. Procedures to create information with minimum user input.

Algorithm Step-by-step procedures used in a computer program.

Alphanumeric terminal Device that can show text and numbers but not graphics.

ANSI (Pronounced "an-see") American National Standards Institute, responsible for many standards, including computer standards.

APT Automatically Programmed Tools, a widely used language for numerically controlled machine tools.

ASCII (Pronounced "as-kee") American Standard Code for Information Interchange, a code of numerical equivalents for keyboard and control characters. Used by most computer systems for alphanumeric data. *See also* EBCDIC.

ASM Analytic Solid Modeling. Consistent mathematical representation of curves, surfaces, and solids.

Aspect ratio Ratio of height to width on a workstation display surface, viewport, or window.

Assembler language Computer language at the elementary level of the direct instructions used by the processor. *See* High-level language.

Assembly drawing Drawing showing how two or more parts fit together.

Attributes Specification of characteristics of graphical entities. For example, the color assigned to objects in a CAD drawing file.

B-Rep Boundary Representation. Computer-based description of surfaces, used for modeling.

B-size paper 11 × 17 inch paper

Background Color of the display surface before other information is output. Also, a process which runs at low priority.

BASIC Beginners All-purpose Symbolic Instruction Code. A high-level language, usually interpreted, most commonly used on microcomputers.

Baud rate Rate, bits per second, of data transmission over a serial line.

Bench mark Performance test to compare computer systems.

Bill of materials List of parts and so forth making up an assembly.

Binary Data represented by 0's (zeros) and 1's only. Also refers to files intended to be read only by machine and not printed.

Bit Single digit, 0 or 1, of binary code.

Bit map Graphics display where the state of each pixel is described by a binary representation.

Buffer Area in RAM used for temporary data storage, such as the "type-ahead" buffer for a keyboard.

Button Freestanding selection keyboard or device located on a puck or mouse which sends a choice to the processor.

Byte Group of 8 bits (usually) making up a character.

C-size paper 17 × 22 inch paper

CAD Computer-Aided Drafting. Also, Computer-Aided Design.

CADD Computer-Aided Design and Drafting.

CAE Computer-Aided Engineering.

CAM Computer-Aided Manufacturing.

CGM Computer Graphics Metafile. System for passing graphical images from and between different computer systems.

CG-VDI Computer Graphics-Virtual Device Interface.

Choice Logical graphical input which selects from a list of options.

CIM Computer-Integrated Manufacturing.

Clipping Elimination of graphical material which lies beyond defined boundaries.

COM Computer Output Microfilm. Graphic output system for high-quality images.

Compiler Program that converts a group of high-level language statements into machine language prior to execution. *See* Translate.

Core An early graphics programming standard.

CPU Central Processor Unit. Part of the computer that controls all functions.

Cross hair Manually adjusted horizontal and vertical line on a display for pointing to a location.

CRT Cathode Ray Tube. An evacuated glass tube with an electron gun, deflection devices, and phosphor-coated screen for information display.

CSA Canadian Standards Association.

CSG Constructive Solid Geometry. Solid modeling system based on combining solid primitives such as cylinders and bricks with Boolean operations.

Cursor Graphical pointing device.

D-size paper 22 × 34 inch paper.

Data base Organized collection of data.

DDA Digital Differential Analyzer. Part of raster graphics hardware that bit-maps lines.

Default Predefined value or setting. Essential for saving users' time.

Device coordinates System-dependent numbers corresponding to actual locations on workstation surfaces.

Digitize Convert graphical information into digital form. A *digitizer* is a device to automate the process.

Disk Rotating medium for semipermanent data storage which permits random access.

Display Graphics output surface such as a screen or plotter paper.

DMA Direct Memory Access. Communication between workstation and processor for high-performance graphics workstations.

Draw Basic graphics operation of making a line.

DVST Direct-View Storage Tube. *See* Storage tube.

E-size paper 34 × 44 inch paper.

EBCDIC (Pronounced eb-see-dick) Extended Binary Coded Decimal Interchange Code. A system of character representation originally developed for punched cards.

Editor Program to create or change files.

Escape function Command preceded by the ASCII code ESC, usually used for changing the state of graphics display devices.

Execution The process of actually running a program. May be preceded by compiling and linking.

Expert system Highly automated interactive computer system which uses knowledge of the problem to aid the user in arriving at a solution.

File Collection of records which are machine-accessible.

Fill Insertion of lines, solid color, or patterns within a closed curve.

Firmware Computer instructions built into ROM for the purpose of providing faster execution.

Floating point Arithmetic performed on numbers represented by mantissa and exponent. Provides an expanded range of representable values.

Font Set of alphanumeric characters matched in style and size.

FORTRAN FORmula TRANslation, a high-level language used by the engineering and scientific communities.

Function key Preprogrammed key which sends any string of characters with one key stroke.

Geometric modeling Process of formally describing the graphical attributes of an object.

GKS Graphical Kernel System, a standard for graphics subroutines.

Graphics tablet Surface from which *x-y* coordinates can be generated for "locator" input. Same as digitizer.

Grid Network of points in CAD program to assist in alignment of drawing primitives. Also see Node.

Group technology Data base system utilized for identifying components by similar characteristics.

Hardcopy Output plotted or printed on paper.

Hardware Physical components making up the computer system.

Hatching Process of filling polygons with line patterns.

High-level language Language with English-like structure, such as FORTRAN or BASIC.

Highlighting Visual enhancement of a graphic segment, usually by making it brighter.

IGES Initial Graphics Exchange Specification. A system for passing graphical images between different computer systems.

Image processing Storing and manipulating actual pictures.

Interactive graphics Graphics with computer input and output display feedback.

Interpolation Fitting more points between given points.

Interpreter Program that changes high-level language to machine language one instruction at a time during execution.

I/O Input/output, any process which transfers data to or from the processor.

ISO International Standards Organization.

Joystick, joydisk Workstation control for interactive positioning of the cursor.

Keyboard Input device for alphanumeric data.

Layer Logical division in a CAD drawing file. Also called "view" or "overlay."

Light pen Stylus which is held up to the graphics screen to indicate a point.

Linker Program to combine binary files into a code for execution.

Locator Input device for returning the *x-y* coordinates of a point selected by the cursor.

Machine language Result of translation of assembler or high-level language into code for execution by the processor.

Macro Stored sequence of commands that is initiated by a short command. Saves time in applications when a given series of commands is used repeatedly.

Mainframe Large computer which supports a large number of simultaneous processes. Usually requires a special room.

Marker Stored special symbol such as a box or star, usually to show data points.

Menu List of operations from which a choice may be made. May appear on the screen or on a graphics tablet.

Metafile Device-independent file of graphics instructions. Special routines are used for creating and translating metafiles.

Microcomputer Small computer based on a single integrated circuit. Often a self-contained, single-user system.

Minicomputer Medium-size computer positioned between the mainframe and the microcomputer. Often a multiuser system.

Mirroring Graphics function which reflects an image across a plane.

Modeling transformation Graphics transformation that changes the points matrix.

Mouse Input device moved on desk top to implement locator function.

Move Basic operation of starting a line.

NAPLPS North American Presentation-Level Protocol Syntax. A standard for transmission of graphical data over telephone lines.

NBS National Bureau of Standards. U.S. government agency.

NCGA National Computer Graphics Association. A professional society.

Node A point of grid in an image. Also a computer network hardware connection point.

Normalized coordinates Intermediate coordinates between world coordinates and device coordinates. Often written in the range 0 to 1.

NURB NonUniform Rational B-spline. A curved surface representation.

Off-line Process under control of a different CPU.

On-line Process under control of the same CPU.

Operating system Program that controls the CPU and peripherals of a computer.

Packing Putting data into a form which conserves storage.

Pan Change the location of a graphics window without changing scale of the image.

Pascal A high-level programming language widely used by computer scientists.

PC Parametric Cubic, a type of geometry used in modeling curves, surfaces, and solids. Also Personal Computer. Also Printed Circuit (board).

Pan Change the location of a graphics window without changing scale of the image.

Peripheral Device auxiliary to the CPU, such as tape drive or terminal.

PHIGS Programmer's Hierarchical Interactive Graphics Standard, a standard for graphics subroutines.

Pick Graphic input process of selecting part of an image.

Pixel Picture element. The smallest cell on a raster-scan image. The pixel may have color and brightness.

Plotter Device for creating graphics hardcopy, often a pen plotter.

Polyline Output primitive that draws one or more lines.

Polymarker Output primitive that draws one or more markers.

Primitive Fundamental graphic entity, such as text and lines.

Program Set of instructions to define a job to be done on the computer.

Prompt Output telling that the system is ready for some user action.

Properties Attributes assigned to drawing objects.

Puck Input device moved on the surface of a graphics tablet to implement locator.

RAM Random Access Memory. An area of the computer for temporary storage of data.

Raster CRT display defined by equally spaced horizontal lines, like television.

Refresh Rapid redrawing to replace image on short-persistence screen.

Repaint Redraw the display image.

ROM Read Only Memory. Similar to RAM, except an area than cannot be written on.

Rotation Motion described by concentric circles around a fixed point.

Rubberband Display of a line segment with one end fixed while the other end moves under cursor control.

Scale Operation of making larger or smaller.

Segment Graphic entity composed of a group of primitives. A segment admits manipulation as a group.

SIGGRAPH Special Interest Group on Graphics. An organization within ACM.

Software Programs and documentation for a computer system, distinguished from "hardware."

Solid modeling Representation of real-world objects in a computer data base.

Spline Continuous interpolating curve fit passing through points.

Storage tube CRT display without circuitry for refreshing which holds an image. The first low-cost graphics display. Also DVST for direct view storage tube.

String Sequence of characters.

Stroke Graphic input of coordinate pairs.

Stylus Pencil-like input device for locating a point on a graphics tablet.

Symbol Named group of primitives for fast retrieval in a CAD system.

Tablet *See* Graphics tablet.

Terminal Peripheral device for entering and retrieving data in a system.

Translation (1) Conversion of program code from a high-level language to machine language. See Compiler. (2) Motion of an object such that lines remain parallel.

Valuator Graphical input device which controls a single floating-point number.

Vector Straight line segment with length and direction.

Viewport Subspace of the display surface.

Virtual Coordinates, devices, which are idealized or generic.

Voxel Volume element as distinct from pixel, which is an area element.

Window In interactive graphics, the display of an image with user-defined bounds, usually used to enlarge a detail. Also, the correspondence between world coordinates and device coordinates.

Wireframe Model defined with line segments.

Word Group of bits for data storage, instruction. Often 32 bits or 4 bytes.

Workstation Input or output system with significant built-in processing capability intended for single user, but usually having communications with other systems.

World coordinates Location of points in user space.

Zoom Enlargement or decrease of image about some selected point.

ASCII Character Codes

The American Standard Code for Information Interchange (ASCII) assigns numbers to all of the characters on the keyboard and to 32 "control" characters. Examples of control characters are BS (backspace), CR (return), and ESC (escape). Some of the others originally developed for teletype machines (system dependent) now find use in controlling graphics workstations.

Serial transmission of characters is accomplished with a fixed-length, binary code formed from the ASCII numbers. There are both seven-bit and eight-bit versions of this code, where the seven or eight bits taken together are known as a "byte." It happens that seven bits provide for 2^7 or 128 different characters to be described. While seven of the bits encode the character, the optional eighth bit (the leftmost or highest) may be used for checking the validity of the other seven. The validity or "parity" check is system dependent and should be set in establishing communications links between computers and workstations using serial transmission. Parity may be specified as "odd," "even," or "none." Odd parity means that the eighth bit is set true if the sum of the other seven bits which are 1 is even. Even parity means that the eighth bit is set true if the sum of the other bits is odd. Thus, the *total* number of true bits is odd for odd parity and even for even parity. None means that parity is not checked.

Example: According to the table, the ASCII decimal equivalent of *W* is 87. The seven-bit binary code for this byte is formed by combining the high bits 1Ø1 and the low bits Ø111 to give 1Ø1Ø111. The binary code may be verified by the process

$$
\begin{array}{rcr}
1 \times 2^6 &=& 64 \\
\emptyset \times 2^5 &=& 0 \\
1 \times 2^4 &=& 16 \\
\emptyset \times 2^3 &=& 0 \\
1 \times 2^2 &=& 4 \\
1 \times 2^1 &=& 2 \\
\underline{1} \times 2^0 &=& \underline{1} \\
5 & & 87
\end{array}
$$

The sum of true bits, 5, is odd. Hence the eight-bit code is 11Ø1Ø111 if parity is *even* and Ø1Ø1Ø111 if parity is *odd* or *none*.

	High bits							
Low bits	ØØØ	ØØ1	Ø1Ø	Ø11	1ØØ	1Ø1	11Ø	111
ØØØØ	NUL *0*	DLE *16*	SP *32*	0 *48*	@ *64*	P *80*	‘ *96*	p *112*
ØØØ1	SOH *1*	DC1 *17*	! *33*	1 *49*	A *65*	Q *81*	a *97*	q *113*
ØØ1Ø	STX *2*	DC2 *18*	" *34*	2 *50*	B *66*	R *82*	b *98*	r *114*
ØØ11	ETX *3*	DC3 *19*	# *35*	3 *51*	C *67*	S *83*	c *99*	s *115*
Ø1ØØ	EOT *4*	DC4 *20*	$ *36*	4 *52*	D *68*	T *84*	d *100*	t *116*
Ø1Ø1	ENQ *5*	NAK *21*	% *37*	5 *53*	E *69*	U *85*	e *101*	u *117*
Ø11Ø	ACK *6*	SYN *22*	& *38*	6 *54*	F *70*	V *86*	f *102*	v *118*
Ø111	BEL *7*	ETB *23*	’ *39*	7 *55*	G *71*	W *87*	g *103*	w *119*
1ØØØ	BS *8*	CAN *24*	(*40*	8 *56*	H *72*	X *88*	h *104*	x *120*
1ØØ1	HT *9*	EM *25*	) *41*	9 *57*	I *73*	Y *89*	i *105*	y *121*
1Ø1Ø	LF *10*	SUB *26*	* *42*	: *58*	J *74*	Z *90*	j *106*	z *122*
1Ø11	VT *11*	ESC *27*	+ *43*	; *59*	K *75*	[*91*	k *107*	{ *123*
11ØØ	FF *12*	FS *28*	, *44*	< *60*	L *76*	\ *92*	l *108*	| *124*
11Ø1	CR *13*	GS *29*	- *45*	= *61*	M *77*	] *93*	m *109*	} *125*
111Ø	SO *14*	RS *30*	. *46*	> *62*	N *78*	∧ *94*	n *110*	~ *126*
1111	SI *15*	US *31*	/ *47*	? *63*	O *79*	_ *95*	o *111*	DEL *127*

References

1. Artwick, Bruce A., *Applied Concepts in Microcomputer Graphics,* Prentice-Hall, Englewood Cliffs, NJ, 1984.
2. Ashfahl, C. Ray, *Robots in Manufacturing Automation,* Wiley, New York, 1985.
3. Barr, Paul C., et al., *CAD: Principles and Applications,* Prentice-Hall, Englewood Cliffs, NJ, 1985.
4. Beatty, John C., and Kellogg S. Booth, ed., *Tutorial: Computer Graphics,* IEEE Computer Society Press, Silver Spring, MD, 1982.
5. Belie, Robert G., Putting Computer Graphics to the Test, *Computers in Mechanical Engineering,* Vol. 3, No. 1, pp. 12–19, 1984.
6. Berger, Marc, *Computer Graphics with Pascal,* Benjamin/Cummings, Menlo Park, CA, 1986.
7. Besant, C. B., *Computer-Aided Design and Manufacture,* Halsted, New York, 1980.
8. Brown, Maxine D., *Understanding PHIGS,* Megatek Corporation, San Diego, 1985.
9. Carter, James R., *Computer Mapping: Progress in the 80's,* Association of American Geographers, Washington, DC, 1984.
10. Casale, Malcolm S., Free-Form Solid Modeling with Trimmed Surface Patches, *IEEE Computer Graphics and Applications,* Vol. 7, No. 1, pp. 33–43, 1987.
11. Casale, Malcolm S., and E. L. Stanton, An Overview of Analytic Solid Modeling, *IEEE Computer Graphics and Applications,* Vol. 5, No. 2, pp. 45–56, 1985.
12. Chasen, Sylvan H., *Geometric Principles and Procedures for Computer Graphic Applications,* Prentice-Hall, Englewood Cliffs, NJ, 1978.
13. *Choosing the Right Chart,* Integrated Software Systems Corporation, San Diego, 1981.
14. Conrac Division, Conrac Corporation, *Raster Graphics Handbook,* Van Nostrand Reinhold, New York, 1985.
15. Cook, Robert D. *Concepts and Applications of Finite Element Analysis,* 2nd ed., Wiley, New York, 1981.
16. deBoor, Carl, *A Practical Guide to Splines,* Springer-Verlag, New York, 1978.

17. Demel, John T., and Michael J. Miller, *Introduction to Computer Graphics,* Brooks/Cole Engineering Division, Monterey, CA, 1984.
18. Doctor, L. J., and J. G. Torborg, Display Techniques for Octree-Encoded Objects, *IEEE Computer Graphics and Applications,* Vol. 1, No. 3, pp. 29–38, 1981.
19. Draper, N. R., and H. Smith, *Applied Regression Analysis,* Wiley, New York, 1966.
20. Encarnação, J., and E. G. Schlechtendahl, *Computer-Aided Design: Fundamentals and System Architectures,* Springer-Verlag, New York, 1983.
21. Falk, David S., et al., *Seeing the Light: Optics in Nature, Photography, Color, Vision and Holography,* Harper & Row, New York, 1986.
22. Faux, I. D., and M. J. Pratt, *Computational Geometry for Design and Manufacture,* Halsted, New York, 1979.
23. Foley, James D., and Andries Van Dam, *Fundamentals of Interactive Computer Graphics,* Addison-Wesley, Reading, MA, 1982.
24. Gardan, Yvon, *Numerical Methods for CAD, Mathematics and CAD,* Vol. 1, MIT Press, Cambridge, MA, 1986.
25. Gere, James M., and William Weaver, Jr., *Matrix Algebra for Engineers,* Van Nostrand, Princeton, NJ, 1965.
26. Goetsch, David L., *Computer-Aided Drafting,* Prentice-Hall, Englewood Cliffs, NJ, 1985.
27. Groover, Mikell P., and Emory W. Zimmers, Jr., *CAD/CAM: Computer-Aided Design and Manufacturing,* Prentice-Hall, Englewood Cliffs, NJ, 1984.
28. Harris, Dennis, *Computer Graphics and Applications,* Chapman & Hall, London, 1984.
29. Hearn, Donald, and M. Pauline Baker, *Computer Graphics,* Prentice-Hall, Englewood Cliffs, NJ, 1986.
30. Hopgood, F. R. A., et al., *Introduction to the Graphical Kernel System (GKS),* 2nd ed., Academic Press, London, 1986.
31. Hordeski, Michael F., *CAD/CAM Techniques,* Reston Publishing Co., Reston, VA, 1986.
32. James, M. L., et al., *Applied Numerical Methods for Digital Computation,* Harper & Row, New York, 1985.
33. Judd, Deane B., and Gunter Wyszecki, *Color in Business, Science, and Industry,* 3rd ed., Wiley, New York, 1975.
34. Meriam, J. L., and L. G. Kraige, *Engineering Mechanics,* 2nd ed., Wiley, New York, 1986.
35. Mitchell, William J., *Computer-Aided Architectural Design,* Van Nostrand Reinhold, New York, 1977.
36. Mortenson, Michael E., *Geometric Modeling,* Wiley, New York, 1985.
37. Mufti, Aftab A., *Elementary Computer Graphics,* Reston Publishing Co., Reston, VA, 1983.
38. Myers, Roy E., *Microcomputer Graphics,* Addison-Wesley, Reading MA, 1982.
39. Newman, William M., and Robert F. Sproull, *Principles of Interactive Computer Graphics,* 2nd ed., McGraw-Hill, New York, 1979.
40. Pao, Y. C., *Elements of Computer-Aided Design and Manufacturing,* Wiley, New York, 1984.
41. Park, Chan S., *Interactive Microcomputer Graphics,* Addison-Wesley, Reading, MA, 1985.
42. *Patran Users Guide,* PDA Engineering, Costa Mesa, CA, 1984.

43. Plastock, Roy A., and Gordon Kalley, *Computer Graphics*, Schaum's Outline Series, McGraw-Hill, New York, 1986.
44. Preparata, F. P., and M. I. Shamos, *Computational Geometry: An Introduction*, Springer-Verlag, New York, 1985.
45. Pusztai, Joseph, and Michael Sava, *Computer Numerical Control*, Reston Publishing Co., Reston, VA, 1983.
46. Raker, Daniel, and Harbert Rice, *Inside AutoCAD*, New Riders Publishing, Thousand Oaks, CA, 1985.
47. Reddy, J. N., *An Introduction to the Finite Element Method*, McGraw-Hill, New York, 1984.
48. Requicha, A. A. G., and H. B. Volker, An Introduction to Geometric Modeling and Its Applications, *Advances in Information Science Systems*, J. Tou, ed., Plenum, New York, pp. 293–328, 1981.
49. Rogers, David F., *Procedural Elements for Computer Graphics*, McGraw-Hill, New York, 1985.
50. Rogers, David F., and J. Alan Adams, *Mathematical Elements for Computer Graphics*, McGraw-Hill, New York, 1976.
51. Ryan, Daniel L., *Computer-Aided Graphics and Design*, 2nd ed., Marcel Dekker, New York, 1985.
52. Schweikert, D. G., An Interpolating Curve Using a Spline in Tension, *Journal of Mathematics and Physics*, Vol. 45, pp. 312–317, 1966.
53. Scott, Joan E., *Introduction to Interactive Computer Graphics*, 2nd ed., Wiley, New York, 1981.
54. Shahinpoor, Moshen, *A Robot Engineering Textbook*, Harper & Row, New York, 1987.
55. Sproull, Robert F., W. R. Sutherland, and Michael K. Ullner, *Device-Independent Graphics, with Examples from IBM Personal Computers*, McGraw-Hill, New York, 1985.
56. Thalman, David, and Nadia Magnenat-Thalman, *Computer Animation: Theory and Practice*, Springer-Verlag, New York, 1985.
57. Tufte, Edward R., *The Visual Display of Quantitative Information*, Graphics Press, Chesire, CT, 1983.
58. Van Deusen, Edmund, ed., *Graphics Standards Handbook*, CC Exchange, Laguna Beach, CA, 1985.
59. Yakowitz, Sidney, and Ferenc Szidarovsky, *An Introduction to Numerical Computations*, Macmillan, New York, 1986.
60. Zienkiewicz, O. C., *The Finite Element Method*, McGraw-Hill, New York, 1977.

Index